Parasitology Research Monographs

Volume 13

Series Editor
Heinz Mehlhorn
Department of Parasitology
Heinrich Heine University
Düsseldorf
Germany

This book series "Parasitology Research Monographs" presents carefully refereed volumes on selected parasitological topics. Parasites have an increasing impact on animal and human health in the present times of globalization and global warming. Parasites may be agents of diseases and- often at the same time- vectors of other agents of disease such as viruses, bacteria, fungi, protozoa and/or worms. The growth in knowledge of parasitic physiology, cell structure, biotechnological and genetic approaches, ecology, therapeutic capabilities, vaccination, immunology, diagnosis, transmission pathways and many other aspects of parasitology is increasing dramatically, even in the face of the breakthroughs that have already been made. Reflecting these most recent achievements and the importance of parasites as a threat to human and animal health, the series' broad scope concentrates on particularly hot topics that were recently covered by review articles in the journal "Parasitology Research" or in other journals. These reviews offer compact but intense insights into the ongoing research and into the methods and technologies used to control parasites. The volumes in the series build on these topics, and the volume editors are well-known experts in their respective fields. Each volume offers 10 to 20 comprehensive reviews covering all relevant aspects of the topic in focus.

More information about this series at http://www.springer.com/series/8816

Christina Strube • Heinz Mehlhorn
Editors

Dog Parasites Endangering Human Health

Springer

Editors
Christina Strube
Institute for Parasitology, Centre for Infection Medicine
University of Veterinary Medicine Hannover
Hannover, Niedersachsen, Germany

Heinz Mehlhorn
Department of Parasitology
Heinrich Heine University
Düsseldorf, Nordrhein-Westfalen, Germany

ISSN 2192-3671 ISSN 2192-368X (electronic)
Parasitology Research Monographs
ISBN 978-3-030-53229-1 ISBN 978-3-030-53230-7 (eBook)
https://doi.org/10.1007/978-3-030-53230-7

This Springer imprint is published by the registered company Springer Nature Switzerland AG.
The registered company address is: Gewerbestrasse 11, 6330 Cham, Switzerland

Preface

Peculiar rapture animals—later called dogs (Canidae)—accompanied the various waves of "early humans" leaving the "paradise" in the African Rift Valley region and wandering together from there to the top ends of the continents of our times. While at first some of these dog precursors benefited from feeding on the food remnants of the early humans, mankind later benefited from round-the-clock presence of these dogs, which helped them to hunt animals and to protect their families from aggressive, free-roaming animals like the precursors of the lions of our days and other related carnivores.

However, since the times of the early humans and dog precursors, host-specific parasites like cestodes and trematodes have developed constantly at the same time as their peculiar life cycles. Thus, it is not astonishing that during the long periods of common co-evolution a number of dog parasites also became able to infect humans or that parasitic cycles have even been developed between both humans and dogs and their related parasites.

This book offers insights into the recent status of some important worldwide spread of dog parasites belonging to protozoans and helminths which are transmitted either via fecally contaminated food or by bloodsucking insects and ticks.

In this book, 12 internationally renowned groups of scientists—belonging to the fields of human and animal parasitology as well as to the different fields of biology—contributed peculiar insights into the important dog parasites based on their knowledge obtained during long periods of research in these fields.

Hannover, Germany — Christina Strube
Düsseldorf, Germany — Heinz Mehlhorn

Contents

1 **Introduction** 1
Heinz Mehlhorn and Christina Strube

2 **The Zoonotic Dog Roundworm *Toxocara canis*, a Worldwide Burden of Public Health** 5
Patrick Waindok, Marie-Kristin Raulf, Andrea Springer, and Christina Strube

3 **Canine Leishmaniosis** 27
Gad Baneth

4 **The Challenges with Canine *Giardia*** 45
Dwight D. Bowman

5 **Dogs and Their Role in the Eco-epidemiology of Chagas Disease** 73
Ricardo E. Gürtler and Marta V. Cardinal

6 ***Echinococcus* Species: Tiny Tapeworms** 107
Heinz Mehlhorn

7 **An Update on the Status of Hydatidosis/Echinococcosis in Domestic Animals, Wildlife and Humans in Australia** 123
David J. Jenkins

8 ***Dipylidium caninum*** 141
Heinz Mehlhorn

9 ***Ancylostoma caninum* and Other Canine Hookworms** 147
John M. Hawdon and Kira A. Wise

10 **Strongyloidiasis: Really a Zoonosis?** 195
Adrian Streit

11 *Dirofilaria* spp. Do They Get Under Your Skin (Or into Your Heart)? 227
Alice Vismarra, Marco Genchi, Chiara Cattabiani, and Laura Kramer

12 Guinea Worm Infection in Dogs: A Case of Reverse Zoonosis that Impedes *Dracunculus medinensis* Eradication 239
M. Teresa Galán-Puchades

13 Ticks, Dogs, and Humans: An Endangered Community 261
Heinz Mehlhorn

14 Trematodes Attacking Dogs and Humans 273
Heinz Mehlhorn

Index 285

About the Editors

Christina Strube is a full professor of veterinary parasitology at the University of Veterinary Medicine Hannover (Germany), where she teaches parasitology courses and supervises PhD students. She has authored more than 130 peer-reviewed publications and received three awards in the field of veterinary medicine for her outstanding research work. Her main research interests include the One Health topics "zoonotic parasites" and "tick-borne diseases," as well as the epidemiology of parasitic infections in pets, livestock, and wildlife.

Heinz Mehlhorn studied at the Rheinische Friedrich-Wilhelms University of Bonn (Germany). He was chairman of the parasitological institutes in Düsseldorf and Bochum. As a former president of the World Society of Protozoology and long-standing member of the steering committee of the World Society of Parasitologists, he runs ongoing lectures in parasitology in several countries and still teaches courses for medical students in Düsseldorf. He has published 42 books on parasitological problems in German, French, English, and Spanish. He holds the patents on 12 antiparasitic drugs, which are the basis of products sold by the university spin-off company Alpha-Biocare GmbH (Neuss, Germany), which produces medications, repellents, and wound care products for humans and animals.

Chapter 1
Introduction

Heinz Mehlhorn and Christina Strube

Abstract Dogs are attendants of humans for thousands of years and have today even closer contacts as pet animals than they had in the past, when dogs just protected humans and their homes and helped to hunt animals. During these passed times the contacts between dogs and humans had been much less closer than those of our days. Thus, today several dog parasites may endanger severely health of humans. The present book describes in 14 chapters the most important parasites of dogs, which also may harm humans. Thus, this book helps to protect the community of dogs and humans.

Keywords *Canis lupus familiaris* · Wolves · Autosomes · Chromosomes

1.1 Dogs: The Longtime Faithful Companions of Mankind

I. Abilities

The domestication of the recent dogs (*Canis lupus familiaris*) started around 100,000 years BC spreading from wolves of these days. Due to their abilities to adapt to the needs in the human world the dogs reached at present a population of around 500 millions. The success of the dogs to become one of the most important partners of humankind is mainly based on some peculiar abilities, which helped them and humans to survive in many situations:

H. Mehlhorn (✉)
Department of Parasitology, Heinrich Heine University, Düsseldorf, Nordrhein-Westfalen, Germany
e-mail: mehlhorn@uni-duesseldorf.de

C. Strube
Institute for Parasitology, Centre for Infection Medicine, University of Veterinary Medicine Hannover, Hannover, Niedersachsen, Germany
e-mail: Christina.Strube@tiho-hannover.de

C. Strube, H. Mehlhorn (eds.), *Dog Parasites Endangering Human Health*, Parasitology Research Monographs 13,
https://doi.org/10.1007/978-3-030-53230-7_1

1. *Hearing*	*Dogs*: 15 up to 50,000 Hz *Humans*: 20 up to 20,000 Hz
2. *Field of vision*	*Dogs*: ~240° *Humans*: ~180°
3. *Smelling abilities*	*Dogs*: They possess 125–220 million sensory cells *Humans*: They possess ~five million sensory cells
4. *Touching sensory cells*	The receptors of the so-called vibri (=stiff) hairs around the mouth are very sensitive and help them to avoid unwished contact of their nose to obstacles.

II. Genome

Dogs possess 38 pairs of autosomes and two sexual chromosomes (X and Y). The genome contains 2,528,446,953 base pairs and around 193,009 genes.

III. Life Span

Small-sized dog breeds have larger life spans (~18–20 years), while large dogs reach mostly less than 8–9 years.

IV. Breeds

The cynological systematic is taken care by the FCI (Féderation Cynologique International), which registers actually 344 breeds, that are divided into 10 groups.

V. Acceptance of Dogs in Human Cultures

Depending on the human cultures/religions dogs have found different degrees of acceptance respectively rejection. While in Christian Europe and related countries dogs are mainly kept for personal and house protection as well as for hunting, people in Islamic countries often reject contact with dogs, since (especially in former times) dogs had been considered as "not clean." In some countries (e.g., especially in some regions of China) dog meat is eaten. In any way mainly the high graded abilities of dogs, e.g., protection of persons and houses, detection of drugs, and personal contacts for lonely persons, have led to the fact that millions of humans are permanently in contact to dogs, so that it is needed to keep these people (and especially kids) free from agents of disease, which could be transmitted from dogs to humans.

Literature

Boyko AR (2011) The domestic dog: man's best friend in the genomic era. In: Genome biology, vol 12, p 216 ff

Lindblad-Toh K, Wade CM, Mikkelsen TS et al (2005) Genome sequence, comparative analysis and haplotype structure of the domestic dog. Nature 438(7069):803–819

Pang JF, Kluetsch C, Zou XJ et al (2009) mtDNA indicate a single origin for dogs south of the Yangtze River less than 16300 years ago from numerous wolves. Mol Biol Evol 26:2849–2864
Stafford K (2006) The welfare of dogs. In: Animal welfare, vol 4. Springer, Dordrecht
Thalmann O, Shapiro B (2013) Complete mitochondrial genomes of ancient canids suggest a European origin of domestic dogs. Science 342:871–874

Chapter 2
The Zoonotic Dog Roundworm *Toxocara canis*, a Worldwide Burden of Public Health

Patrick Waindok, Marie-Kristin Raulf, Andrea Springer, and Christina Strube

Abstract The zoonotic dog roundworm *Toxocara canis* infects not only a broad range of canids as definitive hosts, but also affects humans as paratenic hosts after accidental infection. In humans, clinical apparent toxocarosis (or toxocariasis) is classified into four different syndromes: visceral *larva migrans* (VLM), ocular *larva migrans* (OLM), neurotoxocarosis (NT) and covert toxocarosis. Symptoms range from abdominal pain to irreversible blindness or meningitis and cognitive disorders. Nevertheless, toxocarosis is a neglected zoonotic disease, which is often undiagnosed. Until today, the parasite remains enigmatic in many aspects. This chapter aims to summarise key characteristics of the biology of *T. canis*, risk factors that lead to human infection, clinical presentation and relevance of human toxocarosis and possibilities of control and prevention measures. Further research efforts as well as raising public awareness for this zoonotic dog parasite are essential to improve the control of this challenging zoonotic infection.

Keywords Toxocarosis · Toxocariasis · VLM · OLM · Neurotoxocarosis · Parasitic zoonosis

---

P. Waindok · A. Springer · C. Strube (✉)
Institute for Parasitology, Centre for Infection Medicine, University of Veterinary Medicine Hannover, Hannover, Niedersachsen, Germany
e-mail: Christina.Strube@tiho-hannover.de

M.-K. Raulf
Institute for Parasitology, Centre for Infection Medicine, University of Veterinary Medicine Hannover, Hannover, Niedersachsen, Germany

Immunology Unit and Research Center for Emerging Infections and Zoonoses, University of Veterinary Medicine Hannover, Hannover, Niedersachsen, Germany

C. Strube, H. Mehlhorn (eds.), *Dog Parasites Endangering Human Health*, Parasitology Research Monographs 13,
https://doi.org/10.1007/978-3-030-53230-7_2

2.1 Introduction

The earliest known record of an ascarid infection in canids dates back to the Pleistocene, as eggs of *Toxocara* sp. were found in a 1.2 million-year-old coprolite of the extinct hyena *Pachycrocuta brevirostris* (Perri et al. 2017). Today, ascarid infections in wild and domestic canids are frequently caused by the worldwide distributed zoonotic dog roundworm *Toxocara canis* (Werner 1782). Wild and domestic canids may shed high amounts of *Toxocara* eggs into the environment, including recreational areas, gardens or playgrounds.

Humans can become infected by ingestion of embryonated *T. canis* eggs or of undercooked meat of paratenic hosts containing infective third-stage larvae, which results in larval migration and persistence in different tissues and organs (Beaver et al. 1952; Brunaska et al. 1995). The resulting disease, toxocarosis (or toxocariasis), is one of the most reported zoonotic helminth infections worldwide (Rubinsky-Elefant et al. 2010). It is considered by the US American Centres for Disease Control (CDC) as one of the five neglected parasitic infections with priority for public health action. The effects on human health may range from non-specific symptoms such as lethargy and abdominal pain to irreversible blindness, meningitis as well as behavioural and cognitive disorders. This chapter aims to summarise biological characteristics of *T. canis*, its life cycle and transmission to humans and describes salient aspects of human toxocarosis, diagnostic options and efforts to prevent human toxocarosis.

2.2 Life Cycle and Larval Migration in the Definitive Host

Definitive hosts shed *Toxocara* eggs into the environment with their faeces. These eggs are unembryonated when laid and thus not immediately infective. Development of the infective third-stage larva (L3, Fig. 2.1) inside the egg may require weeks to months, as this process is affected by temperature and humidity in the environment. Embryonation starts at around 10 °C and accelerates with rising temperatures, with optimal conditions for embryonation and survival of eggs in soil ranging from 23 to 35 °C (Rocha et al. 2011). The larva accomplishes two moults inside the egg (Fig. 2.1), and the infective L3 remains within the eggshell until ingestion by a vertebrate host (Araujo 1972; Brunaska et al. 1995; Overgaauw 1997a). Furthermore, the development and the survival of eggs increases with increasing humidity, and *Toxocara* eggs may survive for up to 4 years in warm and humid environments (Deutz et al. 2005; Etewa et al. 2016). In contrast, exposure to sunshine as well as desiccation may lead to developmental arrest and egg degeneration (Etewa et al. 2016).

Although definitive hosts can ingest L3 persisting in tissues of paratenic hosts, infection via environmental contamination is regarded as the epidemiological important route (Schnieder et al. 2011). Larvae of *T. canis* hatch from ingested,

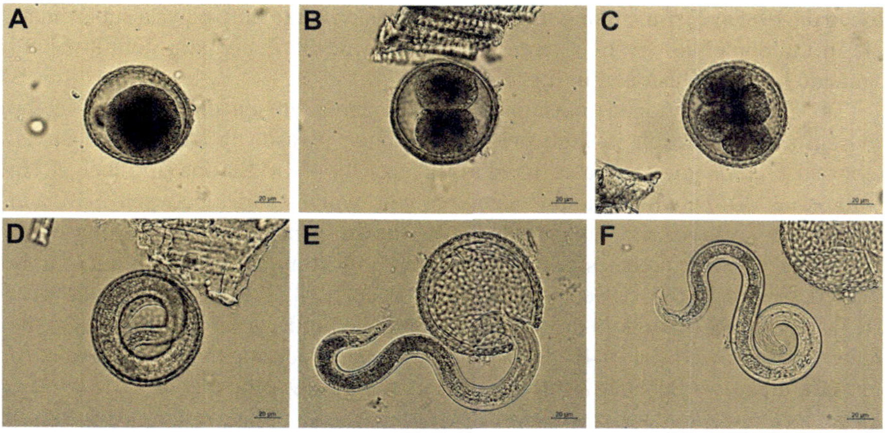

Fig. 2.1 Development of *Toxocara canis* eggs to infective third-stage larvae. (**a**) One-cell stage, (**b**) two-cell stage, (**c**) early morula, (**d**) embryonated egg containing third-stage larva, (**e**) hatching of third-stage larva, (**f**) hatched third-stage larva. (**e**) and (**f**) usually occur within the host. Bar represents 20 μm

embryonated eggs in the duodenum, penetrate the intestinal wall, enter lymph vessels and migrate via the portal circulation to the liver (Webster 1958a, b). Most larvae leave the liver via the caval vein, pass the heart by passive haematogenous transport and reach the lungs via the pulmonary artery (Webster 1958a, b). The subsequent migration route depends on several factors such as the immune status and age of the host and the number of ingested larvae. The predominant route in young puppies is tracheal migration. Larvae penetrate the alveoli and migrate via bronchioles and trachea to the pharynx, get swallowed and reach the intestine, where they mature to adult worms (Sprent 1958). This so-called tracheal migration route also occurs upon infection of older dogs with a low number of larvae (Greve 1971).

In addition to tracheal migration, larvae may re-enter the circulatory system in the lungs and get distributed passively to somatic tissues (Sprent 1958; Webster 1958a, b). The majority of these larvae accumulate in the skeletal muscles and persist in a hypobiotic stage, encapsulated in granulomas (Webster 1958a).

Prenatal transmission, also known as transplacental or intrauterine transmission, is of major importance in the epidemiology of *T. canis*, and occurs either by reactivation of arrested larvae or from new infections of the bitch during gestation (Schnieder et al. 2011). As from approximately day 40 of the pregnancy, larvae penetrate the syncytiotrophoblast and cytotrophoblast, and remain in the liver of the foetus until birth (Koutz et al. 1966; Lloyd et al. 1983; Scothorn et al. 1965). Within hours after the birth, the larvae start to continue migration to the lungs and undergo tracheal migration (Lloyd et al. 1983).

In pregnant bitches, larvae also invade the mammary glands, resulting in lactogenic transmission (Overgaauw 1997b). However, in experimentally infected bitches, only 1.5%–4.5% of larvae were transmitted lactogenically (Burke and

Roberson 1985a, b), illustrating that this transmission route only plays a subordinate role in the epidemiology of *T. canis*. The life cycle of *T. canis* in definitive and paratenic hosts is illustrated in Fig. 2.2.

The various mentioned transmission routes and high tenacity of eggs in the environment may result in high prevalence of the parasite in canid populations. Reported *T. canis* infection rates in red foxes reach 49.4% in Europe (Ilic et al. 2016; Poulle et al. 2017), whilst the prevalence in wild wolves varies between 2.8% and 36.6% (Abdybekova and Torgerson 2012; Bindke et al. 2019; Hermosilla et al. 2017). Detection frequencies in faecal samples from stray dogs in Poland and Italy were 16.8% and 20.6%, respectively (Liberato et al. 2018; Szwabe and Blaszkowska 2017). Regarding household dogs, studies conducted in Belgium, Germany and the Netherlands showed that 4.4% to 6.1% of examined pet dogs excreted *Toxocara* eggs (Barutzki and Schaper 2011; Claerebout et al. 2009; Nijsse et al. 2015).

2.3 Transmission of Infective *Toxocara* Stages to Humans and Larval Migration Through the Human Body

The ubiquitous distribution of *T. canis*, the high reproduction rate and the high tenacity of eggs result in soil contamination of human environments (Fakhri et al. 2018; Traversa et al. 2014). Fakhri et al. (2018) evaluated the global prevalence of *Toxocara* eggs in soil in public places with a total sample size of 42,797 samples from 40 different countries. According to the meta-analysis, *Toxocara* eggs were detected in 21% of soil samples originating from human surroundings (Fakhri et al. 2018). Substantial differences in the contamination rates were detected between the six WHO regions (Africa, America, South East Asia, Europe, Eastern Mediterranean region and Western Pacific), with the highest pooled prevalence in the Western Pacific region (35%) and the lowest in North America (13%) (Fakhri et al. 2018). This clearly illustrates that the extent of soil contamination depends on various environmental conditions and anthropogenic factors. Tropical and subtropical conditions with year-round temperatures >18 °C, high precipitation and therefore a high mean humidity were associated with higher contamination rates, and, consequently, *Toxocara* eggs were found more frequently at lower latitudes (0–20°) than higher latitudes (41–60°) (Fakhri et al. 2018). Moreover, many countries in tropical and subtropical regions are socioeconomically disadvantaged, and this burden, together with different cultural aspects, affects environmental hygiene and behaviour of residents (McMichael 2000; Montgomery and Elimelech 2007). Heavily contaminated environments coupled with warm climatic conditions and poor hygienic standards provide ideal transmission opportunities. Infection usually occurs by ingestion of *Toxocara* eggs with contaminated soil, e.g. by geophagia or pica, or by passing contaminated hands to the mouth. In consequence, children are particularly at risk due to a weakly developed sense of hygiene.

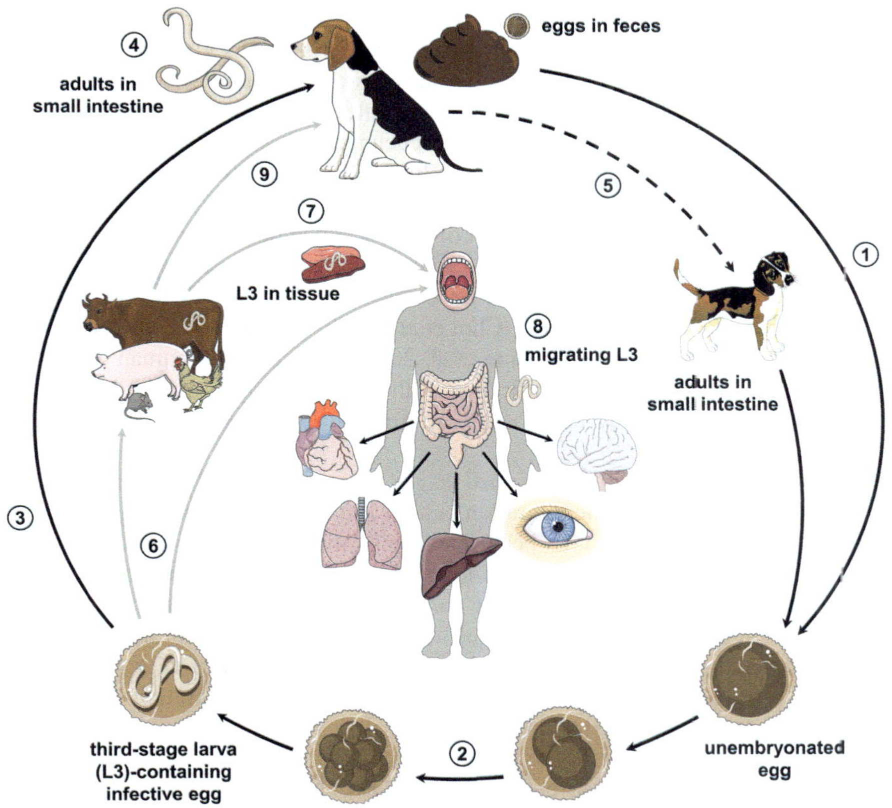

Fig. 2.2 Life cycle of *Toxocara canis*. Adult worms located in the small intestine of the definite host excrete eggs that are shed to the environment via faeces (1). Eggs embryonate within 2–8 weeks, resulting in infective third-stage larvae (L3) (2). Upon ingestion of embryonated eggs by the definitive host (3), the L3 hatch and initiate migration by penetrating the gut wall and invading the blood circulation. Larvae reach the liver and lungs via the cardiovascular system. After entering the pulmonary alveoli, larvae either moult, are coughed up and swallowed (so-called tracheal migration), or enter the circulatory system again to continue migration to somatic tissues. In puppies, larvae primarily follow the tracheal migration route and after re-entering the small intestine, they mature to adult worms and therefore complete their lifecycle (4). In older, immuno-competent dogs, L3 tend to re-enter the circulatory system to eventually arrest in somatic tissues. Such persisting larvae are reactivated in the last trimester of gestation and may be transmitted to the pups via transplacental and lactogenic infection, where they finally develop into reproductive adult worms in the small intestine (5). Humans and a variety of animal species serve as paratenic hosts for *T. canis* after ingestion of embryonated eggs (6). Additionally, humans may acquire infection via consumption of undercooked meat of infected paratenic hosts (7). In paratenic hosts, larvae hatch and penetrate the gut wall with subsequent somatic migration to various tissues such as liver, heart, lungs, eyes and brain (8), where they enter an arrested state. Their life cycle can be completed when definitive hosts ingest paratenic hosts or their tissues (9). This figure was modified from Servier Medical Art (http://smart.servier.com). Servier Medical Art by Servier is licenced under a Creative Commons Attribution 3.0 Unported License

Furthermore, the superficial layer of *Toxocara* eggs consists of craterlike irregular depressions, covered by a web-like albuminous coat with sticky glycoproteins (Uga et al. 2000). This facilitates adherence to surfaces of vegetables, herbs and fruits. Accordingly, contamination of different vegetables has been described, such as lettuce, carrots, potato, zucchini, spinach, cucumber or cress (Abougrain et al. 2010; Adanir and Tasci 2013; Fallah et al. 2012; Klapec and Borecka 2012), some of which are commonly consumed raw.

Human infection due to consumption of raw or undercooked meat of *Toxocara*-infected paratenic hosts is less common. Nevertheless, several case reports describe an infection via raw meat products from cattle, swine, lamb, rabbit and poultry, including chicken, duck and ostrich (Choi et al. 2008; Hoffmeister et al. 2007; Noh et al. 2012; Sturchler et al. 1990; Taira et al. 2004). Direct dog-to-human transmission is not considered to be of significance. Although *Toxocara* eggs may be present in dog fur, they are usually unembryonated and not viable (Keegan and Holland 2013).

After ingestion of infective *Toxocara* stages, a broad range of different animals can act as paratenic hosts. The migration pattern and the influence of migrating and persisting *Toxocara* larvae have been intensively studied in a variety of animals serving as models for human infection (Strube et al. 2013). In paratenic hosts, the migration route of *T. canis* larvae mainly corresponds to the somatic migration in the definitive host. Larvae hatch in the intestine, penetrate the gut wall and enter the mesenteric lymph nodes via the circulatory system. The majority of migrating larvae reach the liver in the first 48 h of infection and continue their migration by passive haematogenous transport to the lungs (Strube et al. 2013). From there, they re-enter the systemic circulation and are distributed throughout the entire body. In somatic tissues, the majority of larvae enter a state of hypobiosis and remain viable for several years (Sprent 1952).

2.4 Human Toxocarosis

Human toxocarosis may range from asymptomatic infections to severe organ injuries, depending on the intensity of infection, the duration of larval migration, the location of larvae and the induced immune response. Nevertheless, toxocarosis rarely fatal and most infections are self-limiting. The disease is classified into four different syndromes, based on the distribution of larvae and the occurring symptoms. Larval migration through major organs causes a systemic disease called visceral *larva migrans* (VLM) syndrome whereas pathological effects in the eye and the optic nerve are termed ocular *larva migrans* (OLM) syndrome. Further forms of the disease are covert toxocarosis, a less severe form of VLM (Taylor et al. 1987, 1988), and cerebral toxocarosis or neurotoxocarosis (NT), characterised by larval migration and persistence in the central nervous system (CNS) (Fan et al. 2015; Finsterer and Auer 2007).

2.4.1 Visceral Larva Migrans *Syndrome*

The term visceral *larva migrans* (VLM) was introduced by Beaver (1952) who detected *Toxocara* larvae in hepatic eosinophilic granulomas in children. The VLM syndrome is a clinically apparent systemic form caused by a high infection dose or repeated exposure. The symptoms are presumably induced by an immediate hypersensitivity response to degenerating larvae in the affected tissues (Despommier 2003). Classical VLM is associated with various symptoms such as general malaise and fatigue, anorexia, fever and abdominal complaints attributed to hepatomegaly and splenomegaly, as well as shortness of breath, wheezing or coughing (Carvalho and Rocha 2011; Gillespie 1987). Severe VLM is relatively uncommon, whereas the mild VLM syndrome with less severe symptoms is frequent (Kayes 1997).

VLM is associated with leukocytosis including high eosinophilia (Despommier 2003; Fan et al. 2013), and many patients suffer from IgG/IgE hyper-γ-globulinaemia (Fan et al. 2013; Maizels 2013). In most visceral tissues, invading larvae are encapsulated into granulomas consisting of polymorphonuclear leukocytes (primarily eosinophils), macrophages and a fibrotic capsule (Hoeppli et al. 1949).

VLM is typically associated with liver pathology, in fact, most cases of granulomatous hepatitis in children may be attributed to *Toxocara* infection (Musso et al. 2007). Furthermore, serum IgG antibodies against *T. canis* were elevated in patients diagnosed with liver abscesses, which disappeared in most cases within 6 months after anthelmintic treatment (Ha et al. 2016).

The clinical picture of VLM may also include lung involvement resulting in respiratory complaints such as a chronic cough, wheezing, dyspnoea or haemoptysis (Mazur-Melewska et al. 2015). The migration of *T. canis* larvae through pulmonary tissues leads to mechanical damage resulting in transient focal haemorrhages. Furthermore, infiltration of eosinophilic granulocytes and T cells may cause pulmonary eosinophilia, an idiopathic hypereosinophilic syndrome, acute bronchitis and pneumonia (Lee et al. 2015; Mazur-Melewska et al. 2015; Yoshikawa et al. 2011). Although involvement of the lung is usually asymptomatic or associated with mild symptoms, VLM is suspected to promote chronic conditions such as asthma or pulmonary fibrosis (Cooper 2008). Experimental research in mice as a model organism for toxocarosis has suggested that infection with *Toxocara* contributes to the development of allergic diseases, including asthma (Pinelli et al. 2008). However, based on epidemiological studies, there is not yet a general consensus about the association between *Toxocara* seropositivity and asthma in humans (Li et al. 2014).

Toxocarosis may be accompanied by cutaneous manifestations, e.g. chronic pruritus and prurigo (Gavignet et al. 2008; Kazmierowski et al. 1978). Furthermore, chronic urticaria has commonly been reported amongst patients seropositive for toxocarosis (Demirci et al. 2003; Wolfrom et al. 1996), as well as miscellaneous eczema, panniculitis and vasculitis (Gavignet et al. 2008).

2.4.2 *Covert or Common Toxocarosis*

Many helminth infections remain asymptomatic or induce only mild or non-specific clinical signs, and thus often remain undiagnosed. *Toxocara* is no exception as epidemiological surveys report a high *Toxocara*-seroprevalence in many countries worldwide. However, only a small number of VLM cases have been described. This has led to the suggestion that a moderate variation of VLM exists, which was termed "covert" or "common" toxocarosis (Taylor et al. 1987). The clinical diagnosis of covert toxocarosis is challenging as the disease comprises various combinations of non-specific symptoms including fever, weakness, lethargy, sleepiness, anorexia, headache, wheezing, abdominal pain, nausea, vomiting, behavioural disorders, pulmonary symptoms, limb pain, cervical lymphadenitis, hepatomegaly, pruritus and rash (Glickman et al. 1987; Kayes 1997; Taylor et al. 1987). Most of these symptoms overlap with the characteristics of VLM, but both forms can be delimited by the severity of disease and the number of co-occurring symptoms, as covert toxocarosis is characterised by a less severe course and smaller number of symptoms.

2.4.3 *Ocular* Larva Migrans *Syndrome*

The ocular *larva migrans* (OLM) syndrome is characterised by larval invasion into the eye (Ashton 1960). The symptomatology of OLM is diverse, ranging from asymptomatic to severe. Migrating larvae induce an eosinophilic inflammatory immune response (Stewart et al. 2005), which may be accompanied by marked vitritis, secondary cataracts and mild anterior chamber reactions (Sabrosa and de Souza 2001). The extent of eosinophilia determines the fibrotic granulomatous response, a strong clinical indication of OLM. Importantly, invasion of a single larva into the eye may be sufficient to induce the inflammatory cascade, wherefore visual impairments occur in most cases of OLM. The granulomas may create visual distortions, heterotopia and/or detachment of the macula (Sabrosa and de Souza 2001; Stewart et al. 2005). Major causes of vision loss are vitreous inflammation as well as macular oedema and macular detachment resulting from uveitis (Sabrosa and de Souza 2001; Taylor 2001), whilst the formation of fibrous traction bands may lead to breaks in the atrophic retina, resulting in rhegmatogenous retinal detachments (Sabrosa and de Souza 2001). Other clinical presentations include retinitis, keratitis and chorioretinitis, which may be followed by choroidal neovascular membrane formation in prolonged *Toxocara* infections (Despommier 2003).

2.4.4 Neurotoxocarosis

In 1951, Beautyman and Woolf (1951) detected a *Toxocara* larva in the left thalamus of a child who died of poliomyelitis. This was the first description of *Toxocara* larvae invasion into the central nervous system (CNS). Today, the accumulation and persistence of *Toxocara* larvae within the CNS are termed cerebral or neurotoxocarosis (NT). In contrast to other tissues, larvae in the CNS are not encapsulated into granulomas, but their migratory tracks result in areas of necrosis and inflammatory infiltration with secondary granuloma formation (Abdel Razek et al. 2011; Pawlowski 2001; Springer et al. 2019). Overall, the pathogenesis of NT is complex and poorly understood compared to other forms of disease provoked by *Toxocara*. Due to the cryptic nature of NT and non-specific clinical signs, the diagnosis remains challenging and the frequency of NT is probably underestimated (Holland and Hamilton 2013; Hotez and Wilkins 2009). Most knowledge on human NT is derived from few clinical cases or studies correlating *Toxocara* seropositivity with neurological disorders.

Migration of larvae induces damage to the brain tissue resulting, e.g. in cerebral lesions, fibrosis of the arachnoidea, bilateral subdural haematoma, subarachnoid haemorrhages and cerebral meningoradiculitis (Fan et al. 2015). Furthermore, NT is accompanied by eosinophilia in both blood and cerebrospinal fluid (CSF) (Finsterer and Auer 2007). Recruitment of eosinophilic granulocytes to the site of inflammation is a common hallmark of NT, and eosinophilic meningoencephalitis, eosinophilic meningitis, and eosinophilic encephalitis are repeatedly recorded (Fan et al. 2015). Furthermore, cerebral vasculitis is common in murine and human NT (Springer et al. 2019; Xinou et al. 2003), but the observed vasculitis-associated infarctions may also be caused by an acute inflammatory response to degenerating larvae or a hypersensitivity reaction to anthelmintic therapy (Xinou et al. 2003). Furthermore, injury of the spinal cord due to migrating larvae can manifest as myelitis, Viliuisk encephalomyelitis, (meningo)encephalomyelitis, lower motor neuron disease and spinal abscesses (Fan et al. 2015).

Neurotoxocarosis has also been implicated as a possible cause of various neurological deficits and neuropsychological disorders. The association between seizures and a positive *Toxocara* serology supports the hypothesis of *T. canis* as a causative agent of epilepsy (Quattrocchi et al. 2012). Moreover, larval invasion into the CNS may induce psychiatric and neurological disorders that could contribute to schizophrenia. This hypothesis is supported by experiments showing that levels of dopamine decreased over time in the brains of *T. canis*-infected mice (Othman et al. 2010), but the role of the dopamine system in NT remains to be elucidated comprehensively. The association between schizophrenia and *Toxocara* seropositivity was the focus of some studies and case reports published in the last decades, but the results remain contradictory (Cong et al. 2014). Thus, it remains unclear whether *T. canis* infection may contribute to mentioned neurological diseases or if patients with neuropsychological disorders exhibit a higher risk of *Toxocara* infection due to abnormal behaviour, e.g. a tendency to eat inappropriate things, and decreased self-care (El-Sayed and Ismail 2012).

Furthermore, *Toxocara* infection has been associated with mental confusion and cognitive impairments, possibly indicative of dementia (Richartz and Buchkremer 2002; Salvador et al. 2010). In addition, *Toxocara* seropositivity has been linked to deficits in the development of speech as well as poor reading achievement, distractibility and lower intelligence in kindergarten children (Jarosz et al. 2010; Walsh and Haseeb 2012). These cognitive dysfunctions and learning difficulties can lead to detrimental long-term consequences, especially in school children. Experimental infections in mice further support the link between toxocarosis and memory function, as they required significantly longer to relocate a water tube in a spatial learning task and a food reward in a maze than uninfected control mice (Cox and Holland 2001; Janecek et al. 2017). In rodents, parenchymal damage and haemorrhagic lesions induced by *T. canis* larvae frequently result in neurodegenerative processes and demyelination, e.g. in the anterior commissure and fornix, important structures regarding memory function (Heuer et al. 2015; Springer et al. 2019). Therefore, *Toxocara* is regarded as a possible cause of dementia, but studies concerning *Toxocara* infection status amongst dementia patients are rare. Nevertheless, *T. canis* as a causative agent of dementia may be underestimated due to the cryptic nature of NT in cases of low parasitic burdens (Holland and Hamilton 2013).

2.5 Immune Recognition and Evasion of *Toxocara* Larvae in the Paratenic Host

Larvae of *T. canis* are capable to enter a long-term developmental arrest in humans and other paratenic hosts. For example, larvae remained viable for up to 9 years in tissues of experimentally infected monkeys (Beaver et al. 1952). Thus, interactions between *T. canis* and immune cells or immune-related compounds have been examined in numerous studies, by using in vitro cell culture assays or rodents as model organisms for paratenic hosts. *Toxocara* infection provokes a Th2-polarised adaptive immune response that is predominantly mediated by $CD4^{+}$ T cells (Del Prete et al. 1991). These $CD4^{+}$ Th2 T cells secrete cytokines such as IL-4 and IL-5, thereby mediating the differentiation of B cells and eosinophils, respectively (Maizels 2013). These cells generate reactive oxygen species like superoxides and free radicals to eliminate invading pathogens (Klion and Nutman 2004). However, eosinophils seem to lack the ability to eliminate *T. canis* larvae. This was shown in transgenic mice exhibiting lifelong eosinophilia due to overexpression of IL-5. *T. canis* larvae were unharmed during hypereosinophilia whilst the nematode *Nippostrongylus brasiliensis* was eliminated (Dent et al. 1999). Co-incubation of *T. canis* larvae and eosinophilic granulocytes showed that eosinophils rapidly adhere to the parasite surface, but larvae evade from surrounding cells within 24 h (Fattah et al. 1986). *Toxocara* larvae are sheathed in a 10–20 nm thick, labile surface coat ("fuzzy coat"), a mucin- and glycan-rich assembly consisting, amongst others, of trisaccharide glycan side-chains and mucin polypeptides (Loukas et al. 2000a; Page et al. 1992a, b; Schabussova et al. 2007). Interestingly, immune reactions directed against *Toxocara* provoke the physical shedding of the fuzzy coat. Thus, vital larvae evade immune recognition, whilst eosinophils and antibodies remain attached to the shed surface (Badley et al. 1987).

Escaping the T cell- and eosinophil-mediated attack is not the only ability that promotes larval survival in paratenic hosts. *Toxocara* spp. release excretory–secretory (TES) products which include, amongst others, proteases, cell adhesion molecules, lectins, SCP/TAPS proteins and mucins, which may be involved in host invasion and in parasite–host interactions such as immune evasion or modulation (Maizels 2013). TES components were first described by SDS-PAGE analysis and are classified by their molecular weight (Maizels 2013). The components TES-32 and TES-70 are both members of the C-type lectin family (Loukas et al. 1999, 2000b). Soluble C-type lectins are involved in numerous physiological and immunogenic functions and regulate many essential processes such as homeostasis, coagulation, angiogenesis and inflammation (Brown et al. 2018). TES-120 contains a set of mucin-like glycoproteins, participating in the formation of the fuzzy coat (Page et al. 1992a). Recent studies investigated TES derived from in vitro cultured *T. canis* larvae using a proteomic approach (da Silva et al. 2018; Sperotto et al. 2017). Various proteins were identified in the secretome of *Toxocara* larvae, including collectins, heat shock proteins (HSPs), troponins, exonucleases and cathepsins. Several of the secreted proteins are homologous to host immunomodulatory proteins, such as HSPs, which prevent protein aggregation and are suggested to stimulate IgG and IgM responses (Schmitt et al. 2007). Furthermore, TES contains macrophage migration inhibitory factor-like proteins, functioning as modulators of the cellular and humoral immune response (Filbey et al. 2019). Overall, genomic and transcriptomic approaches led to the identification of many TES proteins. However, the role of particular and specific components within the complex mixture of proteins remains to be elucidated.

2.6 Diagnosis

Various anamnestic factors such as age, report of geophagy, contact with dogs, clinical signs as well as laboratory results should be taken into consideration during *Toxocara* diagnosis. A prominent hallmark of VLM is an increased number of circulating and persistent tissue eosinophils, but eosinophilia may be absent in cases of covert toxocarosis (Taylor et al. 1988). The pulmonary migration of larvae may result in increased eosinophil numbers in the bronchoalveolar lavage fluid. As the OLM syndrome is usually only caused by few larvae, blood eosinophil levels are usually within the physiological range (Glickman and Schantz 1981), unless larvae previously migrated through visceral tissues (Magnaval et al. 2001). In NT case reports, eosinophilia was present in 41% of blood samples and 24% of CSF samples.

However, the low incidence of around 90 cases does not allow general conclusions regarding eosinophilia in NT (Fan et al. 2015).

Further haematological hallmarks, especially in VLM, include leucocytosis, anaemia, hypoalbuminaemia and increased isohaemagglutinin titres (McGuinness and Leder 2014). Markers of inflammation may also be increased (Fillaux and Magnaval 2013). However, this seems to be rather uncommon in covert toxocarosis, as an elevated erythrocyte sedimentation rate as an indicator for inflammatory processes was observed in only 11% of patients with this disease form (Fillaux and Magnaval 2013). The VLM syndrome is further characterised by hyper-γ-globulinaemia of immunoglobulin M (IgM) and IgG classes (Fillaux and Magnaval 2013; Glickman et al. 1987).

These parameters may be indicative of an infection with *Toxocara*, but are non-specific, as other parasitic infections such as ascariasis, schistosomiasis, strongyloidiasis or fasciolosis may induce similar signs (McGuinness and Leder 2014). Therefore, serological tests are frequently employed to demonstrate the presence of anti-*Toxocara* antibodies in serum, and, in cases of NT, in CSF. The most commonly used immunodiagnostic method is the enzyme-linked immunosorbent assay (ELISA) detecting IgG or IgE antibodies (Fillaux and Magnaval 2013). Several commercial ELISA kits, mostly detecting IgG, are available. However, the variability of positive cut-off values and the quality of used antigens, e.g. somatic antigens from crude *T. canis* larvae as well as native or recombinant TES antigens, result in inconsistent data. Furthermore, false-positive results are common, especially in areas of polyparasitism, where other parasite infections can cross-react in tests for *Toxocara* (McGuinness and Leder 2014). Therefore, ELISA results should be confirmed by immunoblotting (Western Blot) to exclude cross-reactivity with other pathogens (Fillaux and Magnaval 2013; McGuinness and Leder 2014). The Western Blot is often based on fractionated, native antigens and has a higher sensitivity and specificity than the ELISA (Fillaux and Magnaval 2013; Roldan and Espinoza 2009; Zarnowska and Jastrzebska 1994). In summary, ELISAs provide fast and relatively inexpensive results, whereas the Western Blot is cost- and labour-intensive but enhances sensitivity and specificity, especially for values close to the cut-off.

In addition, imaging techniques such as ultrasonography, computed tomography (CT) and magnetic resonance imaging (MRI) may be useful to visualise granulomas in different tissues, supporting the diagnosis in cases of suspected toxocarosis. Larval migration in the hepatopulmonary phase may induce perihepatic lymphadenopathy and hepatomegaly with small hypoechoic diffuse and ill-marginated ovaloid focal liver lesions (Baldisserotto et al. 1999; Lim 2008; Sakai et al. 2006). In CT, pulmonary lesions appear as multifocal subpleural nodules with halos or ground glass opacities and ill-defined margins (Sakai et al. 2006).

In OLM, the use of transducers with subsequent optical coherence tomography, fluorescein angiography, CT or ocular ultrasonography can indicate the location of larvae as well as pathological alterations such as peripheral or posterior pole granulomas and chronic endophthalmitis (McGuinness and Leder 2014).

CT and MRI are suitable to detect and localise lesions caused by migrating *Toxocara* larvae in the CNS. These include, amongst others, diffuse cortical, subcortical or white matter lesions with contrast-enhancing hyperintense areas (Abdel Razek et al. 2011; Finsterer and Auer 2007; Salvador et al. 2010), microhaemorrhages or cortical necrosis (Helsen et al. 2011; Jabbour et al. 2011; Xinou et al. 2003). Larvae in the spinal cord result in swelling and enlargement of the involved spinal segment, and solitary or multiple hyperintense lesions with focal nodular enhancement may be observed after gadolinium injection (Sanchez et al. 2018).

Direct microscopic diagnosis of *Toxocara* larvae by histopathological examination of biopsy specimens is possible and has led to a conclusive diagnosis in some cases (Sanchez et al. 2018), but this is not a routinely used approach as tissue sampling is invasive and the probability of obtaining a tissue sample containing larvae is low (Fillaux and Magnaval 2013). Similarly, a PCR approach by sequencing ITS-1 and ITS-2 sequences of nuclear *Toxocara* rDNA, even if it is commonly conducted in research, is not used in clinical practice (Fillaux and Magnaval 2013).

2.7 Treatment

Even in cases of mild *Toxocara* infections, chemotherapy with specific anthelmintic drugs is highly recommended, particularly to prevent the invasion of *Toxocara* larvae into the brain or eyes (Pawlowski 2001). A number of experiments have been conducted in laboratory animals to test the efficacy of anthelmintic substances, including diethylcarbamazine, ivermectin, levamisole, nitazoxanide and benzimidazoles, against migrating and tissue persisting *Toxocara* larvae. However, experiments showed variable and contradictory results due to the use of different animal models, time points of investigation, inoculation and treatment doses, the application form and duration of treatment as well as the evaluation method (Othman 2012). Thus, the obtained data are of limited use regarding human toxocarosis. Overall, there is neither a recommended treatment scheme nor a consensus about a specific therapy. However, albendazole, thiabendazole and mebendazole have been effectively administered to patients with toxocarosis. Albendazole appears to be the preferred substance due to high tolerability and good tissue penetration when metabolised (Caumes 2003). In contrast to mebendazole and diethylcarbamazine, albendazole has a greater ability to cross the blood–brain barrier, thus reaching neuroinvasive larvae during NT (Othman 2012). An overview of commonly used regimens in the treatment of toxocarosis is listed in Table 2.1.

It should be kept in mind that anthelmintic therapy may induce allergic reactions as disintegrating larvae release high amounts of antigens (Pawlowski 2001). Thus, corticosteroids and non-steroidal anti-inflammatory drugs should be administered during chemotherapy against *Toxocara* (Despommier 2003; Ma et al. 2018). Especially the treatment of OLM requires aggressive anti-inflammatory therapy with corticosteroids, e.g. prednisone or dexamethasone, combined with prolonged

Table 2.1 Common treatment regimens of human toxocarosis

Anthelmintic drug	Dose	Duration of treatment	Comment	References
Albendazole	5–15 mg/kg daily	5–14 days	Common treatment	Reviewed by Magnaval and Glickman (2006)
Thiabendazole	25–50 mg/kg daily	3–7 days	Not recommended; relatively high rate of side effects	Reviewed by Magnaval and Glickman (2006)
Mebendazole	20–25 mg/kg daily	7–21 days	Large variations in plasma concentrations	Magnaval (1995)
Diethylcarbamazine	1–4 mg/kg daily	21 days	Dose-dependent side effects	Pawlowski (2001) Magnaval (1995)
Ivermectin	12 mg	Single dose	Low efficacy	Magnaval (1998)

The chemotherapy of toxocarosis in humans is related to the severity and location of symptoms. A combined treatment with anti-inflammatory compounds is frequently implemented. This table only provides an overview and does not represent a treatment recommendation

albendazole treatment to prevent dangerous ocular inflammations (Barisani-Asenbauer et al. 2001).

2.8 Control and Prevention

Control and prevention of toxocarosis are complicated by high environmental contamination and the numerous transmission modes. To raise public awareness, the CDC has included toxocarosis in the list of the five neglected parasitic diseases, whose control and prevention is a priority for public health action (CDC 2020; Woodhall et al. 2014). Unfortunately, awareness regarding *Toxocara* is still insufficient in many communities. Thus, improving public knowledge on toxocarosis and its prevention through educational programmes is essential, including teaching of adequate hygienic standards, such as washing of hands after soil contact, washing of vegetables and thorough cooking of meats and offal (Fan et al. 2013; Lloyd et al. 1983; Moreira et al. 2014). However, the successful implementation of hygiene measures remains difficult in many parts of the world as, e.g. sanitary facilities are limited in certain regions.

A further approach is to substantially reduce the abundance of infective eggs in public environments. Interventions to prevent dog-mediated pollution include the removal of faeces by dog owners in public areas (Atenstaedt and Jones 2011). Preventing environmental contamination by wild carnivores as well as stray dogs entails considerable effort and costs, e.g. fencing of public areas or the periodic change of sand in playgrounds. Likewise, the restriction of dogs from playgrounds, public squares and gardens is considered to be useful, but the efficiency is questionable as some studies report similar densities of *Toxocara* eggs in- and outside of restricted areas (Kirchheimer and Jacobs 2008).

Cleaning of contaminated surfaces mainly plays a role in settings with high animal densities, e.g. animal shelters or breeding facilities. The efficacy of various disinfectants against *T. canis* eggs under laboratory conditions remains contradictory and seems to be dependent on exposure time and concentration. The use of common disinfectants under routine conditions with short contact times did not exhibit sufficient ovicidal effects (Ursache et al. 2019). Nevertheless, such disinfectants or anionic detergents may decrease the adherence of *Toxocara* eggs to surfaces, making them easier to remove. As not all surfaces or materials can be treated with chemicals, temperature is an appropriate alternative in removing contaminations. For example, pasteurisation for 20 min at 73 °C, temperatures of 60 °C for 1 h and steam sterilisation kill the developmental stages (Azam et al. 2012; van Knapen et al. 1979).

Practicing veterinarians play a pivotal role in raising awareness on the zoonotic potential of *Toxocara* spp. in the general population, especially in pet owners. They should aim to increase compliance with effective deworming regimens. The US American Companion Animal Parasite Council (CAPC) and the European Scientific Counsel for Companion Animal Parasites (ESCCAP) provide independent, research-based guidelines assisting both veterinarians and pet owners in the successful control of worm infections in dogs (CAPC 2016; ESCCAP 2017). Anthelmintics should only be administered following veterinary advice as local epidemiological circumstances and each pet's individual risk, e.g. due to hunting and raw meat diets, must be considered to define an appropriate treatment regimen (ESCCAP 2017).

In addition to companion animals, populations of stray or feral dogs also provide a continuous reservoir of infection and may be an important source of environmental egg contamination (Morgan et al. 2013). In some countries, control programmes have been developed to reduce stray and feral dog populations by implementing measures to reduce reproduction (Macpherson 2013). Other programmes focus on combating zoonotic agents such as *Leishmania donovani*, *Echinococcus* spp. and rabies by therapeutic treatment (Macpherson 2013). However, control of *T. canis* in wild animals such as foxes is challenging and currently not possible on a large scale, as several treatments per year would be necessary (Holland 2017; Morgan et al. 2013). The development of effective strategies in Europe is urgently required as fox populations continuously expand towards urban settlements, thereby possibly increasing the risk for human infection (Deplazes et al. 2011).

2.9 Conclusions

Toxocarosis is one of the most prevalent zoonotic helminthoses worldwide, due to the ubiquitous distribution of *T. canis* and its definitive hosts living in close proximity to humans, in combination with multiple transmission routes and a high tenacity of eggs in the environment. A wide variety of risk factors, including contact with contaminated environments, poor sanitation and hygiene as well as lack of education, have been reported. On this account, the greatest disease burden is linked to socioeconomically disadvantaged populations, and especially to children. Although asymptomatic in many cases, severe and detrimental symptoms such as pulmonary fibrosis, visual impairments, meningitis and behavioural and cognitive disorders may result from *T. canis* infection. As many symptoms are non-specific, a reliable diagnosis should be based on the sum of occurring symptoms, laboratory and serological tests as well as anamnestic information regarding risk factor exposure.

Although treatment of the different disease forms in humans is possible, control efforts should focus on the prevention of transmission to humans. Important aspects are to raise global awareness on risk factors for infection and promoting individual hygiene as well as implementing control programmes to reduce prevalence in dogs. Here, veterinarians play an important role in advising pet owners on effective anthelmintic treatment regimes.

Further experimental investigations of molecular mechanisms regarding parasite–host interactions and pathogenesis of human toxocarosis are crucial to broadening the fundamental knowledge on *T. canis* and toxocarosis, which may provide a basis for the design of novel diagnostic and therapeutic approaches.

References

Abdel Razek AA, Watcharakorn A, Castillo M (2011) Parasitic diseases of the central nervous system. Neuroimaging Clin N Am 21(4):815–841

Abdybekova AM, Torgerson PR (2012) Frequency distributions of helminths of wolves in Kazakhstan. Vet Parasitol 184(2–4):348–351

Abougrain AK, Nahaisi MH, Madi NS, Saied MM, Ghenghesh KS (2010) Parasitological contamination in salad vegetables in Tripoli-Libya. Food Control 21(5):760–762

Adanir R, Tasci F (2013) Prevalence of helminth eggs in raw vegetables consumed in Burdur, Turkey. Food Control 31(2):482–484

Araujo P (1972) Findings related to the 1st ecdysis of *Ascaris lumbricoides*, *A. suum* and *Toxocara canis* larvae. Rev Inst Med Trop Sao Paulo 14(2):83–90

Ashton N (1960) Larval granulomatosis of the retina due to *Toxocara*. Br J Ophthalmol 44:129–148

Atenstaedt RL, Jones S (2011) Interventions to prevent dog fouling: a systematic review of the evidence. Public Health 125(2):90–92

Azam D, Ukpai OM, Said A, Abd-Allah GA, Morgan ER (2012) Temperature and the development and survival of infective *Toxocara canis* larvae. Parasitol Res 110(2):649–656

Badley JE, Grieve RB, Rockey JH, Glickman LT (1987) Immune-mediated adherence of eosinophils to *Toxocara canis* infective larvae: the role of excretory-secretory antigens. Parasite Immunol 9(1):133–143

Baldisserotto M, Conchin CF, Soares Mda G, Araujo MA, Kramer B (1999) Ultrasound findings in children with toxocariasis: report on 18 cases. Pediatr Radiol 29(5):316–319

Barisani-Asenbauer T, Maca SM, Hauff W, Kaminski SL, Domanovits H, Theyer I, Auer H (2001) Treatment of ocular toxocariasis with albendazole. J Ocul Pharmacol Ther 17(3):287–294

Barutzki D, Schaper R (2011) Results of parasitological examinations of faecal samples from cats and dogs in Germany between 2003 and 2010. Parasitol Res 109(Suppl 1):S45–S60

Beautyman W, Woolf AL (1951) An ascaris larva in the brain in association with acute anterior poliomyelitis. J Pathol Bacteriol 63(4):635–647

Beaver PC, Snyder CH, Carrera GM, Dent JH, Lafferty JW (1952) Chronic eosinophilia due to visceral *larva migrans*; report of three cases. Pediatrics 9(1):7–19

Bindke JD, Springer A, Janecek-Erfurth E, Boer M, Strube C (2019) Helminth infections of wild European gray wolves (*Canis lupus* Linnaeus, 1758) in Lower Saxony, Germany, and comparison to captive wolves. Parasitol Res 118(2):701–706

Brown GD, Willment JA, Whitehead L (2018) C-type lectins in immunity and homeostasis. Nat Rev Immunol 18(6):374–389

Brunaska M, Dubinsky P, Reiterova K (1995) *Toxocara canis*: ultrastructural aspects of larval moulting in the maturing eggs. Int J Parasitol 25(6):683–690

Burke TM, Roberson EL (1985a) Prenatal and lactational transmission of *Toxocara canis* and *Ancylostoma caninum*: experimental infection of the bitch at midpregnancy and at parturition. Int J Parasitol 15(5):485–490

Burke TM, Roberson EL (1985b) Prenatal and lactational transmission of *Toxocara canis* and *Ancylostoma caninum*: experimental infection of the bitch before pregnancy. Int J Parasitol 15 (1):71–75

CAPC (2016) CAPC guidelines, ascarid. https://capcvet.org/guidelines/ascarid. Accessed 06 Feb 2020

Carvalho EA, Rocha RL (2011) Toxocariasis: visceral *larva migrans* in children. J Pediatr 87 (2):100–110

Caumes E (2003) Treatment of cutaneous *larva migrans* and *Toxocara* infection. Fundam Clin Pharmacol 17(2):213–216

CDC (2020) Parasites – neglected parasitic infections (NPIs). https://www.cdc.gov/parasites/npi/index.html. Accessed 06 Feb 2020

Choi D, Lim JH, Choi DC, Paik SW, Kim SH, Huh S (2008) Toxocariasis and ingestion of raw cow liver in patients with eosinophilia. Korean J Parasitol 46(3):139–143

Claerebout E, Casaert S, Dalemans AC, De Wilde N, Levecke B, Vercruysse J, Geurden T (2009) Giardia and other intestinal parasites in different dog populations in Northern Belgium. Vet Parasitol 161(1–2):41–46

Cong W, Zhang XX, Zhou N, Yu CZ, Chen J, Wang XY, Li B, Qian AD, Zhu XQ (2014) *Toxocara* seroprevalence among clinically healthy individuals, pregnant women and psychiatric patients and associated risk factors in Shandong Province, Eastern China. PLoS Negl Trop Dis 8(8): e3082

Cooper PJ (2008) *Toxocara canis* infection: an important and neglected environmental risk factor for asthma? Clin Exp Allergy 38(4):551–553

Cox DM, Holland CV (2001) Relationship between three intensity levels of *Toxocara canis* larvae in the brain and effects on exploration, anxiety, learning and memory in the murine host. J Helminthol 75(1):33–41

da Silva MB, Urrego AJ, Oviedo Y, Cooper PJ, Pacheco LGC, Pinheiro CS, Ferreira F, Briza P, Alcantara-Neves NM (2018) The somatic proteins of *Toxocara canis* larvae and excretory-secretory products revealed by proteomics. Vet Parasitol 259:25–34

Del Prete GF, De Carli M, Mastromauro C, Biagiotti R, Macchia D, Falagiani P, Ricci M, Romagnani S (1991) Purified protein derivative of *Mycobacterium tuberculosis* and excretory-secretory antigen(s) of *Toxocara canis* expand in vitro human T cells with stable and opposite (type 1 T helper or type 2 T helper) profile of cytokine production. J Clin Invest 88 (1):346–350

Demirci M, Yildirim M, Aridogan BC, Baysal V, Korkmaz M (2003) Tissue parasites in patients with chronic urticaria. J Dermatol 30(11):777–781

Dent LA, Daly CM, Mayrhofer G, Zimmerman T, Hallett A, Bignold LP, Creaney J, Parsons JC (1999) Interleukin-5 transgenic mice show enhanced resistance to primary infections with *Nippostrongylus brasiliensis* but not primary infections with *Toxocara canis*. Infect Immun 67(2):989–993

Deplazes P, van Knapen F, Schweiger A, Overgaauw PA (2011) Role of pet dogs and cats in the transmission of helminthic zoonoses in Europe, with a focus on echinococcosis and toxocarosis. Vet Parasitol 182(1):41–53

Despommier D (2003) Toxocariasis: clinical aspects, epidemiology, medical ecology, and molecular aspects. Clin Microbiol Rev 16(2):265–272

Deutz A, Fuchs K, Auer H, Kerbl U, Aspock H, Kofer J (2005) *Toxocara*-infestations in Austria: a study on the risk of infection of farmers, slaughterhouse staff, hunters and veterinarians. Parasitol Res 97(5):390–394

El-Sayed NM, Ismail KA (2012) Relationship between *Toxocara canis* infection and schizophrenia. Rawal Med J 37(02):155–161

ESCCAP (2017) ESCCAP guidelines, GL1: worm control in dogs and cats, 3rd edn. https://www.esccap.org/guidelines/. Accessed 06 Feb 2020

Etewa SE, Abdel-Rahman SA, Abd El-Aal NF, Fathy GM, El-Shafey MA, Ewis AM (2016) Geohelminths distribution as affected by soil properties, physicochemical factors and climate in Sharkyia governorate Egypt. J Parasit Dis 40(2):496–504

Fakhri Y, Gasser RB, Rostami A, Fan CK, Ghasemi SM, Javanian M, Bayani M, Armoon B, Moradi B (2018) *Toxocara* eggs in public places worldwide – a systematic review and meta-analysis. Environ Pollut 242(Pt B):1467–1475

Fallah AA, Pirali-Kheirabadi K, Shirvani F, Saei-Dehkordi SS (2012) Prevalence of parasitic contamination in vegetables used for raw consumption in Shahrekord, Iran: influence of season and washing procedure. Food Control 25(2):617–620

Fan CK, Liao CW, Cheng YC (2013) Factors affecting disease manifestation of toxocarosis in humans: genetics and environment. Vet Parasitol 193(4):342–352

Fan CK, Holland CV, Loxton K, Barghouth U (2015) Cerebral toxocariasis: silent progression to neurodegenerative disorders? Clin Microbiol Rev 28(3):663–686

Fattah DI, Maizels RM, McLaren DJ, Spry CJ (1986) *Toxocara canis*: interaction of human blood eosinophils with the infective larvae. Exp Parasitol 61(3):421–431

Filbey KJ, Varyani F, Harcus Y, Hewitson JP, Smyth DJ, McSorley HJ, Ivens A, Nylen S, Rottenberg M, Loser S, Maizels RM (2019) Macrophage migration inhibitory factor (MIF) is essential for type 2 effector cell immunity to an intestinal helminth parasite. Front Immunol 10:2375

Fillaux J, Magnaval JF (2013) Laboratory diagnosis of human toxocariasis. Vet Parasitol 193 (4):327–336

Finsterer J, Auer H (2007) Neurotoxocarosis. Rev Inst Med Trop Sao Paulo 49(5):279–287

Gavignet B, Piarroux R, Aubin F, Millon L, Humbert P (2008) Cutaneous manifestations of human toxocariasis. J Am Acad Dermatol 59(6):1031–1042

Gillespie SH (1987) Human toxocariasis. J Appl Bacteriol 63(6):473–479

Glickman LT, Schantz PM (1981) Epidemiology and pathogenesis of zoonotic toxocariasis. Epidemiol Rev 3:230–250

Glickman LT, Magnaval JF, Domanski LM, Shofer FS, Lauria SS, Gottstein B, Brochier B (1987) Visceral *larva migrans* in French adults: a new disease syndrome? Am J Epidemiol 125 (6):1019–1034

Greve JH (1971) Age resistance to *Toxocara canis* in ascarid-free dogs. Am J Vet Res 32 (8):1185–1192

Ha KH, Song JE, Kim BS, Lee CH (2016) Clinical characteristics and progression of liver abscess caused by *Toxocara*. World J Hepatol 8(18):757–761

Helsen G, Vandecasteele SJ, Vanopdenbosch LJ (2011) Toxocariasis presenting as encephalomyelitis. Case Rep Med 2011:503913

Hermosilla C, Kleinertz S, Silva LM, Hirzmann J, Huber D, Kusak J, Taubert A (2017) Protozoan and helminth parasite fauna of free-living Croatian wild wolves (*Canis lupus*) analyzed by scat collection. Vet Parasitol 233:14–19

Heuer L, Beyerbach M, Lühder F, Beineke A, Strube C (2015) Neurotoxocarosis alters myelin protein gene transcription and expression. Parasitol Res 114(6):2175–2186

Hoeppli R, Feng L, Li F (1949) Histological reactions in the liver of mice due to larvae of different Ascaris species. Peking Nat Hist Bull 18(2):119–131

Hoffmeister B, Glaeser S, Flick H, Pornschlegel S, Suttorp N, Bergmann F (2007) Cerebral toxocariasis after consumption of raw duck liver. Am J Trop Med Hyg 76(3):600–602

Holland CV (2017) Knowledge gaps in the epidemiology of *Toxocara*: the enigma remains. Parasitology 144(1):81–94

Holland CV, Hamilton CM (2013) The significance of cerebral toxocariasis: a model system for exploring the link between brain involvement, behaviour and the immune response. J Exp Biol 216(Pt 1):78–83

Hotez PJ, Wilkins PP (2009) Toxocariasis: America's most common neglected infection of poverty and a helminthiasis of global importance? PLoS Negl Trop Dis 3(3):e400

Ilic T, Becskei Z, Petrovic T, Polacek V, Ristic B, Milic S, Stepanovic P, Radisavljevic K, Dimitrijevic S (2016) Endoparasitic fauna of red foxes (*Vulpes vulpes*) and golden jackals (*Canis aureus*) in Serbia. Acta Parasitol 61(2):389–396

Jabbour RA, Kanj SS, Sawaya RA, Awar GN, Hourani MH, Atweh SF (2011) *Toxocara canis* myelitis: clinical features, magnetic resonance imaging (MRI) findings, and treatment outcome in 17 patients. Medicine 90(5):337–343

Janecek E, Waindok P, Bankstahl M, Strube C (2017) Abnormal neurobehaviour and impaired memory function as a consequence of *Toxocara canis*- as well as *Toxocara cati*-induced neurotoxocarosis. PLoS Negl Trop Dis 11(5):e0005594

Jarosz W, Mizgajska-Wiktor H, Kirwan P, Konarski J, Rychlicki W, Wawrzyniak G (2010) Developmental age, physical fitness and *Toxocara* seroprevalence amongst lower-secondary students living in rural areas contaminated with *Toxocara* eggs. Parasitology 137(1):53–63

Kayes SG (1997) Human toxocariasis and the visceral larva migrans syndrome: correlative immunopathology. Chem Immunol 66:99–124

Kazmierowski JA, Chusid MJ, Parrillo JE, Fauci AS, Wolff SM (1978) Dermatologic manifestations of the hypereosinophilic syndrome. Arch Dermatol 114(4):531–535

Keegan JD, Holland CV (2013) A comparison of *Toxocara canis* embryonation under controlled conditions in soil and hair. J Helminthol 87(1):78–84

Kirchheimer R, Jacobs DE (2008) *Toxocara* species egg contamination of soil from children's play areas in southern England. Vet Rec 163(13):394–395

Klapec T, Borecka A (2012) Contamination of vegetables, fruits and soil with geohelmints eggs on organic farms in Poland. Ann Agric Environ Med 19(3):421–425

Klion AD, Nutman TB (2004) The role of eosinophils in host defense against helminth parasites. J Allergy Clin Immunol 113(1):30–37

Koutz FR, Groves HF, Scothorn MW (1966) The prenatal migration of *Toxocara canis* larvae and their relationship to infection in pregnant bitches and in pups. Am J Vet Res 27(118):789–795

Lee KH, Kim TJ, Lee KW (2015) Pulmonary toxocariasis: initial and follow-up CT findings in 63 patients. Am J Roentgenol 204(6):1203–1211

Li L, Gao W, Yang X, Wu D, Bi H, Zhang S, Huang M, Yao X (2014) Asthma and toxocariasis. Ann Allergy Asthma Immunol 113(2):187–192

Liberato C, Berrilli F, Odorizi L, Scarcella R, Barni M, Amoruso C, Scarito A, Filippo MMD, Carvelli A, Iacoponi F, Scaramozzino P (2018) Parasites in stray dogs from Italy: prevalence, risk factors and management concerns. Acta Parasitol 63(1):27–32

Lim JH (2008) Toxocariasis of the liver: visceral *larva migrans*. Abdom Imaging 33(2):151–156

Lloyd S, Amerasinghe P, Soulsby E (1983) Periparturient immunosuppression in the bitch and its influence on infection with *Toxocara canis*. J Small Anim Pract 24(4):237–247

Loukas A, Mullin NP, Tetteh KK, Moens L, Maizels RM (1999) A novel C-type lectin secreted by a tissue-dwelling parasitic nematode. Curr Biol 9(15):825–828

Loukas A, Doedens A, Hintz M, Maizels RM (2000a) Identification of a new C-type lectin, TES-70, secreted by infective larvae of *Toxocara canis*, which binds to host ligands. Parasitology 121 (Pt 5):545–554

Loukas A, Hintz M, Linder D, Mullin NP, Parkinson J, Tetteh KK, Maizels RM (2000b) A family of secreted mucins from the parasitic nematode *Toxocara canis* bears diverse mucin domains but shares similar flanking six-cysteine repeat motifs. J Biol Chem 275(50):39600–39607

Ma G, Holland CV, Wang T, Hofmann A, Fan CK, Maizels RM, Hotez PJ, Gasser RB (2018) Human toxocariasis. Lancet Infect Dis 18(1):e14–e24

Macpherson CN (2013) The epidemiology and public health importance of toxocariasis: a zoonosis of global importance. Int J Parasitol 43(12–13):999–1008

Magnaval JF (1995) Comparative efficacy of diethylcarbamazine and mebendazole for the treatment of human toxocariasis. Parasitology 110(Pt 5):529–533

Magnaval JF (1998) Apparent weak efficacy of ivermectin for treatment of human toxocariasis. Antimicrob Agents Chemother 42(10):2770

Magnaval JF, Glickman LT (2006) Management and treatment options for human toxocariasis. In: Holland C, Smith HV (eds) *Toxocara*: the enigmatic parasite. CABI, Oxfordshire, pp 113–126

Magnaval JF, Glickman LT, Dorchies P, Morassin B (2001) Highlights of human toxocariasis. Korean J Parasitol 39(1):1–11

Maizels RM (2013) *Toxocara canis*: molecular basis of immune recognition and evasion. Vet Parasitol 193(4):365–374

Mazur-Melewska K, Jonczyk-Potoczna K, Kemnitz P, Mania A, Figlerowicz M, Sluzewski W (2015) Pulmonary presentation of *Toxocara* sp. infection in children. Pneumonol Alergol Pol 83 (4):250–255

McGuinness SL, Leder K (2014) Global burden of toxocariasis: a common neglected infection of poverty. Curr Trop Med Rep 1(1):52–61

McMichael AJ (2000) The urban environment and health in a world of increasing globalization: issues for developing countries. Bull World Health Organ 78(9):1117–1126

Montgomery MA, Elimelech M (2007) Water and sanitation in developing countries: including health in the equation. Environ Sci Technol 41(1):17–24

Moreira GM, Telmo Pde L, Mendonca M, Moreira AN, McBride AJ, Scaini CJ, Conceicao FR (2014) Human toxocariasis: current advances in diagnostics, treatment, and interventions. Trends Parasitol 30(9):456–464

Morgan ER, Azam D, Pegler K (2013) Quantifying sources of environmental contamination with *Toxocara* spp. eggs. Vet Parasitol 193(4):390–397

Musso C, Castelo JS, Tsanaclis AM, Pereira FE (2007) Prevalence of *Toxocara*-induced liver granulomas, detected by immunohistochemistry, in a series of autopsies at a Children's Reference Hospital in Vitoria, ES, Brazil. Virchows Arch 450(4):411–417

Nijsse R, Ploeger HW, Wagenaar JA, Mughini-Gras L (2015) *Toxocara canis* in household dogs: prevalence, risk factors and owners' attitude towards deworming. Parasitol Res 114(2):561–569

Noh Y, Hong ST, Yun JY, Park HK, Oh JH, Kim YE, Jeon BS (2012) Meningitis by *Toxocara canis* after ingestion of raw ostrich liver. J Korean Med Sci 27(9):1105–1108

Othman AA (2012) Therapeutic battle against larval toxocariasis: are we still far behind? Acta Trop 124(3):171–178

Othman AA, Abdel-Aleem GA, Saied EM, Mayah WW, Elatrash AM (2010) Biochemical and immunopathological changes in experimental neurotoxocariasis. Mol Biochem Parasitol 172 (1):1–8

Overgaauw PA (1997a) Aspects of *Toxocara* epidemiology: human toxocarosis. Crit Rev Microbiol 23(3):215–231

Overgaauw PA (1997b) Aspects of *Toxocara* epidemiology: toxocarosis in dogs and cats. Crit Rev Microbiol 23(3):233–251

Page AP, Rudin W, Fluri E, Blaxter ML, Maizels RM (1992a) *Toxocara canis*: a labile antigenic surface coat overlying the epicuticle of infective larvae. Exp Parasitol 75(1):72–86

Page AP, Rudin W, Maizels RM (1992b) Lectin binding to secretory structures, the cuticle and the surface coat of *Toxocara canis* infective larvae. Parasitology 105(Pt 2):285–296

Pawlowski Z (2001) Toxocariasis in humans: clinical expression and treatment dilemma. J Helminthol 75(4):299–305

Perri AR, Heinrich S, Gur-Arieh S, Saunders JJ (2017) Earliest evidence of *Toxocara* sp. in a 1.2-million-yr-old extinct hyena (*Pachycrocuta brevirostris*) coprolite from Northwest Pakistan. J Parasitol 103(1):138–141

Pinelli E, Brandes S, Dormans J, Gremmer E, van Loveren H (2008) Infection with the roundworm *Toxocara canis* leads to exacerbation of experimental allergic airway inflammation. Clin Exp Allergy 38(4):649–658

Poulle ML, Bastien M, Richard Y, Josse-Dupuis E, Aubert D, Villena I, Knapp J (2017) Detection of *Echinococcus multilocularis* and other foodborne parasites in fox, cat and dog faeces collected in kitchen gardens in a highly endemic area for alveolar echinococcosis. Parasite 24:29

Quattrocchi G, Nicoletti A, Marin B, Bruno E, Druet-Cabanac M, Preux PM (2012) Toxocariasis and epilepsy: systematic review and meta-analysis. PLoS Negl Trop Dis 6(8):e1775

Richartz E, Buchkremer G (2002) Cerebral toxocariasis: a rare cause of cognitive disorders. A contribution to differential dementia diagnosis. Nervenarzt 73(5):458–462

Rocha S, Pinto RM, Floriano AP, Teixeira LH, Bassili B, Martinez A, Costa SO, Caseiro MM (2011) Environmental analyses of the parasitic profile found in the sandy soil from the Santos municipality beaches, SP, Brazil. Rev Inst Med Trop Sao Paulo 53(5):277–281

Roldan WH, Espinoza YA (2009) Evaluation of an enzyme-linked immunoelectrotransfer blot test for the confirmatory serodiagnosis of human toxocariasis. Mem Inst Oswaldo Cruz 104 (3):411–418

Rubinsky-Elefant G, Hirata CE, Yamamoto JH, Ferreira MU (2010) Human toxocariasis: diagnosis, worldwide seroprevalences and clinical expression of the systemic and ocular forms. Ann Trop Med Parasitol 104(1):3–23

Sabrosa NA, de Souza EC (2001) Nematode infections of the eye: toxocariasis and diffuse unilateral subacute neuroretinitis. Curr Opin Ophthalmol 12(6):450–454

Sakai S, Shida Y, Takahashi N, Yabuuchi H, Soeda H, Okafuji T, Hatakenaka M, Honda H (2006) Pulmonary lesions associated with visceral *larva migrans* due to *Ascaris suum* or *Toxocara canis*: imaging of six cases. Am J Roentgenol 186(6):1697–1702

Salvador S, Ribeiro R, Winckler MI, Ohlweiler L, Riesgo R (2010) Pediatric neurotoxocariasis with concomitant cerebral, cerebellar, and peripheral nervous system involvement: case report and review of the literature. J Pediatr 86(6):531–534

Sanchez SS, Garcia HH, Nicoletti A (2018) Clinical and magnetic resonance imaging findings of neurotoxocariasis. Front Neurol 9:53

Schabussova I, Amer H, van Die I, Kosma P, Maizels RM (2007) O-methylated glycans from *Toxocara* are specific targets for antibody binding in human and animal infections. Int J Parasitol 37(1):97–109

Schmitt E, Gehrmann M, Brunet M, Multhoff G, Garrido C (2007) Intracellular and extracellular functions of heat shock proteins: repercussions in cancer therapy. J Leukoc Biol 81(1):15–27

Schnieder T, Laabs EM, Welz C (2011) Larval development of *Toxocara canis* in dogs. Vet Parasitol 175(3–4):193–206

Scothorn MW, Koutz FR, Groves HF (1965) Prenatal *Toxocara canis* infection in pups. J Am Vet Med Assoc 146:45–48

Sperotto RL, Kremer FS, Aires Berne ME, Costa de Avila LF, da Silva PL, Monteiro KM, Caumo KS, Ferreira HB, Berne N, Borsuk S (2017) Proteomic analysis of *Toxocara canis* excretory and secretory (TES) proteins. Mol Biochem Parasitol 211:39–47

Sprent JF (1952) On the migratory behavior of the larvae of various *Ascaris* species in white mice. I Distribution of larvae in tissues. J Infect Dis 90(2):165–176

Sprent JF (1958) Observations on the development of *Toxocara canis* (Werner, 1782) in the dog. Parasitology 48(1–2):184–209

Springer A, Heuer L, Janecek-Erfurth E, Beineke A, Strube C (2019) Histopathological characterization of *Toxocara canis*- and *T. cati*-induced neurotoxocarosis in the mouse model. Parasitol Res 118(9):2591–2600

Stewart JM, Cubillan LD, Cunningham ET Jr (2005) Prevalence, clinical features, and causes of vision loss among patients with ocular toxocariasis. Retina 25(8):1005–1013

Strube C, Heuer L, Janecek E (2013) *Toxocara* spp. infections in paratenic hosts. Vet Parasitol 193 (4):375–389

Sturchler D, Weiss N, Gassner M (1990) Transmission of toxocariasis. J Infect Dis 162(2):571

Szwabe K, Blaszkowska J (2017) Stray dogs and cats as potential sources of soil contamination with zoonotic parasites. Ann Agric Environ Med 24(1):39–43

Taira K, Saeed I, Permin A, Kapel CM (2004) Zoonotic risk of *Toxocara canis* infection through consumption of pig or poultry viscera. Vet Parasitol 121(1–2):115–124

Taylor MR (2001) The epidemiology of ocular toxocariasis. J Helminthol 75(2):109–118

Taylor MR, Keane CT, O'Connor P, Girdwood RW, Smith H (1987) Clinical features of covert toxocariasis. Scand J Infect Dis 19(6):693–696

Taylor MR, Keane CT, O'Connor P, Mulvihill E, Holland C (1988) The expanded spectrum of toxocaral disease. Lancet 1(8587):692–695

Traversa D, Frangipane di Regalbono A, Di Cesare A, La Torre F, Drake J, Pietrobelli M (2014) Environmental contamination by canine geohelminths. Parasit Vectors 7:67

Uga S, Matsuo J, Kimura D, Rai SK, Koshino Y, Igarashi K (2000) Differentiation of *Toxocara canis* and *T. cati* eggs by light and scanning electron microscopy. Vet Parasitol 92(4):287–294

Ursache AL, Mircean V, Dumitrache M, Andrei S, Stefanut L, Cozma V, Catana R, Cernea M (2019) Is routine disinfection efficient in preventing contamination with *Toxocara canis* eggs? J Helminthol 94:e60

van Knapen F, Franchimont JH, Otter GM (1979) Steam sterilisation of sandpits infected with *Toxocara eggs*. Br Med J 1(6174):1320

Walsh MG, Haseeb MA (2012) Reduced cognitive function in children with toxocariasis in a nationally representative sample of the United States. Int J Parasitol 42(13–14):1159–1163

Webster GA (1958a) On prenatal infection and the migration of *Toxocara canis* Werner, 1782 in dogs. Can J Zool 36(3):435–440

Webster GA (1958b) A report on *Toxocara canis* Werner, 1782. Can J Comp Med Vet Sci 22 (8):272–279

Wolfrom E, Chene G, Lejoly-Boisseau H, Beylot C, Geniaux M, Taieb A (1996) Chronic urticaria and *Toxocara canis* infection. A case-control study. Ann Dermatol Venereol 123(4):240–246

Woodhall DM, Eberhard ML, Parise ME (2014) Neglected parasitic infections in the United States: toxocariasis. Am J Trop Med Hyg 90(5):810–813

Xinou E, Lefkopoulos A, Gelagoti M, Drevelegas A, Diakou A, Milonas I, Dimitriadis AS (2003) CT and MR imaging findings in cerebral toxocaral disease. Am J Neuroradiol 24(4):714–718

Yoshikawa M, Koyama N, Hontsu S, Yamamoto Y, Mikasa K, Kimura H (2011) Lessons from eight cases of adult pulmonary toxocariasis: abridged republication. Respirology 16 (6):1014–1015

Zarnowska H, Jastrzebska M (1994) Excretory-secretory larval antigens of *Toxocara canis*: physico-chemical characteristics and specificity assayed by Western blot technique. Acta Parasitol 39(1)

Chapter 3
Canine Leishmaniosis

Gad Baneth

Abstract Canine leishmaniosis (CanL) is one of the most important and widespread zoonotic disease associated with the domestic dog as a reservoir. Canine *Leishmania infantum* infection presents frequently as a severe chronic clinical disease affecting the skin, hemolymphoid system, kidneys, and other internal organs, and associated with hyperglobulinemia, hypoalbuminemia, and anemia. High rates of subclinical infection are found in dogs living in endemic areas, however, the disease is also common in dogs imported into non-endemic countries. Both dogs with clinical and subclinical infections are infectious to sand flies that transmit the infection to humans and other animals. Infection can also be transmitted transplacentally, venereally, and through blood transfusions. Coinfections with tick-borne diseases have been shown to increase the risk of progression from infection to clinical disease and the risk of mortality. The diagnosis of the disease is done mainly by quantitative serology, aided by cytology of hemolymphoid organs and skin, and PCR. The currently recommended treatment of CanL includes long-term treatment with allopurinol combined for the first month with meglumine antimoniate or miltefosine. Treatment is aimed at decreasing the parasite load and allowing the dog's immune system to recover, and is associated, if successful, with clinical cure and substantial decrease in antileishmanial antibody levels and parasite load, however, sometimes not with the parasite's elimination. Treated dogs should be followed up by repeated serology, serum biochemistry, and urinalysis for more than 1 year after treatment initiation. A CanL staging system is helpful in evaluating the disease's severity, deciding on treatment, and forecasting the dog's prognosis. Prevention of the disease is done by the use of topical insecticides with pyrethroids and vaccination.

Keywords *Leishmania infantum* · Leishmaniosis · Zoonosis · Allopurinol · Meglumine antimoniate · Miltefosine · Exfoliative dermatitis · LeishVet

---

G. Baneth (✉)
Koret School of Veterinary Medicine, The Hebrew University of Jerusalem, Rehovot, Israel
e-mail: gad.baneth@mail.huji.ac.il

C. Strube, H. Mehlhorn (eds.), *Dog Parasites Endangering Human Health*, Parasitology Research Monographs 13,
https://doi.org/10.1007/978-3-030-53230-7_3

3.1 Introduction

The leishmaniases are a group of diseases caused by protozoa parasites of the genus *Leishmania* (order: Kinetoplastida, family: Trypanosmatidae). More than 30 species of *Leishmania* that infect mammals have been described. Most of them are zoonotic and transferred from animals to humans, and only a few are anthroponotic and transmitted from human to human. Phlebotomine sand flies are the main vectors of leishmaniosis and are responsible for the transmission of this disease. Three forms of the disease are reported in humans, cutaneous leishmaniosis affecting the skin, mucocutaneous leishmaniosis affecting the junctions between skin and mucous membranes mainly in the face, and visceral leishmaniosis affecting the visceral organs. Animals infected with the same species of *Leishmania* that affect humans are often infected subclinically, and if they manifest clinical disease, it may be different from the disease found in people due to the same species, for instance, *Leishmania infantum* causes visceral leishmaniosis in humans affecting the internal organs, and there is also a form which affects only the skin with no visceral manifestations. On the other hand, the same species in dogs causes disease that affects both the skin and the visceral organs in most clinical cases. Out of 200 countries and territories reporting to the world health organization (WHO), 98 were endemic for leishmaniosis in 2018. This includes 68 countries that are endemic for both visceral and cutaneous leishmaniosis, 9 countries that are endemic only for visceral leishmaniosis, and 21 countries that are endemic only for cutaneous leishmaniosis (WHO 2018a). An estimated 700,000–1 million new cases of human leishmaniosis occur annually, of these, an estimated 50,000–90,000 are new visceral leishmaniosis cases (WHO 2018b). The majority of these visceral leishmaniosis cases are caused by the anthroponotic species *Leishmania donovani* in India and East Africa, whereas the zoonotic *L. infantum*, with the dog as the main reservoir, is associated with most human visceral leishmaniosis cases in other areas of the world including South America, Southern Europe, North Africa, and the Middle East. Visceral leishmaniosis caused by *L. infantum* in the Mediterranean basin was traditionally predominantly a disease of young children and the name of the causative agent of this disease reflects the predilection to infants. Malnutrition has been recognized as a risk factor for infantile leishmaniosis, and may explain why this disease is more prevalent among children in poor countries as compared with affluent ones, despite high prevalence rates in the dog populations. Immune-compromised people and HIV+ patients are an important risk group for human visceral leishmaniosis in Southern Europe.

Dogs are infected with several species of *Leishmania* which may cause clinical disease including *L. infantum*, *L. braziliensis*, *L. mexicana*, *L. tropica*, *L. major*, and others (Dantas-Torres et al. 2012; Baneth et al. 2017). This chapter focuses on canine infection with the main species for which the dog is considered a reservoir, *L. infantum*, and the disease termed canine leishmaniosis (CanL) is referred to in this chapter as the disease inflicted by *L. infantum*.

CanL is a major zoonotic disease endemic in more than 70 countries. It is enzootic in Southern Europe, Northern Africa, the Middle East, Central Asia, China, South America, and has sporadically emerged also in the United States and Canada (Baneth et al. 2008; Solano-Gallego et al. 2011; Lun et al. 2015; Toepp et al. 2017; Ribeiro et al. 2018). The importation of infected dogs makes CanL an important concern also in non-endemic countries where it is a veterinary and public health problem (Schäfer et al. 2019). Dogs are the main reservoir for human visceral leishmaniosis and the disease is usually fatal if not treated in people. Female phlebotomine sand flies of several species are the vectors of *L. infantum* with *Phlebotomus* spp. serving as vectors in Europe, Asia, and Africa, and *Lutzomyia* spp. in South and Central America (Baneth et al. 2008).

3.2 Epidemiology

Leishmania infantum infection is mainly subclinical in dogs and also in immune-competent humans. Population studies in *Leishmania*-endemic areas have shown that a proportion of the canine population develops clinical disease, another fraction has persistent asymptomatic infection, while yet another fraction is resistant to the infection or intermittently resolves it without developing clinical disease. A study from the *Leishmania*-endemic island of Mallorca, Spain, found that about 13% of the dogs in a municipal pound had clinical signs of leishmaniosis, 26% were seropositive, and 63% were PCR-positive in at least one tissue evaluated (Solano-Gallego et al. 2001) (Fig. 3.1). Seroprevalence rates found in studies carried out in the Mediterranean basin range between 10 and 37% of the dogs in disease foci. Surveys employing the polymerase chain reaction (PCR) method for the detection of leishmanial DNA in canine tissues, or combining serology and DNA detection, have revealed even higher infection rates approaching 70% in some foci. Dogs with clinical signs as well as subclinically infected dogs are infectious to sand flies and may transmit infection (Molina et al. 1994; Borja et al. 2016; Rocha et al. 2020).

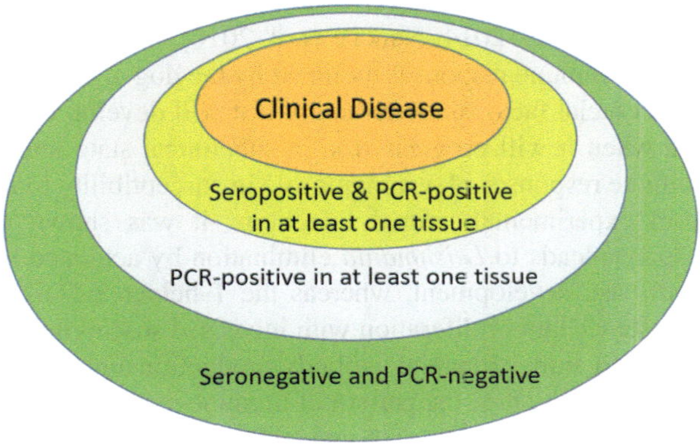

Fig. 3.1 Disease and infection in a canine population in a *Leishmania infantum endemic* area. Only a small part of the population displays clinical disease, while a larger proportion is subclinically infected and seropositive, and a yet larger part of the population is infected, PCR-positive in its tissues, and seronegative

Therefore, dogs with clinical disease are just the tip of the iceberg representing a much higher prevalence of infection in the population of canines in endemic foci of the disease where transmission takes place (Baneth et al. 2008). It has been estimated based on seroprevalence studies from Italy, Spain, France, and Portugal that 2.5 million dogs in these countries are infected (Moreno and Alvar 2002). The number of infected dogs in South America is also estimated in millions with high infection rates in some areas of Brazil, Paraguay, and Venezuela (Baneth et al. 2008; Marcondes and Day 2019).

3.3 Pathogenesis

Leishmania spp. have two life stages and complete their life cycle by development in two hosts, a female sand fly that harbors the flagellated extracellular promastigote stage, and a mammal where the intracellular amastigote stage develops. Dogs are infected by *L. infantum* promastigotes deposited in the skin during the bites of infected female sand fly vectors. The promastigotes invade host cells and replicate as intracellular amastigotes. The parasite disseminates in macrophages from the skin to the draining lymph node, and from there to the spleen, bone marrow, additional lymph nodes, and other internal organs. It then spreads again to the skin of the dog and is present in lesional as well as healthy looking dermal tissues (Ordeix et al. 2017). The disease incubation period prior to the appearance of clinical signs may last months to years, during which the parasite disseminates throughout the host's body (Foglia Manzillo et al. 2013).

Although natural transmission of *L. infantum* takes place by the bite of sand flies, the transmission of *L. infantum* through blood products has been reported in dogs that received blood transfusions from infected donors (Owens et al. 2001). Other proven non-vectorial modes of transmission include vertical transplacental transmission from dam to its offspring and venereal transmission (Silva et al. 2009; Boggiatto et al. 2011; Svobodova et al. 2017). Direct transmission from dog to dog by contact or bite without involvement of an hematophagous vector has been suspected in some cases of infection in areas where vectors of the disease are apparently absent (Karkamo et al. 2014; Naucke et al. 2016).

The immune responses mounted by the dog at the time of infection and thereafter are a crucial factor in determining if it will develop a lasting infection and whether and when it will progress from a subclinical state into a clinical disease. Specific immune responses play a major role in susceptibility to infection. Based on findings from experimental animal infections, it was shown that the T-helper 1 (Th1) response leads to *Leishmania* elimination by activated macrophages and resistance to disease development, whereas the T-helper 2 (Th2) response leads to parasite persistence and proliferation with increased susceptibility to disease associated with increased antileishmanial antibody production and hyperglobulinemia (Baneth et al. 2008). Despite this, the pattern of immune responses in infected dogs is a mixture of Th1 and Th2 responses without clear delineation between them, and it is thought that the imbalance between these responses is what tilts the infected animal toward resistance or predisposition to clinical disease (Baneth et al. 2008). During clinical disease, dogs become increasingly immune-suppressed and may develop decreased CD_4+ lymphocyte counts and a decrease in the CD_4+/CD_8+ ratio (Papadogiannakis et al. 2010). Moreover, it has been demonstrated that the infectiousness of dogs with leishmaniosis to sand flies increases with the decrease in CD_4+ counts (Guarga et al. 2000). Due to prolonged antigen exposure, dogs with chronic clinical disease develop a T cell exhaustion which is characterized by increased expression of cell surface receptors including the programmed death-1 (PD-1) receptor and decreased lymphocyte proliferation when stimulated with *L. infantum* antigen (Solano-Gallego et al. 2017; Toepp and Petersen 2020). The decreased proliferation of T lymphocytes in chronically diseased dogs is associated with a decrease in gamma interferon production, a Th1 cytokine that is imperative in the activation of macrophages to kill *Leishmania* amastigotes, and an increase in the levels of IL10, a regulatory cytokine which decreases the proliferation of Th1 lymphocytes and prevents production of reactive oxygen species in macrophages reducing their capability to kill parasites. During the disease process, B lymphocytes are activated via Th2 dominated responses to produce more antibodies as plasma cells, and these antibodies, in addition to causing hyperglobulinemia, also bind to parasite antigen and complement C3 to form circulating immune complexes (Toepp and Petersen 2020).

Immune-mediated mechanisms are responsible for much of the pathological findings in CanL. Circulating immune complexes and antinuclear antibodies have been detected in animals with CanL (Ginel et al. 2008; Parody et al. 2019). Immune complex glomerulonephritis associated with the deposition of immune complexes in the kidneys is a hallmark of the disease. Renal pathology is present, even if not manifested clinically, in the majority of dogs with this disease (Zatelli et al. 2003).

Resistance or susceptibility to CanL is influenced by the dog's genetics. The presence of overt CanL among Ibizan hounds in the Balearic Islands of Spain is rare and significantly lower than among other breeds, and it has been shown that this breed produces a predominantly cellular immune response against *L. infantum* infection (Solano-Gallego et al. 2000). Other breeds that originate from regions that are not enzootic for leishmaniosis such as the Boxer, Rottweiler, and German shepherd are overrepresented in CanL surveys (Miranda et al. 2008; Gharbi et al. 2018). Studies on the polymorphism of the canine Slc11a1 (NRAMP1) gene that encodes an iron transporter protein involved in the control of intraphagosomal replication of parasites and macrophage activation, have implied that susceptible dogs have mutations in this gene (Altet et al. 2002; Sanchez-Robert et al. 2008). A DLA class II DLA-DRB1 genotype, which is a dog major histocompatibility complex (MHC) class II allele, has also been linked to increased risk of being infected in an endemic area in Brazil (Quinnell et al. 2003).

Infection with another vector-borne infectious disease has been shown to affect the likelihood of developing CanL and the progression to clinical disease (Mekuzas et al. 2009; Toepp et al. 2019). In a study of 223 dogs from Natal in Brazil, dogs infected with three or more tick-borne diseases were 11 times more likely to be associated with progression to clinical disease than dogs with no tick-borne disease, and dogs with exposure to both *Leishmania* infection and tick-borne diseases were five times more likely to die (Toepp et al. 2019). A significant association was found between CanL and canine ehrlichiosis caused by *Ehrlichia canis* (Mekuzas et al. 2009; Attipa et al. 2018) and dogs with this coinfection may develop a higher skin *Leishmania* parasite load (Andrade et al. 2014).

3.4 Clinical Findings

The typical history reported by owners of dogs with CanL includes the appearance of skin abnormalities, ocular lesions, or epistaxis. These are frequently accompanied by lethargy and weight loss. Dogs of all breeds and both sexes are infected with leishmaniosis and the age distribution of the disease is bimodal with a peak of prevalence at 2–4 years and a secondary peak from the age of 7 years (Miranda et al. 2008). On physical examination, the main clinical signs associated with CanL are dermal lesions, lymphadenomegaly, splenomegaly, abnormal nails growth (onychogryphosis), muscle atrophy, and poor body condition. Additional findings may include epistaxis, renal failure, decreased appetite, tongue lesions (Fig. 3.2), polyuria and polydipsia, vomiting, melena, and lameness. Fever has only been described in about 20% of cases or less as CanL usually presents as a chronic disease (Meléndez-Lazo et al. 2018). 16–80% of the dogs with clinical leishmaniosis have ocular or periocular lesions including keratoconjunctivitis and uveitis (Solano-Gallego et al. 2011).

About 78% of the dogs with clinical disease due to CanL are admitted for veterinary care with skin abnormalities (Meléndez-Lazo et al. 2018). The dermal lesions associated with CanL include exfoliative dermatitis which can be generalized

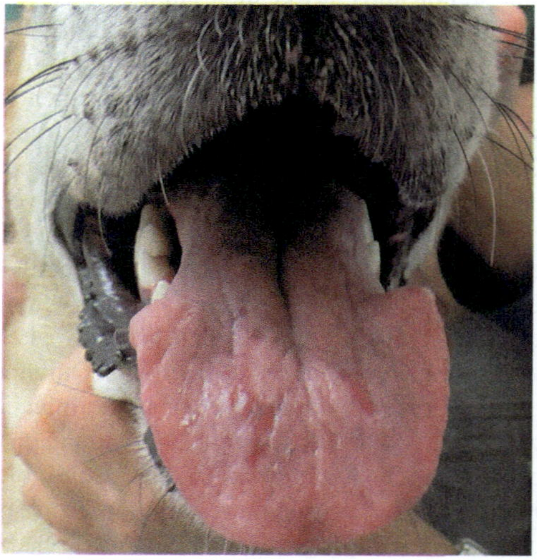

Fig. 3.2 Indented lesions on the surface of the tongue of a dog with CanL

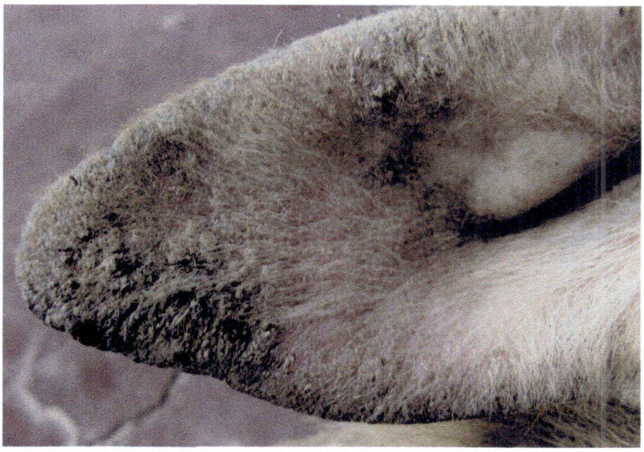

Fig. 3.3 Exfoliative dermatitis affecting the ear pinna of a dog with CanL

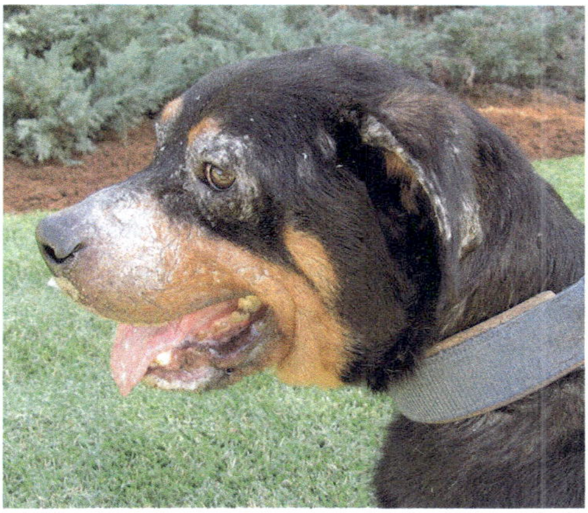

Fig. 3.4 A typical presentation of CanL in a Rottweiler with exfoliative dermatitis on the face, around the eyes, and on the ears

or localized over the face, ears, and limbs (Figs. 3.3 and 3.4). Deep cutaneous ulceration is frequently found with bleeding from the pinna, nose, or other local ulceration sites (Fig. 3.5). A form of nodular dermatitis has been reported, especially in boxers, and a mild form of papular dermatitis involving small skin papules in otherwise apparently healthy dogs has been described (Solano-Gallego et al. 2011; Lombardo et al. 2014).

Despite the typical picture of CanL associated with dermal abnormalities, epistaxis, lymphadenomegaly, ocular abnormalities, neurological abnormalities, or renal failure, may be the only presenting clinical findings in CanL without obvious skin disease, and therefore, this disease should be considered among the differential diagnoses for these conditions.

About 63% of the dogs admitted for veterinary care due to CanL are anemic, usually with mild-to-moderate non-regenerative anemia, and 25% have lymphopenia (Meléndez-Lazo et al. 2018). The most common serum biochemistry findings in dogs with clinical CanL are serum hyperproteinemia with

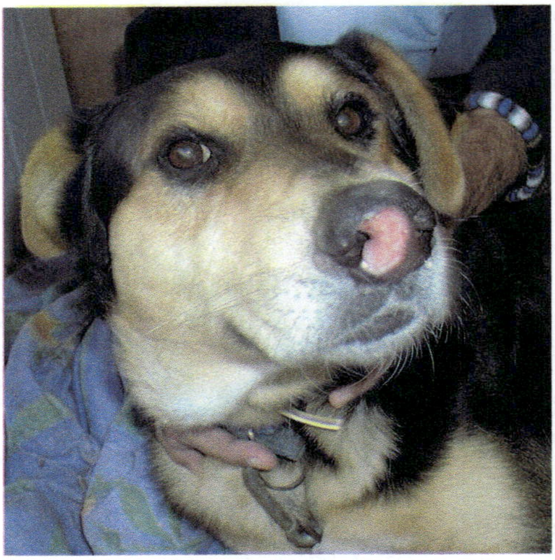

Fig. 3.5 Deep ulceration of the nasal planum in a dog with CanL

hyperglobulinemia and hypoalbuminemia, causing a decreased albumin/globulin ratio (Baneth et al. 2008; Meléndez-Lazo et al. 2018). Acute-phase proteins such as C-reactive protein (CRP) and ferritin increase in CanL whereas the negative acute-phase protein paraoxonase 1 (PON1) decreases in clinical disease. The levels of these acute-phase proteins may be used to assess the severity of CanL and follow-up response to treatment (Ceron et al. 2018). Grossly elevated activities of liver enzymes or azotemia are found in only a minority of dogs with CanL. However, proteinuria is frequently present in dogs with CanL and renal failure with azotemia due to immune-complex glomerulonephritis eventually develops and is considered the main natural cause of death in dogs with CanL (Koutinas and Koutinas 2014). Marked hyperglobulinemia with no apparent cause should also be investigated for CanL in dogs from endemic regions for CanL, or dogs that have travelled to such areas.

3.5 Diagnosis

Leishmaniosis is an example of an infection which does not equal clinical disease due to the high prevalence of subclinical infection. This makes CanL a diagnostic challenge for veterinary practitioners, clinical pathologists, and public health officials in endemic countries as well as in non-endemic regions where imported infection is a concern. There are several tests that can be used to diagnose this infection, and each of them has advantages and shortcomings. The diagnosis of CanL is performed for several indications. The most common indication is the confirmation of disease in a dog with compatible clinical signs. Other indications for diagnosis of infection include: evaluation of blood donors to prevent

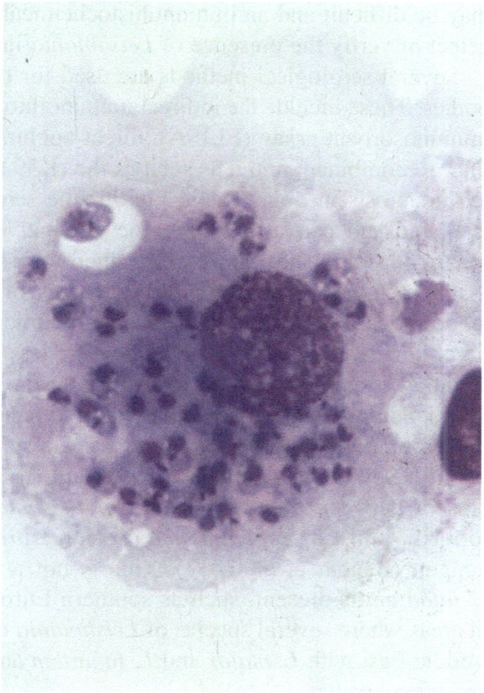

Fig. 3.6 *Leishmania infantum* amastigotes in the cytoplasm of a canine macrophage from a lymph node aspirate. Note the basophilic staining nucleus and smaller kinetoplast in the amastigotes

transmission of infection by blood products, testing of dogs before importation or exportation, epidemiological surveys in endemic areas, monitoring the health status of dogs that live in endemic areas or have travelled to them, and monitoring of dogs that have been diagnosed for CanL and are being treated against it to follow their response and progress.

Cytology of tissues for the detection of *Leishmania* amastigotes is a simple test to assess needle aspirates from the bone marrow, spleen, lymph nodes or skin, or touch impressions from the skin and other tissues. These are stained by Romanoswsky-type stains such as May Grunwald-Giemsa or a quick commercial stain, and viewed by light microscopy. Amastigotes are detected in the cytoplasm of macrophages and more rarely in neutrophils, and may also be found free as a result of cell damage in the process of smear preparation. Amastigotes found in macrophages are about 1–4 μm long by 1–2 μm wide and contain a basophilic staining nucleus and a prominent basophilic staining kinetoplast (Fig. 3.6). However, detection of amastigotes by cytology is frequently unrewarding due to the possibility of a low number of detectable parasites present even in dogs with a full-blown clinical disease (Moreira et al. 2007). In addition, other inclusions in macrophage cytoplasm may look like amastigotes and therefore it is imperative to detect a nucleus and a kinetoplast in the suspected amastigotes. *Leishmania* parasites may also be viewed in histopathologic formalin-fixed, paraffin-embedded biopsy sections of the skin or other infected organs. Definite identification of parasites within tissue macrophages may be difficult and an immunohistochemical staining method can be employed to detect or verify the presence of *Leishmania* in the tissue (Paltrinieri et al. 2016).

Several serological methods are used for the detection of anti-*Leishmania* antibodies. These include the indirect immunofluorescence assays (IFA), enzyme-linked immunosorbent assay (ELISA), direct agglutination test (DAT), and western blotting. Recombinant antigens such as the rK39 are also used for detection of visceral leishmaniosis in dogs. Rapid kits for the serological evaluation of CanL are also available and can usually provide qualitative positive or negative results. Good sensitivities and specificities are gained with quantitative serological methods such as the IFA and ELISA for the diagnosis of clinical CanL cases in experienced laboratories. However, while dogs with clinical disease are almost always seropositive, dogs with subclinical infection are less frequently seropositive and therefore serology is not the optimal assay for testing subclinical infection (Mettler et al. 2005). Serological cross-reactivity with trypanosomatids such as *Trypanosoma cruzi* is a problem in areas where canine trypanosomiasis is common, especially in some areas of South and Central America. Furthermore, serologic cross-reactivity is also present between different species of *Leishmania* that infect dogs, and antibodies formed against *L. tropica*, *L. major*, and *L. braziliensis* cross-react with *L. infantum* antigen (Baneth et al. 2017). This is not a major problem in areas where only *L. infantum* is present, such as southern Europe, however, it may pose difficulties in areas where several species of *Leishmania* cause clinical disease in dogs such the Middle East with *L. major* and *L. infantum* and South America with *L. braziliensis* and other species.

Detection of parasite DNA in tissues by PCR allows sensitive and specific diagnosis. Several different PCR assays with various target sequences of genomic or kinetoplast DNA (kDNA) have been developed for CanL. kDNA PCR assays are considered more sensitive than genomic DNA assays which target parts of the *Leishmania* ribosomal operon DNA such as the internal transcribed spacer (ITS) PCR, however, ITS-PCR is able to pinpoint the infecting *Leishmania* species while kDNA PCR is only indicative of *Leishmania* infection without species identification (Talmi-Frank et al. 2010). PCR can be performed on DNA extracted from tissues, blood, or even from histopathologic specimens. The lymph nodes, bone marrow, and spleen are considered the most sensitive tissues for PCR detection of *L. infantum*, while PCR on blood is not sensitive and may be negative also in cases of overt clinical disease (Solano-Gallego et al. 2011, 2016). Conjunctival swab PCR is a sensitive noninvasive technique for sampling dogs, which can be used in surveys and when aspirations of bone marrow, lymph nodes, or spleen are undesirable (Strauss-Ayali et al. 2004; Di Muccio et al. 2012).

The LeishVet association recommends to use quantitative serology as the main diagnostic test in the case of a dog with suspected clinical signs of CanL, or hematological and serum biochemistry abnormalities compatible with the disease. Moderate to high anti-leishmanial antibody levels with clinical findings compatible with the disease are usually sufficient to reach a diagnosis of CanL (Table 3.1). Cytology and PCR are ancillary tests that may aid in the diagnostic process in the case of doubtful serology (Solano-Gallego et al. 2011). Precaution and exceptions to

Table 3.1 Recommendation for a diagnostic plan for dogs suspected of canine leishmaniosis and presenting clinical signs or laboratory findings compatible with the disease

Step	Description
Information from dog owners	Take a thorough history from the dog's owner. Where does the dog live? Is it mostly indoors or outdoors? Are there other animals diagnosed with CanL in the area? Has the dog been imported from an endemic area for CanL or has travelled to one? Is the dog protected with topical insecticides against sand flies and is the protection applied regularly? Has it been vaccinated against CanL and when? What suspected clinical signs has the owner seen and when? Has the dog been diagnosed with other vector-borne diseases or other medical conditions?
Physical examination	Pay special attention to findings that are compatible with CanL. Is there lymph node enlargement or splenomegaly? Are the mucous membranes pale? Is there exfoliation of skin in the ear tips, face, or anywhere in the body? Evaluate the eyes and conjunctivas for any abnormalities. Is there any sign of nasal discharge or epistaxis and are the foot nails elongated or brittle?
Laboratory tests	Complete blood count (CBC); serum biochemistry panel with special focus on albumin, globulins, albumin/globulin ratio, and creatinine levels; urinalysis and the urine protein/creatinine ratio (UPC) if proteinuria is evident.
Diagnostic tests	1. *Serology* for CanL. A quantitative assay such as IFA or ELISA as a first choice test is recommended. A reliable commercial fast kit can be used initially and if positive submit the serum for a quantitative test. 2. *Cytology* of an enlarged lymph node, skin lesion, spleen, or bone marrow. A negative cytology does not rule out CanL. 3. *PCR*, preferably of lymph node, spleen, conjunctiva, or bone marrow, may be used as an adjunct step to serology or in cases of doubtful serological results. A negative blood PCR does not rule out CanL.

the recommendation of diagnosis based on quantitative serology and clinical findings should be taken in particular situations, for instance, in areas endemic for canine trypanosomiasis.

3.6 Clinical Staging and Additional Tests Needed in the Diagnosis of CanL

Dogs suspected of CanL with compatible clinical findings should be evaluated clinically by a thorough physical examination, complete blood count (CBC), serum biochemistry, and urinalysis (Table 3.1). Dogs with CanL will typically be hyperglobulinemic, hypoalbuminemic, anemic, and will frequently have proteinuria due to glomerular loss of albumin, even if they are not azotemic with high levels of serum creatinine and urea. Quantification of proteinuria by the urine protein/creatinine (UPC) ratio is needed when proteinuria is detected to evaluate the magnitude of protein loss through the kidney.

A clinical staging system proposed by the LeishVet association for CanL divides the disease into four clinical stages based on clinical signs, clinicopathological abnormalities, and level of anti-leishmanial antibodies (Solano-Gallego et al. 2017). This system is helpful for decisions on the most suitable therapy for each patient and for consideration of a prognosis. The clinical stage may change if the dog deteriorates or improves. LeishVet stage I includes dogs that have papular dermatitis or enlargement of a solitary lymph node and usually no clinicopathological abnormalities. LeishVet stage II includes most of the dogs diagnosed in veterinary clinics with the disease. At this stage dogs present with typical findings such as exfoliative or ulcerative dermatitis, generalized lymphadenomegaly, possible loss of appetite and weight loss, and they have low to high levels of anti-leishmanial antibodies. Stage II is subdivided to stage IIa in which there is a normal renal profile, and stage IIb in which the UPC is mildly elevated (UPC 0.5–1). Stage III includes dogs with immune-mediated phenomena such as uveitis and immune-complex glomerulonephritis, in addition to the lesions found in previous stages, and have medium to high anti-leishmanial antibody levels with chronic kidney disease of stages I and II as defined by the International Renal Interest Society (IRIS). Stage IV includes dogs with medium to high anti-leishmanial antibodies, and very severe diseases such as the end-stage kidney disease, the nephrotic syndrome, and thromboembolism, in addition to lesions present in the previous stages. The prognosis is good for dogs in LeishVet stage I, good to guarded in stage II, guarded to poor in stage III, and poor in stage IV (Solano-Gallego et al. 2017).

3.7 Treatment and Follow-up

The main drug used for the treatment of CanL is allopurinol that acts by interfering with the purine pathway and the parasite's RNA synthesis. Allopurinol is used for long-term treatment of at least 6 months and frequently a year or more. The pentavalent antimony meglumine antimoniate (Glucantime®) that selectively inhibits leishmanial glycolysis and fatty acid oxidation is frequently used in combination with allopurinol for the first 4 weeks of treatment. Alternatively, miltefosine (Milteforan®), which is an oral anti-leishmanial drug, can be used for the first month of treatment in combination with allopurinol instead of meglumine antimoniate. Other drugs including paromomycin and marbofloxacin have also been shown to have some degree of anti-leishmanial effect (Miró et al. 2017). The treatment of CanL may vary for dogs in different clinical conditions, and the standard treatment regime in Europe for dogs in a stable clinical condition is allopurinol at 10 mg/kg every 12 h per-os (P.O.) in combination with meglumine antimoniate at 100 mg/kg injected subcutaneously every 24 h for 4 weeks, or in combination with miltefosine at 2 mg/kg P.O. every 24 h for 4 weeks (Miró et al. 2017). Treatment of CanL is long term and can be stopped when the following three conditions are all met: (1) disappearance of clinical signs; (2) normalization of the hematology, blood biochemistry profile and urinalysis; and (3) serology should become negative (e.g., below the cut-off of quantitative serological assays) (Solano-Gallego et al. 2011).

Anti-leishmanial treatment may achieve only partial clinical improvement in some dogs with CanL and it is frequently not associated with the complete elimination of the parasite, which is kept under control by the decrease in parasite load following treatment, and the recovery of the immune system function of treated animals. Treated dogs can remain carriers of the disease, be infectious to sand flies, and experience clinical relapses. Owners must receive a thorough and realistic explanation about the disease, its zoonotic potential, and the prognosis for their dog. Dogs in therapy should be treated with topical insecticides to prevent transmission of infection to other animals and to humans.

The side effects of the drugs most commonly used against CanL include gastrointestinal signs for miltefosine, local reaction in the injection site and potential nephrotoxicity for meglumine antimoniate, and formation of xanthine uroliths for allopurinol (Miró et al. 2017). Disease relapse of dogs with CanL during allopurinol treatment has been described and is associated with allopurinol resistance of *L. infantum* isolated from these animals (Yasur-Landau et al. 2016). Relatively little is known about resistance to pentavalent antimonials and miltefosine in dogs, however, resistance to these drugs in human leishmaniosis is well known.

Treatment of dogs in early stages of CanL with the allopurinol combination therapy is usually successful and often results in clinical cure. A longitudinal study of 1 year with 37 dogs that received the allopurinol and meglumine antimoniate combination treatment, of which 32 dogs were in LeishVet stage II and 5 dogs in stage III, indicated that all dogs improved in their clinical and clinicopathological abnormalities within 30 days of treatment, and there was a significant drop in their antibody levels and blood parasite load in the first 6 months of treatment. However, despite the marked clinical improvement of most dogs, only 5 dogs (16%) were eligible for stopping treatment at the end of 1 year of therapy (Solano-Gallego et al. 2016). In a second study with 23 treated dogs in LeishVet stage II followed up for 2–9 year, survival was long, although antibody levels remained positive in most dogs after 1 year of treatment. Three dogs experienced clinical relapse with high antibody levels and parasitemia, eight dogs had immune-mediated lesions, such as uveitis, arthritis and cutaneous vasculitis, and in all of these cases, the dogs had high anti-*Leishmania* antibody levels at diagnosis and during follow-up. Three dogs in this study presented xanthine urolithiasis that was likely associated with the allopurinol treatment (Torres et al. 2011).

Follow up of sick dogs under treatment of CanL varies according to their clinical status. It is recommended for dogs in stable condition to be evaluated 1 month after the beginning of treatment with a physical examination, CBC, serum biochemistry panel and urinalysis, and then again every 3–4 months during the first year of treatment. Later when fully recovered clinically, dogs should be followed up with the same evaluation program every 6–12 months. A quantitative serological test is recommended 6 months after the beginning of treatment, and then every 6–12 months. Quantitative PCR of blood is optional at the same time as serology (Solano-Gallego et al. 2017). A marked increase in the antibody titer of treated dogs, or dogs after the end of therapy, often precedes clinical relapse, and should prompt additional testing and possible repeated treatment, or the introduction of an additional effective drug and increase in allopurinol dose in the case of dogs under treatment.

Ancillary treatment of CanL includes treatment with domperidone (Leishguard®) which is registered in Europe for prophylaxis of the disease (Sabaté et al. 2014). Domperidone is a dopamine D2 receptor antagonist reported to have immunostimulant properties via the stimulation of prolactin secretion which acts as a pro-inflammatory cytokine. It is claimed to reduce the risk of the development of clinical disease by stimulating specific cellular immunity. Additional ancillary treatment includes a dietary supplement of nucleotides and active hexose as an adjunctive therapy for CanL (Segarra et al. 2018). Other drugs and immune-modulators have also been suggested for the treatment of CanL and prevention of progression to clinical disease (Baxarias et al. 2019).

3.8 Prevention

The use of topical insecticides against CanL in collars and spot-on formulations containing pyrethroids has been shown to be effective in reducing disease transmission. Pyrethroid collars and permethrin with imidacloprid or fipronil spot-on drops have been shown to significantly reduce the number of sand fly bites to dogs under experimental transmission and demonstrated decreased transmission of infection in field studies. Commercial vaccines against CanL have been approved in Brazil and Europe and are used to protect dogs, often in combination with topical insecticides. However, some studies have shown that vaccines may decrease the occurrence of clinical disease but do not prevent infection (Miró et al. 2017; Dantas-Torres et al. 2020). Vaccination with some canine vaccines against CanL may elicit a positive antibody response with *L. infantum* antigen detectable by serological tests, especially in the case of recent vaccination (Solano-Gallego et al. 2017).

References

Altet L, Francino O, Solano-Gallego L et al (2002) Mapping and sequencing of the canine NRAMP1 gene and identification of mutations in leishmaniasis-susceptible dogs. Infect Immun 70:2763–2771

Andrade GB, Barreto WT, Santos LL et al (2014) Pathology of dogs in Campo Grande, MS, Brazil naturally co-infected with *Leishmania infantum* and *Ehrlichia canis*. Rev Bras Parasitol Vet 23:509–515

Attipa C, Solano-Gallego L, Papasouliotis K et al (2018) Association between canine leishmaniosis and *Ehrlichia canis* co-infection: a prospective case-control study. Parasit Vectors 11:184

Baneth G, Koutinas AF, Solano-Gallego L et al (2008) Canine leishmaniosis – new concepts and insights on an expanding zoonosis: part one. Trends Parasitol 24:324–330

Baneth G, Yasur-Landau D, Gilad M et al (2017) Canine leishmaniosis caused by *Leishmania major* and *Leishmania tropica*: comparative findings and serology. Parasit Vectors 10:113
Baxarias M, Martínez-Orellana P, Baneth G et al (2019) Immunotherapy in clinical canine leishmaniosis: a comparative update. Res Vet Sci 125:218–226
Boggiatto PM, Gibson-Corley KN, Metz K et al (2011) Transplacental transmission of *Leishmania infantum* as a means for continued disease incidence in North America. PLoS Negl Trop Dis 5: e1019
Borja LS, Sousa OMF, Solcà MDS et al (2016) Parasite load in the blood and skin of dogs naturally infected by *Leishmania infantum* is correlated with their capacity to infect sand fly vectors. Vet Parasitol 229:110–117
Ceron JJ, Pardo-Marin L, Caldin M et al (2018) Use of acute phase proteins for the clinical assessment and management of canine leishmaniosis: general recommendations. BMC Vet Res 14:196
Dantas-Torres F, Solano-Gallego L, Baneth G et al (2012) Canine leishmaniosis in the old and new worlds: unveiled similarities and differences. Trends Parasitol 28:531–538
Dantas-Torres F, dos Santos Nogueira F, Menz I, Tabanez P, da Silva SM, Ribeiro VM, Miró G, Cardoso L, Petersen C, Baneth G, Oliva G, Solano-Gallego L, Ferrer L, Pennisi MG, Bourdeau P, Maia C, Otranto D, Gradoni L, Courtenay O, Costa CHN (2020) Vaccination against canine leishmaniasis in Brazil. Int J Parasitol 50(3):171–176
Di Muccio T, Veronesi F, Antognoni MT et al (2012) Diagnostic value of conjunctival swab sampling associated with nested PCR for different categories of dogs naturally exposed to *Leishmania infantum* infection. J Clin Microbiol 50:2651–2659
Foglia Manzillo V, Di Muccio T, Cappiello S et al (2013) Prospective study on the incidence and progression of clinical signs in naïve dogs naturally infected by *Leishmania infantum*. PLoS Negl Trop Dis 7:e2225
Gharbi M, Jaouadi K, Mezghani D et al (2018) Symptoms of canine leishmaniosis in Tunisian dogs. Bull Soc Pathol Exot 111:51–55
Ginel PJ, Camacho S, Lucena R (2008) Anti-histone antibodies in dogs with leishmaniasis and glomerulonephritis. Res Vet Sci 85:510–514
Guarga JL, Moreno J, Lucientes J et al (2000) Canine leishmaniasis transmission: higher infectivity amongst naturally infected dogs to sand flies is associated with lower proportions of T helper cells. Res Vet Sci 69:249–253
Karkamo V, Kaistinen A, Näreaho A et al (2014) The first report of autochthonous non-vector-borne transmission of canine leishmaniosis in the Nordic countries. Acta Vet Scand 56:84
Koutinas AF, Koutinas CK (2014) Pathologic mechanisms underlying the clinical findings in canine leishmaniasis due to *Leishmania infantum/chagasi*. Vet Pathol 51:527–538
Lombardo G, Pennisi MG, Lupo T et al (2014) Papular dermatitis due to *Leishmania infantum* infection in seventeen dogs: diagnostic features, extent of the infection and treatment outcome. Parasit Vectors 7:120
Lun ZR, Wu MS, Chen YF et al (2015) Visceral leishmaniasis in China: an endemic disease under control. Clin Microbiol Rev 28:987–1004
Marcondes M, Day MJ (2019) Current status and management of canine leishmaniasis in Latin America. Res Vet Sci 123:261–272
Mekuzas Y, Gradoni L, Oliva G et al (2009) *Ehrlichia canis* and *Leishmania infantum* co-infection: a 3-year longitudinal study in naturally exposed dogs. Clin Microbiol Infect 5(Suppl 2):30–31
Meléndez-Lazo A, Ordeix L, Planellas M et al (2018) Clinicopathological findings in sick dogs naturally infected with *Leishmania infantum*: comparison of five different clinical classification systems. Res Vet Sci 117:18–27
Mettler M, Grimm F, Capelli G (2005) Evaluation of enzyme-linked immunosorbent assays, an immunofluorescent-antibody test, and two rapid tests(immunochromatographic-dipstick and gel tests) for serological diagnosis of symptomatic and asymptomatic *Leishmania* infections in dogs. J Clin Microbiol 43:5515–5519

Miranda S, Roura X, Picado A et al (2008) Characterization of sex, age, and breed for a population of canine leishmaniosis diseased dogs. Res Vet Sci 85:35–38

Miró G, Petersen C, Cardoso L et al (2017) Novel areas for prevention and control of canine leishmaniosis. Trends Parasitol 33:718–730

Molina R, Amela C, Nieto J (1994) Infectivity of dogs naturally infected with *Leishmania infantum* to colonized *Phlebotomus perniciosus*. Trans R Soc Trop Med Hyg 88:491–493

Moreira MA, Luvizotto MC, Garcia JF et al (2007) Comparison of parasitological, immunological and molecular methods for the diagnosis of leishmaniasis in dogs with different clinical signs. Vet Parasitol 145:245–252

Moreno J, Alvar J (2002) Canine leishmaniasis: epidemiological risk and the experimental model. Trends Parasitol 18:399–405

Naucke TJ, Amelung S, Lorentz S (2016) First report of transmission of canine leishmaniosis through bite wounds from a naturally infected dog in Germany. Parasit Vectors 9:256

Ordeix L, Dalmau A, Osso M et al (2017) Histological and parasitological distinctive findings in clinically-lesioned and normal-looking skin of dogs with different clinical stages of leishmaniosis. Parasit Vectors 10:121

Owens SD, Oakley DA, Marryott K et al (2001) Transmission of visceral leishmaniasis through blood transfusions from infected English foxhounds to anemic dogs. J Am Vet Med Assoc 219:1076–1083

Paltrinieri S, Gradoni L, Roura X et al (2016) Laboratory tests for diagnosing and monitoring canine leishmaniasis. Vet Clin Pathol 45:552–578

Papadogiannakis E, Andritsos G, Kontos V et al (2010) Determination of CD4+ and CD8+ T cells in the peripheral blood of dogs with leishmaniosis before and after prolonged allopurinol monotherapy. Vet J 186:262–263

Parody N, Cacheiro-Llaguno C, Osuna C et al (2019) Circulating immune complexes levels correlate with the progression of canine leishmaniosis in naturally infected dogs. Vet Parasitol 274:108921

Quinnell RJ, Kennedy LJ, Barnes A et al (2003) Susceptibility to visceral leishmaniasis in the domestic dog is associated with MHC class II polymorphism. Immunogenetics 55:23–28

Ribeiro RR, Michalick MSM, da Silva ME et al (2018) Canine leishmaniasis: an overview of the current status and strategies for control. Biomed Res Int 3296893

Rocha MF, Michalsky ÉM, de Oliveira L-SF et al (2020) Dogs with divergent serology for visceral leishmaniasis as sources of *Leishmania* infection for *Lutzomyia longipalpis* phlebotomine sand flies – an observational study in an endemic area in Brazil. PLoS Negl Trop Dis 14:e0008079

Sabaté D, Llinás J, Homedes J (2014) A single-centre, open-label, controlled, randomized clinical trial to assess the preventive efficacy of adomperidone-based treatment programme against clinical canine leishmaniasis in a high prevalence area. Prev Vet Med 115:56–63

Sanchez-Robert E, Altet L, Utzet-Sadurni M et al (2008) Slc11a1 (formerly Nramp1) and susceptibility to canine visceral leishmaniasis. Vet Res 39:36

Schäfer I, Volkmann M, Beelitz P et al (2019) Retrospective evaluation of vector-borne infections in dogs imported from the Mediterranean region and southeastern Europe (2007-2015). Parasit Vectors 12:30

Segarra S, Miró G, Montoya A et al (2018) Prevention of disease progression in *Leishmania infantum*-infected dogs with dietary nucleotides and active hexose correlated compound. Parasit Vectors 11:103

Silva FL, Oliveira RG, Silva TM et al (2009) Venereal transmission of canine visceral leishmaniasis. Vet Parasitol 160:55–59

Solano-Gallego L, Llull J, Ramos G (2000) The Ibizian hound presents a predominantly cellular immune response against natural *Leishmania* infection. Vet Parasitol 90:37–45

Solano-Gallego L, Morell P, Arboix M et al (2001) Prevalence of *Leishmania infantum* infection in dogs living in an area of canine leishmaniasis endemicity using PCR on several tissues and serology. J Clin Microbiol 39:560–563

Solano-Gallego L, Miró G, Koutinas A et al (2011) The LeishVet group. LeishVet guidelines for the practical management of canine leishmaniosis. Parasit Vectors 4:86
Solano-Gallego L, Di Filippo L, Ordeix L (2016) Early reduction of *Leishmania infantum*-specific antibodies and blood parasitemia during treatment in dogs with moderate or severe disease. Parasit Vectors 9:235
Solano-Gallego L, Cardoso L, Pennisi MG et al (2017) Diagnostic challenges in the era of canine *Leishmania infantum* vaccines. Trends Parasitol 33:706–717
Strauss-Ayali D, Jaffe CL, Burshtain O et al (2004) Polymerase chain reaction using noninvasively obtained samples, for the detection of *Leishmania infantum* DNA in dogs. J Infect Dis 189:1729–1733
Svobodova V, Svoboda M, Friedlaenderova L et al (2017) Canine leishmaniosis in three consecutive generations of dogs in Czech Republic. Vet Parasitol 237:122–124
Talmi-Frank D, Nasereddin A, Schnur LF et al (2010) Detection and identification of old world *Leishmania* by high resolution melt analysis. PLoS Negl Trop Dis 4:e581
Toepp AJ, Petersen CA (2020) The balancing act: immunology of leishmaniosis. Res Vet Sci 130:19–25
Toepp AJ, Schaut RG, Scott BD et al (2017) *Leishmania* incidence and prevalence in U.S. hunting hounds maintained via vertical transmission. Vet Parasitol Reg Stud Rep 10:75–81
Toepp AJ, Monteiro GRG, Coutinho JFV et al (2019) Comorbid infections induce progression of visceral leishmaniasis. Parasit Vectors 12:54
Torres M, Bardagí M, Roura X et al (2011) Long term follow-up of dogs diagnosed with leishmaniosis (clinical stage II) and treated with meglumine antimoniate and allopurinol. Vet J 188:346–3451
WHO (2018a). https://www.who.int/gho/neglected_diseases/leishmaniasis/en/
WHO (2018b). https://www.who.int/news-room/fact-sheets/detail/leishmaniasis
Yasur-Landau D, Jaffe CL, David L et al (2016) Allopurinol resistance in *Leishmania infantum* from dogs with disease relapse. PLoS Negl Trop Dis 10:e0004341
Zatelli A, Borgarelli M, Santilli R et al (2003) Glomerular lesions in dogs infected with *Leishmania* organisms. Am J Vet Res 64:558–561

Chapter 4
The Challenges with Canine *Giardia*

Dwight D. Bowman

Abstract This chapter reviews the prevalence of *Giardia* and *Giardia* Assemblages in people and dogs and the risks that dogs may present as sources of human infection. The review shows that the treatment of dogs with any of the currently available therapeutics given to dogs often do not clear all *Giardia* from dogs or halt *Giardia* from repopulating a negative. Humans sometimes host *Giardia* that is very difficult to clear with therapy. The concern is expressed that the treatment of dogs without clinical signs repeatedly with drugs that are less than perfect is likely selecting for *Giardia* that will be resistant to treatment remaining in the treated dogs and potentially shared with other dogs. If transmission is occurring in people that comes from dogs and if it is demonstrated that people are infected with resistant canine isolates, it will have a negative impact on the human–animal bond. Therefore, it is suggested that veterinarians give careful consideration to the perceived need to treat all dogs testing positive for *Giardia*, because it may in the end have serious impacts on the relationship between pet owners, their dogs, veterinarians, and the practitioners of human medicine.

Keywords *Giardia* · Resistance · Metronidazole · Fenbendazole · Tinidazole · Secnidazole · Ronidazole · Febantel · Oxfendazole · Nitazoxanide · Quinacrine · Chloroquine

4.1 Introduction

In 2014, Tysnes et al. published an excellent opinion piece in *Trends in Parasitology* discussing subclinical *Giardia* in dogs and the associated veterinary conundrum relevant to human infection, and overall, things have not changed to any great extent

---

D. D. Bowman (✉)
Department of Microbiology and Immunology, CVM Cornell University, Ithaca, NY, USA
e-mail: ddb3@cornell.edu

C. Strube, H. Mehlhorn (eds.), *Dog Parasites Endangering Human Health*, Parasitology Research Monographs 13,
https://doi.org/10.1007/978-3-030-53230-7_4

since this publication. So, dogs continue to fairly commonly harbor *Giardia* within their intestinal tracts, and some dogs may have signs that appear in association with the presence of this agent. Many people in the general public are aware that giardiasis can induce in some persons signs and symptoms associated with intestinal distress. Veterinarians, especially with the availability of antigen testing, can now with relative ease diagnose a dog as "*Giardia* positive." The improved diagnosis and awareness lead to questions as to what is to be done with a positive dog. These questions have come in the form of the following queries. Should every dog be treated whether or not it has any associated signs? Does one require a dog to be negative prior to it entering a "doggie daycare" facility or a boarding kennel? Does one suggest that the owners might want to have all dogs in their household tested? Does one suggest that the owners might want to consider having themselves and other family members tested? Can a dog that is positive for *Giardia* be sold? Should all shelters be required to test all dogs for *Giardia*, which are scheduled for adoption, and then, only be allowed to place *Giardia* negative dogs in new homes? Should the presence of *Giardia* in puppies be used as an indicator of poor care within an animal facility? Should all persons caring for dogs in shelters or animal facilities be tested for *Giardia* infections on a regular basis, for example, every 3 months? Should all dogs that are *Giardia* positive be banned from public dog parks? Now that we can use DNA identification to identify the specific source of a fecal sample, are owners going to be accused of spreading a pathogen if a certain canine fecal with *Giardia* is found in the "wrong" yard? These are not esoteric questions, and the author of this chapter has been asked on multiple occasions to respond to each of these and other similar questions. However, at this time, there appears to be no resolution in sight.

It is currently not known to what extent canine giardiasis is endangering human health. *Giardia* occurs in both people and dogs. The common C and D assemblages of dogs usually are not found in people (Bowman and Lucio-Forster 2010). Also, Assemblage B is almost never found in dogs (Bowman and Lucio-Forster 2010). In the case of Assemblage A, the sub-Assemblage AII appears restricted to humans, sub-Assemblage AIII to wild hoofed stock, and sub-Assemblage AI occurs in people and occasionally dogs. Two recent papers reported on the examination of *Giardia* in people and their pets within Germany (Rehbein et al. 2018) and Spain (de Lucio et al. 2017). In Germany there were 38 households, 69 humans, and 31 dogs; and of the humans sampled, 3 (4.3%) had asymptomatic *Giardia* infections, and 13 of the 31 dogs (42%) had *Giardia* with 3 dogs having gastrointestinal signs. Only 2 of the human samples amplified successfully, and these were both Assemblage B. For the dog samples, 2 were Assemblage A, 1 was C, and 2 were D, and one was A/B. The canine sample that was the A/B assemblage shared a household with a human with Assemblage B, the sequencing of the material showed that the canine B and human B were not identical. In Spain, there were 63 households with 179 adults and 55 pet dogs. There were 6 humans (3.4%) and 16 dogs (29.1%), that were *Giardia* positive, and of these samples, 1 human and 3 canine samples were successfully genotyped with the human sample being sub-Assemblage BIV and the canine samples all being Assemblage C. Thus, there is still no clear evidence of pets and owners regularly sharing the same Assemblages or sub-Assemblages.

At the same time, people do get sick during *Giardia* outbreaks, and the signs associated with an infection can persist a long time, even years, after the infection has been cleared with treatment. This was demonstrated after a major outbreak in Bergen, Norway where there was sewage contamination of the drinking water (Robertson et al. 2006), and 1500 people were diagnosed as becoming infected and were still suffering sequelae 6 years later (Robertson et al. 2010). On the other hand, in the developing world, neonatal giardiasis is ubiquitous in children just as it is in puppies in the developed world. In one study of 9439 children with moderate-to-severe diarrhea and 13,129 control children, "*Giardia* was not significantly positively associated with moderate to severe diarrhoea; to the contrary, in univariate analyses *Giardia* was identified significantly more frequently in controls than in patients with moderate-to-severe diarrhoea aged 12–59 months" (Kotloff et al. 2013). The other reality is that *Giardia* in people is typically treated only in those who are symptomatic, while in dogs it has become more and more common to treat all cases, rather than just those with associated clinical signs. Therefore, besides arguing about whether or not *Giardia* in dogs is zoonotic, it seems that thought should be given to what it would mean if it is actually a common zoonosis affecting people. If that is the case, then what would then have to occur to keep people safe and healthy.

4.2 The Ubiquity of Humans as Hosts of *Giardia*

In the two reports discussed above (Robertson et al. 2006; Kotloff et al. 2013), *Giardia* in people was not an uncommon occurrence. In the developing world, although not associated with infant mortality of moderately to severely affected children with diarrhea, *Giardia* was found in the stools of 9.4% of 4029 such children between 0 and 11 months of age, and in 19.4% of 3204 such children aged between 12 and 23 months (The *Giardia* data was not presented for the control children.) Kotloff et al. (2013). In a study in the Netherlands in 2001 (Medema and Schijven 2001), it was found that the number of cysts of *Giardia* coming into 5 sewage treatment plants was steady at around 1000 cysts per liter entering the plants, and the authors calculated that the average number of cysts of *Giardia* produced by the 15.1 million persons was 25 million *Giardia* cysts per inhabitant. Such a number is feasible since they cite that infected individuals produce around 10^9–10^{10} cysts during the course of an infection. In the United States, the Centers for Disease Control and Prevention (CDC) published a report on fecal "accidents" in swimming pools, where in 1999 between 29 May and 6 September, from 47 swimming pools throughout the United States, they collected 293 formed fecal samples and 13 (4.4%) of these samples were positive for *Giardia* (CDC 2001). In 2008, further work on backwash filters from 160 public swimming pools from 2 counties in the metropolitan Atlanta, Georgia area were examined and, of the 160 sampled pools, 10 (6.2%) were positive for *Giardia* cysts (Shields et al. 2008). In this same paper, the authors also cite work showing that in the Netherlands, 7 pools sampled

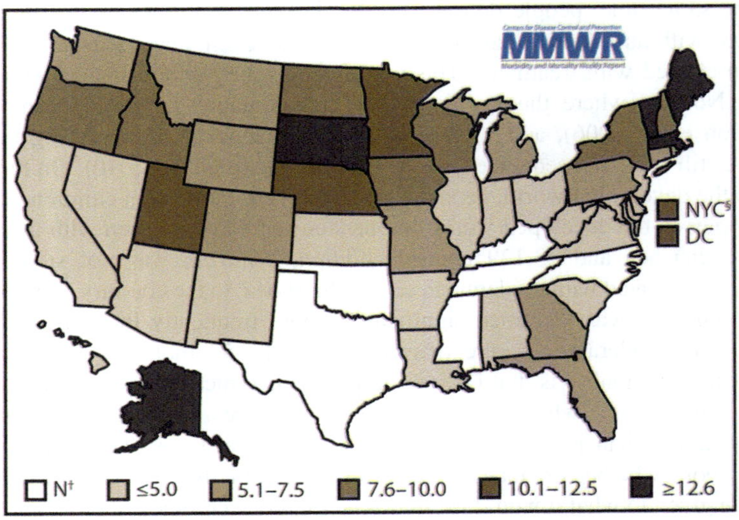

Fig. 4.1 This is Fig. 2 of Painter JE, Gargano JW, Collier SA et al. (2015) *Giardia* surveillance United States 2011–2012 MMWR 64(3):15–25, showing: Incidence rate (Cases per 100,000 population) of giardiasis, by reporting jurisdiction—National Notifiable Diseases Surveillance System, United States, 2012. † Not a reportable disease in these states. § New York State and New York City data are mutually exclusive

for more than a year had a 4.9% *Giardia* prevalence, and that in Italy, one study had found 2/7 tested pools positive, and another found 4/10 pools positive. Also, CDC published the Giardia Surveillance for 2011–2012 in 2015 (Painter et al. 2015). Here it was shown based on reports from all but six US states that cases were highest in the United States in the northwest (9.4 per 100,000) and northeast (8.8 per 100,000) (Fig. 4.1). The CDC data also showed that the highest prevalence of cases around 14–17% occurs in children from 1 to 4 years of age, but that overall, some 6% of those infected are between 20 and 64 year of age (Fig. 4.2). Of course, in people the diagnosis of *Giardia* usually occurs when testing is performed in response to gastrointestinal upset. Two recent studies produced markedly different outcomes likely due to the environment of the sampled individuals. In a study of 104 individuals (1–66 years of age, with a median age of 10) in marginalized rural communities in Palestine that were chosen without any consideration of signalment, the prevalence of *Giardia* was 37% (Al-Jawabreh et al. 2019). In a study in Korea of 8571 patients who had been tested in response to acute diarrhea, only 47 (0.55%) were positive for *Giardia* Assemblage A (Ma et al. 2019). An additional point to consider is when sampling is performed during the year. In the United States, there is a regular increase in the number of cases from June to October (Fig. 4.3), and "The summer peak coincides with increased outdoor activities (e.g., camping and swimming) that likely increase exposure to contaminated water. Transmission associated with outdoor activities is facilitated by the substantial number of *Giardia* cysts that can be shed by a single person, the environmental hardiness of the organism, the extended

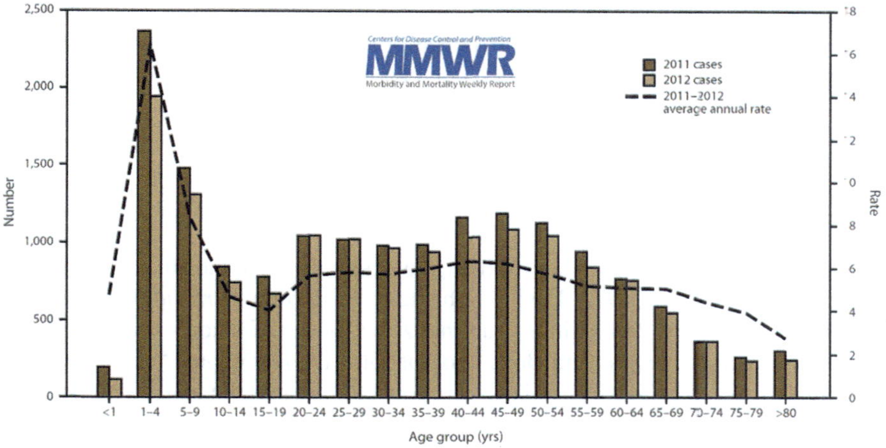

Fig. 4.2 This is Fig. 3 of Painter JE, Gargano JW, Collier SA et al. (2015) *Giardia* surveillance United States 2011–2012 MMWR 64(3):15–25, showing: Number ($N = 31{,}167$; age for 814 patients was unknown) of cases per year and average annual incidence rate (Cases per 100,000 population) of giardiasis, by age group—National Notifiable Diseases Surveillance System, United States, 2011–2012

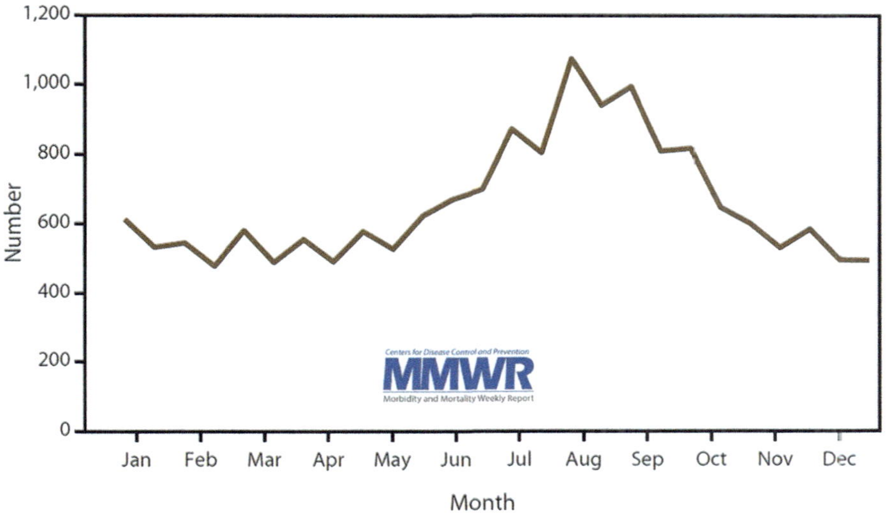

Fig. 4.3 This is Fig. 5 of Painter JE, Gargano JW, Collier SA et al. (2015) *Giardia* surveillance United States 2011–2012 MMWR 64(3):15–25. Figure shows: Number (of total number of cases ($N = 31{,}981$), date of onset for 14,876 patients was unknown) of giardiasis case reports, by date of symptom onset—National Notifiable Diseases Surveillance System, United States, 2011–2012

periods of time that cysts can be shed, and the low infection dose for infection." However, taken altogether, it should be fairly clear that overall a percentage of people are in a good number of cases commonly living with their own *Giardia*.

4.3 The Ubiquity of Dogs as Hosts of *Giardia*

Parasitologists, both veterinary and human, have been aware of the ubiquitous nature of *Giardia* for years. In 1973, Levine in his text, *Protozoan Parasites of Domestic Animals and of Man*, 2nd Edition, summarized a series of studies that presented prevalence data on people and dogs. For people, Levine wrote: "*G. lamblia* is common in man. In 86 surveys of 134,966 people throughout the world summarized by Belding (1965), its prevalence was 2.4–67.5% with a, mean of 10.4%. It was found in 7.4% of 35,299 persons in 24 surveys in the United States, and in 6.9% of 65,295 persons in 20 surveys in the rest of the world. It is about three times as common in children as in adults." The dog form was a separate species at that time, *Giardia canis* Hegner, 1922. Levine summarized the literature on prevalence in dogs: "Catcott (1946) found *G. canis* in 18% of 113 dogs in Ohio, Choquette and Gelinas (1950) in 9% of 155 dogs in Montreal, Canada, Craige (1948) in 5.6% of 160 dogs in California, Bemrick (1961) in 8% of 2063 dog stools in Minnesota; it was much more common in young dogs than in old ones. Levine and Ivens (1965) found its cysts in the feces of 4% of 175 dogs in Illinois. Burrows (1968) found it in 15% of 835 stray dogs in New Jersey, and Thomas (1960) in 0.5% of 1027 German Shepherd dogs received at the US Army Dog Training Center from 44 states. Rijpstra (1967) found it in 17% of 65 dogs in Holland, and Chang (1935) in 29% of 14 dogs in Shantung (=Shandong), China. Hegner and Chu (1930) found it in one of 12 dogs in the Philippines."

More recently, the Companion Animal Parasite Council (CAPC) has been presenting maps for canine and feline parasites based on data collected by two very large diagnostic laboratories, IDEXX Laboratories, Inc., One Idexx Drive, Westbrook, Maine 04092 and ANTECH Diagnostics, VCA Professional Animal Laboratory, Inc.; 12401 West Olympic Boulevard Los Angeles, CA 90064. The 2019 CAPC *Giardia* Prevalence Map (Fig. 4.4) reports the resulting text run on approximately 9.6 million samples. Of these 9.6 million samples, 6.96% (667,005) were positive for *Giardia*. This data is not controlled to remove duplicate or multiple tests on the same animal, such as before and after treatment; however, it seems likely that repeated tests represent only a minimal percentage of the total tests performed. The map does seem to mirror fairly well the areas where there is a higher prevalence in the northeast and northwest, but the canine map seems higher in the southwestern United States. If you examine it more carefully, one will see that the breakdown of colors as percentages also by chance mirror each other. The breaks on the CAPC map are at >0 to 4%; >4 to 7%, and >7%, and the breaks on the CDC map are at

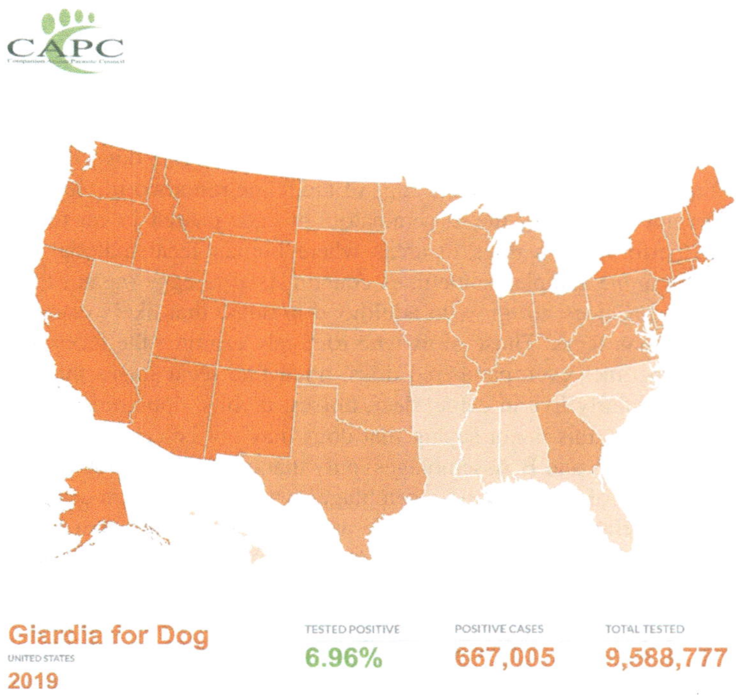

Fig. 4.4 Map from the website of the Companion Animal Parasite Council [CAPC] showing the year of 2019 results from the testing of 9.6 million fecal samples for the presence of *Giardia*. The data for the maps is provided to CAPC by ANTECH Diagnostics® and IDEXX Laboratories. Examination of the online state maps revealed the breaks for prevalence shades were: Light orange: 0.01 to 4.0%, medium orange: >4 to 7%, and dark orange: >7%

0–5, >5 to 7.5, and >7.5 to >12.6 (broken into three distinct categories). Again, it must be remembered that the majority of the canine data represents wellness testing, while the human data probably represents a test performed on symptomatic individuals. It is also likely that the prevalence and transmission in different areas have more to do with the ability of cysts to survive in the environment than the number of potentially infected hosts in a given area. The older texts stated that the parasite is more common in the southern climates (Craig and Faust 1945), but the first significant outbreak in the United States as described by Levine (1973) occurred "during the 1965–1966 ski season at Aspen, Colo.; at least 11% of 1094 skiers surveyed by Moore et al. (1969) had the disease; after the outbreak Gleason et al. (1970) found that 5.0% of 419 persons (about half the permanent population of the town-in the middle-to upper-income brackets) were infected."

4.4 What If Dogs Are Endangering Human Health?

From the above, it should be fairly apparent that *Giardia* commonly occurs in both people and dogs. Also, the canine associated Assemblages C and D only rarely show up in people, and although there have been reports of A and B in dogs, one must also consider that dogs routinely shed the eggs of the feline roundworm, *Toxocara cati*, but they are not hosts, only the disseminators of eggs passed in cat feces. When surveys are performed on dogs in areas where human fecal material is readily available, the canine samples sometimes contain *Ascaris* eggs (again, this roundworm, like *T. cati*, does not grow to adulthood in dogs) that likely came from the ingestion of human feces. Those of us who routinely examine the feces of stray or owned and sometimes even pampered dogs, often find fecal stages that belong to parasites that regularly live in cattle, deer, horses, rabbits, opossums, and rodents. Relative to *Giardia*, this is not saying that dogs may not be capable of serving as regular hosts of the A and B Assemblages, only that we really do not know in most cases what to make of finding these assemblages when they occur in canine feces. It has been stated relative to human disease and the study of resistant isolates that "Most work regarding drug resistance mechanisms in *Giardia* has been done using laboratory induced resistant assemblage A1 isolates, which rarely infects humans, compared to infections with *Giardia* assemblages A2 and B, which are common" (Saghaug et al. 2019) [to which Argüelo-Garcia et al. (2020) add] "and, therefore, are more often associated with treatment failure." However, it is important to discuss what should be done if dogs are a regular source of an agent that is being transmitted to people and causing significant disease. Tysnes et al. (2014) provide an excellent decision tree for veterinarians relative to dealing with the *Giardia* positive dog, the client, and the client's household, and these are part of the daily concern of the working veterinarian. However, if a diagnosis is made, there will always be a desire to treat, and when diarrhea is involved, the human–animal bond is in jeopardy if the condition continues for too long.

An obvious fix would be to clear all dogs of their infections, which is very unlikely to happen solely with the use of drug therapy. A vaccine for dogs was available for several years but has disappeared from the market. The vaccine was purported to minimize the shedding of organisms by vaccinated animals, and thus, it did provide a means by which dogs were allowed to enter facilities without requiring a test or routine testing after they became regular inhabitants or visitors to a facility. Otherwise, dogs can sometimes be treated for weeks or months before they are negative on their *Giardia* tests, and thus, cannot be housed within various daycare or boarding facilities. There has been a report of work to develop an oral vaccine for dogs and cats (Serradell et al. 2016, 2018), but it appears still under development. A vaccine would likely reduce the rate of infection in dogs without selecting for resistance to the drugs that are used to treat people and dogs. This would be of value, because all the drugs used to treat infections in people are the same or in the similar drug classes as those used to treat the presence of *Giardia* in dogs, and this is the concern relative to shared infections, i.e., it could lead to the sharing of resistant isolates with those from treated dogs being transferred to people.

4.5 Drugs Shared in the Treatment of Canine and Human Anti-*Giardia* Therapy

The drugs currently used most commonly to treat giardiasis in people and dogs are the 5-nitroimidazoles, including metronidazole, tinidazole, ornidazole, and secnidazole, and benzimidazoles, including for people albendazole and mebendazole, and in the case of dogs, fenbendazole (Argüelo-Garcia et al. 2020). Other drugs that have been used in people, and sometimes in dogs, include nitazoxanide, furazolidone, quinacrine, chloroquine, and paromomycin (Argüelo-Garcia et al. 2020). In people, metronidazole and albendazole are no longer working as well as they once did, and in the outbreak in the Netherlands, it was necessary to use combination therapies utilizing both agents (Mørch et al. 2008). The aminoglycoside paromomycin is the drug of choice for giardiasis in pregnant women since it is not readily absorbed and so its action is primarily within the intestinal tract (Leitsch 2015). In people, quinacrine has been used, but is a second-line therapeutic that is usually used in combination with metronidazole or albendazole (Leitsch 2015). Quinacrine has side effects in people that include nausea and skin discoloration. Furazolidone in people is used mainly in children because it is available in a liquid formulation, but it must be administered for 7–10 days or the cure rate is reduced.

In dogs, the therapeutics are not perfect. In most trials, efficacy had not been 100% no matter which drug, drug regimen, facility sanitation method, dog bathing, or environmental treatment has been used. There have been many trials with the different products, and often two or three products are compared in the same sets of studies. Not all the drugs are shared between people and dogs. One drug not used in people, such as the very commonly used benzimidazole, fenbendazole, that is approved for treating dogs in the EU and other parts of the world and is routinely used in much of the rest of the world. Another benzimidazole, i.e., albendazole, is one of the most commonly used drugs in people; however, due to its induction of bone marrow suppression in both dogs and cats, it is no longer used to treat parasitic diseases in these domestic animals (Tantrongsup and Scorza 2010). Thus, in this review due to fenbendazole being approved for the treatment of *Giardia* in many countries, the report will only present the data from some of the more recent fenbendazole trials. Also, because albendazole in dogs is no longer considered an appropriate therapeutic in most cases in dogs and cats, it will not be included in the review. Trials with other benzimidazoles and other products will be reviewed.

4.5.1 Nitroimidazoles

These products enter cells, they are reduced to active agents with toxic effects on cellular reactions resulting in the release of nonfunctional end products The modes of actions of the different forms that are used have been reviewed recently by Argüelo-Garcia et al. (2020) and Leitsch et al. (2019).

Metronidazole Metronidazole is used off label in the United States (CAPC) but is licensed in most European countries for dogs and cats (ESCCAP). Metronidazole is the drug used most commonly in dogs for *Giardia*, and metronidazole is also the drug that is used most commonly in people. Metronidazole has not reliably been found to clear all dogs of their infections as can be seen from the tabular review of the data for the therapy of metronidazole in dogs (Table 4.1). The data presented in Table 4.1 only presents the data for metronidazole, and where other drugs were involved in these studies, the results are presented below under the discussion of each of the different compounds.

There have been studies in dogs looking at the efficacy of other members of the nitroimidazole family. These include tinidazole, ornidazole, secnidazole, and ronidazole (which has more complex substitution at C2 and was introduced fairly recently) (Argüelo-Garcia et al. 2020). None of these products have proven fully satisfactory in clearing all dogs of their *Giardia*.

Tinidazole and/or Ornidazole A number of the early studies with nitroimidazoles were performed with tinidazole and ornidazole, with some including quinacrine for comparative purposes. Nesvadba et al. (1980) presented on the treatment of 232 of 814 dogs with intestinal signs that included diarrhea and the presence of *Giardia* as diagnosed by a flotation method. Treatment was with various regimes using orally administered tablets of metronidazole (Flagyl), tinidazole (Fasigyn), and ornidazole (Tiberal). The most successful regimen was twice daily treatments for 5 days followed by once-daily treatments for a further 10 days. Using these treatments, they achieved complete clearance in less than 50% of the dogs treated with any of the regimens. Rohde (1983) in a thesis submitted to the Veterinary Faculty in Zurich treated six normal laboratory beagles that were passing *Giardia* cysts with two regimens of ornidazole followed by treatment with the acridine derivative quinacrine. Initially, the beagles were treated with ornidazole, and then the feces were examined for cysts: re-excretion of cysts began 7 days after a 2-day treatment with 56.9 mg/kg, 16 days after a 2-day treatment with 84.5 mg/kg, and 11 days after a 10-day treatment with 81.5 mg/kg. The subsequent treatment with quinacrine led to an excretion-free interval of 34 days, but a parasitological cure was not obtained, and dogs again became positive. Zimmer and Burrington (1986) reported treating 66 dogs that had *Giardia* (Interestingly, 21 of these dogs also had *Pentatrichomonas* which has no cyst stage and is therefore not routinely seen with the commonly used fecal examination methods that detect the cyst stage of *Giardia*). The dogs were treated for 5 days with tinidazole (44 mg/kg QD), metronidazole (22 mg/kg BID), quinacrine (6.6 mg.kg BID), and both metronidazole (22 mg/kg) and quinacrine

Table 4.1 Summary of published reports on the treatment of dogs for *Giardia* with metronidazole

Country	Number of dogs	Dose	Detection method	Cure rate (#Neg/#Cases) % Cured	References
Americas					
PEI Canada	**1 5-month-old Dalmatian**	**Not stated; do not know how long after treatment fecal flotation performed**	**Cysts—do not know how long after treatment fecal flotation performed**	**0/1—Dog treated before the study began**	**Conboy** (1996)
USA	15 1- to 2-year-old German shepherd or Labrador retrievers	22 mg/kg PO BID for 5 days	Trophozoites and cysts—fecal smear and zinc sulfate centrifugal flotation	10/15 66.6%	Zimmer and Burrington (1986)
USA	**10 Unspecified breed and age**	**73 mg/kg PO SID for 5 days**	**Trophozoites and cysts—direct smear and PVA fixed material for slides**	**10/10; however, only monitored clinical cure—did not reexamine for *Giardia***	**Simmons and Passon** (1981)
USA	**5 [1 7-month-old Malamute and 4 8-week old Afghan puppies]**	**Case 1—dose not given. Case 2 50 mg/kg SID for 7 days**	**Direct smears and flotation examinations**	**4/5—one infected puppy died** **No posttreatment examinations for organisms** **THIS WAS WHEN CANINE PARVO 2C HIT THE United States**	**Roudebush** (1978)
USA	6 Beagles, >3 months old, 5–10 kg, ≥750 cysts/gram by Direct Immunofluorescence Assay, Merifluor® Meridian BIOSCIENCE	3 treated: oral metronidazole suspension (Eradia/Ayradia, Virbac) 25 mg/kg BID for 5 days. 3 placebo-treated controls	Cysts in feces identified using DFA, Merifluor® Meridian BIOSCIENCE Trophozoites determined from washings of mucosae of intestinal sections after necropsy	Cyst counts reduced by >90% from day 3 to 5 (86.8%, 83.7%, 99.97%, 99.97%, 93.5%, from day 1 to day 5, respectively) Trophozoite counts reduced 99.9% in treated dogs (mean 183 in treated dogs vs. 192,106 in controls)	Bowman et al. (2019)

(continued)

Table 4.1 (continued)

Country	Number of dogs	Dose	Detection method	Cure rate (#Neg/#Cases) % Cured	References
Cuba	16 1- to 2-month-old dogs, breed unspecified	30, 40, 50, or 60 mg/kg PO BID for 10 days	ZnSO4 Centrifugal flotation—followed for 2 months with weekly samples beginning 7 days after the last treatment	16/16—all doses provided 100% efficacy 100%	Hernández et al. (1984)
Europe					
France	6 16–17-week-old Beagles	25 mg/kg PO and spiramycin 150,000 IU/kg PO SID 5 days	Cysts (Ovassay) and ELISA (ProSpecT) Before day 0, 3 consecutive Ovassays and 2 ELISAs—if any positive, dog positive ELISA/Cysts days 3, 5, 10, 17, and 18 after the first day of treatment (day 0)	Days 3, 5, 10, 17, 18: Ovassay: 2/6, 4/6, 6/6, 5/6, 1/6 ELISA 5/6, 6/6, 6/6, 1/6, 0/6 100% on day 10	Faure et al. (2018)
Czech Republic	**134 German shepherds; >164 treatments administered (some dogs treated more than once; cannot determine how many)**	**40–50 mg/kg PO SID 3–5 days—some (not specified) of the dogs were treated with metronidazole, but others were treated with ornidazole**	**Feces examined—flotation with Birch sugar Solution—specific gravity 1.300. Feces were examined again a week from the last day of treatment**	**Treated many dogs, but all were positive at least one time after treatment—but time from treatment to sampling was greater than or equal to a week**	**Decock et al.** (2003)
Denmark	1 10-week-old Danish Pointer	100 mg/kg BID for 3 days and ampicillin SID for 3 days	Trophozoites in peritoneum	1/1 100%	Hořejš and Koudela (1994)
Germany	2 (4-month-old German shepherd and 3-month-old king Charles spaniel)	250 mg BID for 5 days	Trophozoites and cysts with modified thiomersal, iodine, formalin sedimentation	2/2 100%	Bastholm and Kristensen (1986)

Italy	11 client-owned breed not specified	50 mg/kg SID 5 days with "low dose" sulfonamide antibiotic	Trophozoites via direct and stained smears	11/11 were *Giardia* negative 20 and 30 days after treatment. 100%	Jungmann et al. (1986)
Austria	6 (3 toy poodles, and a cocker spaniel, spaniel, and a shepherd cross) 2- to 5-months-old, <10 kg	125 mg PO BID 5 days	Cysts and trophozoites in smears and zinc sulfate flotation	6/6 no cysts in some dogs 3 days after the start of treatment and all negative 14 days after treatment 100%	Agresti et al. (1977)
Asia and Australasia					
Korea	7 (2 Shetland sheepdogs, 2 golden retrievers, 1 Samoyed, and 2 border collies) 12- to 36-months old	50 mg/kg PO SID 14 days	Triplicate zinc sulfate flotation days 2, 4, 6, 8, 10, 12, 14 of treatment performed only on IDEXX SNAP + samples [all negative samples were negative by SNAP and flotation]	7/7 on day 14 100%	Chon and Kim (2005)
Japan	1 pet 4-year-old mixed breeds coinfected with *Cryptosporidium*	60 mg/kg PO SID 6 days	Trophozoites via direct smear and sugar flotation	1/1 initially, trophozoites seen in feces. No cysts seen at days 0, 6, 10, 11, 12, 17, 18 100%	Matsubayashi et al. (2004)
Japan	4 (Labrador retriever, Yorkshire terrier, German shepherd, Welsh corgi) 2- to 4-months-old from pet shops or breeding kennel	25 mg/kg PO BID 5 days	Trophozoites and/or cysts via direct smear and formalin ethyl acetate sedimentation	2/4; retreated and 1 cleared of cysts, other did not bu. had clinical improvement 50% 3/4 (75%) after the 2 dogs that did not clear after the first treatment was retreated	Itoh and Muraoka (2001)
Tasmania	**9-month-old Weimaraner**	**Not stated; metronidazole with cortisone**	**Trophozoites in fecal smears**	**Improvement in diarrhea**	**Davies et al.** (1993)

(continued)

Table 4.1 (continued)

Country	Number of dogs	Dose	Detection method	Cure rate (#Neg/#Cases) % Cured	References
India	**16 3-month to 2-year old dogs; breed unspecified**	**30 mg/kg PO BID 3 (12 dogs) to 5 (8 dogs) days**	**Trophozoites in fecal smears**	**Signs have gone—but did not check the dogs again to see if shedding cysts or trophozoites**	**Reddy et al.** (1992)
Japan	9 2- to 3-month old puppies (2 Yorkshire terriers, 3 Maltese, 4 Shetland sheepdogs), 6 adults (Dachshund, Shetland sheepdog, Maltese, Yorkshire terrier, Pomeranian, Shiatzu)	60 mg/kg PO SID 6 days	Trophozoites in fecal smears and cyst counts using formalin ether sedimentation	14/15—all puppies and 5/6 adults were cleared. 93.3%	Sugano et al. (1989)
India	**8, client-owned 3–9-year-old; unspecified breed**	**100 mg/kg TID for 7 days**	**Trophozoites and cysts—direct smear and fecal flotation**	**Clinical recovery, but feces not examined post treatment for any stage of *Giardia***	**Chakrabarti et al.** (1982)
Japan	25 1.5- to 12-months-old (8 Shiba Inus, 5 Maltese, 2 Poodles, 5 Pomeranians, 2 Yorkshire terriers, 2 Pekinese, 1 Dachshund), many coinfected with coccidia and helminths[a]	60–120 mg/kg SID or BID for 6 or 7 days	Trophozoites and cysts in direct fresh and stained smears	25/25 100%	Sugano and Ando (1978)
Australia	**1 18-month-old King Charles Spaniel**	**200 mg BID for 6 days, then after recurrence of diarrhea, 450 mg BID for 10 days**	**None on flotations or smears. Trophozoites in duodenal aspirates and biopsies**	**100% after re-treatment**	**Watson** (1980)

Bold rows are papers where only clinical recovery was scored, not reductions in organisms

[a]Authors claim to have infected four puppies and one adult with stool from a dog with *Giardia*, and thought these dogs passed trophozoites 5–7 days PI and cysts 10–12 days PI—these results are confounded by the fact that the authors point out that these animals were also infected with coccidia and helminths. Therefore, a prior *Giardia* infection cannot be ruled out

(6.6 mg/kg) BID. Only the dogs receiving quinacrine were cleared of their infections, while tinidazole and metronidazole each only cleared two-thirds of the treated dogs. Hořejš and Koudela (1994) reported on the treatment of German Shepherd dogs, 5 adult males, 28 bitches, and 101 puppies (between 1.5 and 12 months of age) in a breeding colony over a period of 18 months with a total of 36.2% of 494 fecal samples being found positive by magnesium sulfate flotation (1.30 sp. gr.). There were cysts seen in 3.4% of 29 samples from males, 7.0% of 157 samples from bitches, and 53.2% of the samples from puppies. There was no temporal association observed between bouts of diarrhea and peaks of *Giardia* cyst shedding or between the shedding of *Giardia* cysts by bitches and the pups in their litters. Also, the majority of pups at 6–12 weeks of age had passed the peak of cyst shedding. When *Giardia* was detected in fecal samples, the dogs were treated with either ornidazole at 25 mg/kg or metronidazole 50 mg/kg for 3–5 days with efficacy being determined by fecal exams within 7 days after the completion of treatment. All treated animals were found to be *Giardia*-positive at least once following treatment over the period of the study. Coprological examination of the persons working in a breeding unit did not reveal any *Giardia* cysts. Itoh and Muraoka (2001) reported on the treatment of 11 dogs with histories of diarrhea or soft feces. *Giardia* was detected in the feces, and 5 of 7 dogs treated with tinidazole (50 mg/kg SID) for 3 days and 2 of 4 dogs treated with metronidazole (25 mg/kg BID) for 5 days were cleared of *Giardia*. When the four positive dogs were retreated each with the same regimen as previously used, only 1 dog remained positive.

Secnidazole There have been two reports examining the effects of secnidazole on *Giardia* in dogs. Karahallı and Ural (2017) followed two groups of six hospitalized dogs that had been diagnosed with Giardia and soft to liquid stools. The treated group received a single 500 mg tablet with a target dose of 30 mg/kg per dog; the other group remained untreated until the trial was completed. During the 10 days prior to treatment, all dogs had fecal consistency scores that were abnormally soft. After treatment, the feces in the treated dogs improved. Three days after treatment, one treated dog was positive, and on days 7 and 10 after treatment, all treated dogs were negative. Prior to the study all the control and treated dogs had cysts counts from 100,000 to 300,000 each and these counts remained high in the control dogs while they fell to zero in the treated dogs during the 10-day follow-up period. Thus, the efficacy of cyst reduction was calculated to be 100%. Cheung et al. (2019) reported on treating *Giardia* that was determined to be Assemblage C and D in two groups of shelter dogs, 14 dogs in Group A were adults or weaned dogs >10 weeks of age, and Group B consisted of 10 younger puppies <10 weeks of age. The dogs in Group A were treated with a single treatment of secnidazole at 30 mg/kg. In this group within 5–13 days after treatment, clinical signs ascribed as being *Giardia* associated resolved and 10 of the 14 dogs were *Giardia* antigen negative; one *Giardia* antigen dog continued to have signs. At the time of the second follow-up examination one dog that had been *Giardia* antigen negative and having normal stools were lost to the study, and at this time 12 of the 13 dogs were antigen negative. The third follow-up examination at 3–4 weeks after treatment looked at the

7 remaining dogs (6 had been returned to the general shelter population), and all these dogs were negative on the antigen test. In Group B, the 10 younger were treated with 30 mg/kg secnidazole and then moved to isolation, and at this time 3 of 9 group-housed littermate puppies with soft stools were antigen positive, and a single individually housed puppy continued to have signs while being antigen negative. At this time, a second dose of secnidazole was administered to each of the 9 littermates. At 3 weeks after the first treatment, only one littermate with clinical signs tested as being *Giardia* antigen positive. At the next follow up about 4 weeks after the first treatment, 6 of the 10 puppies were antigen positive, and the individual puppy was antigen negative and without signs. At 43 days after the first treatment, all 9 were considered to have normal stools, but 4 of the 9 puppies were positive for *Giardia* antigen.

Ronidazole Fiechter et al. (2012) examined the treatment of research dogs in a laboratory animal facility of the university. The isolate of *Giardia* in the infected dogs was identified as Assemblage C. There were 6 dogs with *Giardia* that were moved into the facility that was shampooed before and after oral treatment with ronidazole 30–50 mg/kg BID for 7 days. These dogs all were then negative on the tests for cysts and antigen. However, 33 days after the last treatment, one dog was coproantigen positive and 7 days later 4 dogs were coproantigen positive. The first cysts were detected in this group of dogs 54 days after the last treatment day. In the initially untreated group, 1–5 of the dogs tested positive regularly on at least one of the diagnostic tests performed. These dogs were then treated with ronidazole with the same regimen as the first group of dogs, and they were all still negative for *Giardia* when the study ended 19 days after the last treatment.

4.5.2 Benzimidazoles

The benzimidazoles bind to tubulin and inhibit its function in cells and by so doing inhibit the myriad of functions microtubules perform in cells. Fortunately, they disassociate rapidly from mammalian tubulin, but bind much more tightly to the tubulin of invertebrates and protists. In protists, like *Giardia*, this greatly interferes with their ability to undergo cellular division.

Again because fenbendazole is approved in many places for the treatment of *Giardia* in dogs, and because albendazole is not a treatment of choice for *Giardia* in dogs due to side effects, only recent work on fenbendazole will be reviewed, and albendazole research is not included. With the benzimidazoles, it seems likely that the different molecules will target similar mechanisms within the *Giardia* organism, and so, the development of resistance may be associated with cross-resistance seen with other members of the class.

Fenbendazole There have been two relatively recent studies on the efficacy of fenbendazole in the treatment of *Giardia* in dogs. Saleh et al. (2016) examined the control of laboratory dogs at a University laboratory colony in Virginia, and Faure et al. (2018) examined fenbendazole's efficacy in a clinical field trial comparing its use with that of an oral formulation of metronidazole.

Saleh's group dealt ultimately with 22 newly delivered older laboratory Beagles and an existing group consisting of a Beagle and 11 mixed breed dogs; overall the dogs were between 1.5 and 2.5 years of age. The 22 Beagles that joined the group came from a commercial breeding facility and all 22 had been treated with fenbendazole at 50 mg/kg SID for 3 days. A month after the arrival of the 22 dogs, it was noted that some of the dogs had diarrhea, and samples were collected from 32 of the 34 dogs and examined in several pooled groups using zinc sulfate flotation, and all were found to be possible. At this time all the dogs under observation were treated with metronidazole at 30 mg/kg SID for 7 days. The staff reported that some dogs still had diarrhea, and a decision was made to put in place an integrated *Giardia* control procedure, and the day before the program began, fecal flotations revealed that all of the 34 dogs were positive for *Giardia* cysts by zinc sulfate flotation, and it was ultimately revealed that the *Giardia* in question was Assemblage C. The floors of the runs where the dogs were housed were cleaned routinely using hot water and the washing of the material to a central drain. All dogs then began a 10-day course of oral fenbendazole at 50 mg/kg SID. On day 5 of treatment, all dogs were bathed in a top with shampoo and moved temporarily to part of the facility that had not housed any dogs in a year. At this time the runs were washed with hot water and then disinfected with quaternary ammonium solution, and then the temperature of the room was raised to 26.7 °C for 24 h (on day 6) to dry the runs. The dogs were returned to the runs on day 7 of treatment. The dogs were handed throughout by students, staff, and faculty with no restrictions other than the need to wear a laboratory coat, and no footbaths were utilized. All dogs had access to the outside where they were walked by students in the area surrounding the facility three times per week, and the dogs could also roam freely in an outdoor enclosure. The diagnostic for the purpose of following was the examination of a portion of formalin-fixed tissue dried on a slide and examined with a direct fluorescent assay from Meridian Bioscience. Using this test weekly sampling of the dogs during their stay showed that they all remained negative until the last day they were available for sampling at 115–199 days after the last day of treatment.

The study reported by Faure et al. (2018) utilized 178 of 193 dogs recruited at 44 study sites in France and Germany. To be included in the study, the dogs had to weigh at least 2 kg and be positive on the IDEXX *Giardia* Snap™ Test. Dogs were treated with metronidazole suspension at 25 mg/kg BID for 5 days or 50 mg of fenbendazole suspension at 50 mg/kg SID for 3 days. Of the dogs entering the trial, the final number analyzed for efficacy was 131 due to protocol deviations or no cysts in pretreatment samples being detected on indirect fluorescent antibody assays (IFAs); in the IFA, the detection of a single cyst was considered the equivalent of >750 cysts per gram. The pretreatment cysts per gram ranged from 0 to 1,116,000. This study showed the percentage of cyst per gram reduction for metronidazole was 91.9 and 30.3% for fenbendazole. This study compared only cyst reduction, not whether or not dogs were cyst free at the end of the treatment regimen.

Febantel There have been a good number of studies utilizing the pro-benzimidazole febantel. It appears in several formulations: one with pyrantel pamoate and praziquantel, and in a formulation with only febantel and pyrantel. In the United States, each "medium" tablet contains 68 mg praziquantel, 68 mg pyrantel pamoate, and 340 mg febantel, and in the EU, each tablet contains 50 mg of praziquantel, 144 mg of pyrantel embonate (=pamoate), and 150 mg febantel.

There were a number of early studies looking at the efficacy of febantel in these combination products as to their efficacy against *Giardia* shedding from dogs harboring the parasite. Barr et al. (1998) followed three groups of 5 dogs using zinc-sulfate flotation treated with a febantel, praziquantel, pyrantel combination. All the dogs were checked 3 times between 14 and 8 days before allocation and the day of allocation, and to be included in the studies the dogs had to be positive 3 of the 4 times, with one sample being on the day of allocation. The dogs were divided into three groups by randomly assigning them by weight—the medium weight in the three groups was 86, 9.1, and 10.4 kg. Group 1 dogs were treated with 26.8–35.2 mg febantel/kg SID for 3 days, Group 2 was given the same dose but only once on one day, and the third group was untreated. Four posttreatment samples were collected from all dogs at least 24 h apart within 6 days of the final treatment. All dogs in Group 1 were negative, 60% of the dogs in Group 2 were negative, and all the Group 3 dogs were positive on each of the four tests. Barutzki et al. (2002) treated 38 client-owned dogs where initial and follow up diagnoses were made with the ProspecT *Giardia* Microplate assay. Dogs were treated with about 15 mg/kg febantel (this was a combination product also containing pyrantel and praziquantel) being given on two consecutive days (dosing involved combinations of whole and half tablets). Fecal samples were collected on the third day after treatment for the antigen assay, and 35 of the 38 samples (92.1%) were negative. Schlüshe et al. (2002) treated 42 (34 one time and 8 dogs 2 days in a row) of 50 client-owned dogs with diarrhea that had been diagnosed with *Giardia* by the MIF (merthiolate–iodine–formaldehyde) concentration method. All 50 were shedding cysts and three were also passing trophozoites. In this case, the febantel was in a suspension with pyrantel, and the dogs were treated orally with 15 mg/kg febantel and 14.4 mg/kg pyrantel. Of the 34 dogs treated once, 27 dogs did not shed cysts for several days, and 15 of 27 dogs that could be tested remained negative for 5 days after the first treatment. The animals in the group that was treated twice did not she *Giardia* in the 5 days after the last treatment. The dogs had only been treated just once and were still shedding, were cured after a second treatment. Giangaspero et al. (2002) reported on treating 26 dogs with Febantel with 7 dogs being untreated controls. The treated groups A–D were treated as follows: A (9 dogs) 30–60 mg febantel/kg one time; B (10 dogs) treated at 15–30 mg febantel/kg for two consecutive days; C (7 dogs) for 3 consecutive days; and D (7 dogs) not treated. Fecal samples were collected on 2–6, 10, and 14 days after treatment. Of nine dogs in group A, 7 shed no cysts after treatment, 1 ha cysts on day 14, and another on days 3, 6, and 14; In Group B, all dogs were negative except one that shed on day 2, all dogs in group C remained negative, and in Group D dogs remained positive. Itoh et al. (2002) reported on the treating of 26 clinical canine cases in Japan: 8 with febantel at 30 mg/kg; 7 with albendazole at 25 mg/kg;

and 11 with fenbendazole at 50 ng/kg. The dogs were all cleared of their *Giardia* and associated diarrhea ended. Montoya et al. (2008) looked at the efficacy of febantel in 24 naturally infected dogs in three groups of 8 each (Groups A, B, and C) from a shelter in Spain. Group A received tablets at the dose of 15 mg/kg for 3 consecutive days, Group B the same dose for 5 consecutive days, and Group C served as untreated controls. In Group A one dog remained positive consistently after treatment, while a second dog only shed cysts on days 5 and 10. In Group B, two dogs shed on Day 9 and 1 dog shed on day 5. The dogs in group C remained consistently positive. De Souza et al. (2010) looked at the treatment of naturally infected dogs in Brazil with febantel. This study utilized 21 laboratory Beagles that were treated on three consecutive days with 15 mg febantel/kg. Feces were collected 5–3 days before treatment and after treatment on days 7–9, 14–16, and 21–23. After treatment at 7–9 days, feces of 14 dogs were negative for cysts, but only 3 were negative on the samples examined on days 14–16 after treatment, and all the dogs were positive again on days 21–23 after treatment.

Three additional papers on febantel compared the treatment of dogs with febantel to fenbendazole and mebendazole (Miro et al. 2007), metronidazole, fenbendazole, and oxfendazole (Decock et al. 2003), and with nitazoxanide (Moron-Soto et al. 2017). The last two listed here will be discussed in the sections dealing with oxfendazole and nitazoxanide, respectively. The report by Miro et al. (2007) presented a study on the effects of different benzimidazoles on both helminths and *Giardia*. After performing 1161 fecal samples from dogs in shelters, 321 dogs were noted to be infected with one of a number of parasites. From these positive dogs, 30 dogs were randomly allocated to five groups: *Toxocara canis*, *Toxascaris leonina*, hookworms (Ancylostomidae), Taeniidae, and *Giardia duodenalis*. The group assigned by the parasite was then subdivided into groups of 10 dogs each that were labelled A, B, and C. Dogs in Group A received mebendazole at a dose of 22 mg/kg once daily for 3 days, Group B dogs were treated with fenbendazole at 50 mg/kg once daily for 3 days, and Group C received a single treatment with febantel–pyrantel–praziquantel combination at a dose of 15–5–5 mg kg. After treatment, efficacy was determined for all parasites, not by the number of dogs cleared of their infections, but by posttreatment reductions in eggs/gram (or cysts per gram in the case of *Giardia*). In the case of reductions in *Giardia*, efficacies were not significantly different between treatments. The efficacies calculated at days 9 and 16 after treatment were, respectively, for mebendazole 75 and 85%, for fenbendazole 84 and 78%, and for febantel 72 and 73%.

Oxfendazole This benzimidazole is produced during the hydrolysis of the prodrug febantel as a step after fenbendazole. It is formulated in some oral anthelminthic suspensions for use in cattle. This drug has recently been recognized as a potentially new agent for helminth treatment in people (Gonzalez et al. 2019). It has been shown to have effects against *Giardia* trophozoites in vitro (Morgan et al. 1993), and there have been examinations as to its treatment efficacy for *Giardia* in dogs (Decock et al. 2003; Villeneuve et al. 2000).

Decock et al. (2003) published a comparison of four treatments for *Giardia* in dogs. These treatments were oxfendazole, febantel (in combination with pyrantel and praziquantel), fenbendazole, and metronidazole. The dogs were 30 laboratory Beagles (15 males and 15 females) with 6 dogs in 5 groups making up the controls and the four treated groups. Doses were oxfendazole 11.3 mg/kg SID for 3 days; febantel, praziquantel, and pyrantel at 15-5-5 mg/kg SID for 3 days; fenbendazoles 50 mg/kg SID for 3 days; and metronidazole at 25 mg/kg and spiramycin at 150,000 IU/kg SID for 5 days. Diagnostics that were employed included three consecutive stationary zinc sulfate flotations paired on the second and third tests with an indirect plate ELISA; dogs were considered positive if they were positive on any of the five tests performed. All dogs were positive before treatment, and samples from days 3, 5, 10, 17, and 18 were examined after treatment. After treatment, 100% of the untreated dogs were positive. The efficacy Oxfendazole appeared to be the lowest efficacy of any of the treatments with almost all dogs being found positive on all posttreatment sampling days. The metronidazole/spiramycin combination appeared to have the best overall efficacy, with no dogs in this group being positive 10 days after treatment. However, by 18 days after treatment, all dogs in all groups were positive with the exception of a single dog in the febantel group.

The study by Villeneuve et al. (2000) dealt with four different trials on dogs in breeding kennels. The diagnostic method used was a McMaster method with magnesium sulfate (1.28 specific gravity) where cysts were recorded per gram in groups of 0–5; corresponding to $0 = 0$, $1 < 10$, $2 = 10–100$, $3 = 100–200$, $4 = 200–1000$, and $5 = >5000$ cysts per 5 g of feces. The oxfendazole treatment regimens used in the four trials were as follows: Trial 1, 6 dogs with a mean age of 4.57 months treated with 11.3 mg/kg SID for 3 days; Trial 2, 11 dogs with a mean age of 3.81 months treated with 22.6 mg/kg SID for 3 days; Trial 3, 10 dogs with a mean age of 5.0 months treated with 22.6 mg/kg SID for 3 days and disinfection of housing; and Trial 4, 9 dogs with a mean age of 24.11 months treated with 11.3 mg/kg and disinfection of housing. The authors stated that it was their opinion that when dogs were positive again at 12 days, it was likely from reinfection, so samples were only tested on day12 in Trial 1. In Trial 1, there was a very marked drop from high cyst/gram numbers on day 5 and no cysts present in samples from any dog on day 9, but a percentage of samples were again positive on day 12. In Trial 2, there was a significant drop in cysts per gram in the treated dogs, but no apparent clearing of the infections. In Trial 3, all dogs were cleared of their infections on Days 5, 7, and 9. In Trial 4, as in Trial 3, all dogs were negative by Day 5 and stayed negative through Day 9.

4.5.3 *Thiazolides*

Nitazoxanide is the first of the synthetic salicylanilide derivatives of nitrothiazole. Different analogs of these compounds have activity against a number of parasites and viruses.

Nitazoxanide This product is known to most in the world of parasitology through the product Alinia that is approved for treating cryptosporidiosis in people. It also appeared in Navigator as a suspension for the treatment of horses with equine protozoal myeloencephalitis. There have been two cases where it has been examined for its effects on *Giardia* in dogs.

Lappin et al. (2008) reported on the attempted treatment of 16 dogs with *Giardia* as diagnosed with SNAP*Giardia* and fecal flotation. There were 7 dogs treated with fenbendazole (50 mg/kg SID for 5 days) and 9 with nitazoxanide (25 mg/kg BID for 5 days; 8 dogs). Of the 16 dogs, 8 had unexplained reactions, 3 treated with fenbendazole and 5 with nitazoxanide, and were removed from the study. This left four dogs in each of the two groups. After treatment, using the SNAP*Giardia* and fecal flotation, for fenbendazole there were 2, 1, and 2 dogs, respectively, positive on days 10, 14, and 34 after fenbendazole treatment, and similarly for nitazoxanide there were 4, 4, and 3, dogs infected on these days.

Moron-Soto et al. (2017) utilized five groups each containing seven naturally infected dogs that had *Giardia* and other routine protist and helminth parasites. The four treated groups included three groups treated with different doses of nitazoxanide on days 0 and 14 and one group treated with febantel on days 1, 2, and 3. The nitazoxanide was given to the dogs in Groups 1, 2, and 3 at 37.5 mg/kg, 75 mg/kg, or 150 mg/kg SID on days 0 and 14; the dogs were dosed with the febantel on days 1, 2, and 3, with the febantel, praziquantel, and pyrantel doses being portion of the dog at 15, 5, and 14.4 mg/kg, respectively. For the nitazoxanide treated dogs, for Group 1 the shedding of these dogs was equivalent to the untreated controls. For the groups receiving 37.5 mg/kg and 75 mg/kg the only dogs that shed were 1 in each of the two groups on days 11 and 14 In the case of febantel, a dog passed cysts on days 9, 11, and 14, and 5 dogs passed cysts on days 18, 25, and 28.

4.5.4 *Quinolines*

Two different quinolines have been used in dogs in trials for the treatment of *Giardia* in dogs. These are the acridine derivative quinacrine, = mepacrine, (known to most by the trade name Atabrine) and the related 4-aminoquinoline, chloroquine. These and other related products have historically been used in the prevention and control of malaria.

Quinacrine Croquette (1950) reported on the use of quinacrine to treat dogs with dysentery in Canada that had high numbers of *Giardia*; some of the dogs were felt to also be suffering from distemper. Dogs were treated with 0.2 g TID on day 1, followed by 0.2 g BID for the next 6 days. Smaller dogs received the same regimen, but with only 0.1 g per treatment with the dosages based on earlier work where the drug had been used to treat *Giardia* in people (cited in Croquette 1950). Croquette reported that overall, the responses were positive, and that the dysentery receded within a few days after the initiation of therapy. Rohde (1983) as cited above treated six normal Beagles passing cysts initially with ornidazole, and subsequently treated the same dogs with quinacrine that led to an excretion-free interval of 34 days, but a parasitological cure was not obtained, and dogs again became positive. Zajac et al. (1992) in an attempt to control for giardiasis that was diagnosed in a group of 12 mixed-breed dogs in an ongoing laboratory nutrition study, treated the dogs with either quinacrine or metronidazole and carefully monitored for *Giardia* for a period of 18 months with monthly fecal examinations and the collection of duodenal aspirates by endoscopy. The first treatment occurred 3 weeks after the initial diagnosis of *Giardia* was made with all dogs being treated with quinacrine at 6.6 mg/kg BID for 5 days. Thereafter, whenever a dog was found positive, all dogs were again treated with quinacrine or were treated with metronidazole at 50 mg/kg SID for 5 days [There were two exceptions, one when only individual positive dogs were treated and a case where a dog was cyst positive with a negative duodenal aspirate.]. Overall, fecal samples from all treated dogs were examined again 7–14 days after treatment. Over the course of the 18 months, there were 7 months where no dogs were found positive, i.e., most months after the initial treatment only 1–3 dogs were positive. During the last 3 months of the study, on months 16–18, the numbers of positive dogs were 7, 6, and 1, respectively. Interestingly, after the conclusion of the nutrition trial, 6 of the dogs were treated with corticosteroids, and three again became positive based on fecal examination of on aspirate. During this same period, one of the six dogs that did not receive corticosteroids also changes from negative to positive status.

Chloroquine Ural et al. (2017) reported on a study in Turkey where 20 of 26 dogs of various breed, age, and gender were treated with chloroquine and 6 dogs served as untreated controls. The treated dogs received chloroquine at 2.5 mg/kg BID for 5 consecutive days. The diagnostics used were fecal examinations and an antigen detection assays that were performed 10, 7, and 3 days before treatment and on days 0, 3, 7, and 10 after treatment. On day 0, the mean cysts per gram in the controls were about 208,000 and for the 20 controls was 165,000. On day 10 after treatment, the geometric mean number of cysts in the control dogs was 223,765/g with the mean of the controls being 7.7/g. On day 7 post treatment, there were 15 positive dogs and on day 10 post treatment there were only 8 positive dogs.

4.6 Concerns and Conclusions

It should be obvious from Table 4.1 summarizing the efficacy of metronidazole on *Giardia* in dogs with and without signs along with the text in Sect. 5 dealing with *Giardia* treatment trials in dogs using other products with some efficacy against protozoa, that these drugs often perform less than perfect and very rarely clear >90% of treated dogs of their *Giardia* infections. Overall, most parasitologists from both animal and human medicine agree that if suboptimal doses of a drug are routinely delivered that the chances of selecting for resistant traits in the target pathogen increase. This occurs simply because the ones that survive are the ones that can survive with a good deal of drug present in the environment. This is how drug-resistant *Giardia* isolates are selected in the laboratory. The organisms are grown in cultures, and then, a little drug is added, and any that have survived are moved to another new culture with the same low dose of the drug. Then, after they continue to thrive and multiply at this level, the drug dose is increased again, and the process repeated. Finally, one can have *Giardia* trophozoites that can live in high concentrations of the product that once killed almost every single trophozoite. As was noted above, it has proven difficult for unknown reasons to isolate the canine C and D Assemblages into culture, so these studies have not been performed to any extent with canine isolates, but basically, the same process is being performed routinely in the real world. If you think about it as one thinks about nematodes in sheep or horses, when you treat all the animals, you leave no parasite refugia to provide competing forms that are susceptible to the drug to breed with the resistant forms. Also, in the case of *Giardia*, whether or not sex occurs between organisms still remains an open question. Thus, these organisms maybe even more likely to retain their drug-derived characteristics than are mating nematodes. Of course, the resistant forms may lose in a race with conspecific nonresistant trophozoites, if they have downregulated some multiplicative gene for more resistance or part of the process for attachment to the mucosal surface, trophozoites not yet be exposed to the drug may outcompete them in hosts where there is no drug. We are probably still a long way from answering this question at this time.

Again, the canine Assemblages C, and D, and the Assemblage AI may be zoonotic, but C and D seem to be highly constrained to the dog, and AI is relatively uncommon in dogs (and of course, who gives what to whom?). However, what happens if C and D are found in people somewhere in some regularity? What if researches are able to get C and D routinely into cultures? What if the C and D in cultures are found to be resistant to the drugs that are used to treat *Giardia* in people? There are likely to be no positive outcomes for dogs or veterinarians. It would seem that the rational approach is to follow the rational protocol put forward by Tysnes et al. (2014), and herein it is suggested that veterinarians should be suggesting treatment of a given dog without signs only under very carefully considered conditions. Also, if treatment clears the signs, there is really no reason to treat repeatedly until the dog or all the dogs in the household are negative, because if you look at all the papers summarized in this report, there are going to be many, many dogs that never clear with any of the product currently available.

References

Agresti A, D'Ambrosio G, Gravino E (1977) La Giardiasi ne cane. Indagine epizoologica ed osservazioni cliniche [Italian]. Acta Med Vet 23:175–182

Al-Jawabreh A, Ereqat S, Dumaid K et al (2019) Prevalence of selected intestinal protozoan infections in marginalized rural communities in Palestine. BMC Public Health 19:1667, 11 pp

Argüelo-Garcia R, Leitsch D, Skinner-Adams T et al (2020) Drug resistance in *Giardia*: mechanisms and alternative treatments for Giardiasis. Adv Parasitol 107:201–282

Barr SC, Bowman DD, Frongillo MF et al (1998) Efficacy of a drug combination of praziquantel, pyrantel pamoate, and febantel against giardiasis in dogs. Am J Vet Res 59(9):1134–1136

Barutzki D, Schimmel A, Schaper R (2002) Efficacy of pyrantel embonate, febantel and praziquantel against *Giardia* spp. in naturally infected dogs. In: Olson BE, Olson ME, Wallis PM (eds) *Giardia* the cosmopolitan parasite. CABI, Wallingford, pp 177–180

Bastholm AH, Kristensen TS (1986) *Giardia* i vesica og peritonealhule hos hund. [Danish]. Dansk Vet Tidsskr 69(12):686–690

Belding DL (1965) Giardia lamblia; chapter 5: intestinal, oral, and vaginal flagellates of man, pp 125–126. In: Textbook of clinical parasitology. AppletonCentury-Crofts, Hobe Sound, FL, 1374 pp

Bemrick WJ (1961) A note on the incidence of three species of Giardia in Minnesota. J Parasitol 47 (1):87–88

Bowman DD, Lucio-Forster A (2010) Cryptosporidiosis and giardiasis in dogs and cats: veterinary and public health importance. Exp Parasitol 124(1):121–127

Bowman DD, Nicolas C, Navarro C et al (2019) A pilot study to determine the efficacy of a metronidazole suspension on the reduction of cyst and trophozoite counts in dogs infested with *Giardia*. Abstracts of the 27th conference of the world association for the advancements of veterinary parasitology, July 7–11, 2019 Madison, Wisconsin. Abstract PS01.34, p 206

Burrows TB (1968) Internal parasites of dogs and cats from central New Jersey. Bull NJ Acad Sci 13(2):3–8

Catcott EJ (1946) The incidence of intestinal protozoa in the dog. J Am Vet Med Assoc 108:34–36

CDC (2001) Prevalence of parasites in fecal material from chlorinated swimming pools – United States, 1999. MMWR 50(2):410–412

Chakrabarti A, Pramanik AK, Mitra SK (1982) Malabsorption syndrome in *Giardia canis* infested dogs in Calcutta. Indian J Vet Med 2:42–44

Chang K (1935) A general survey of the protozoa parasitic in the digestive tract of Shan-tung mammals. Peking Nat Hist Bull 9(3):151–159. 1 pl., figs. 1–12

Cheung W, Russo C, Maher S et al (2019) Successful use of secnidazole to manage a giardiosis outbreak in a shelter. Vet Parasitol 274:108911

Chon SK, Kim NA (2005) Evaluation of silymarin in the treatment on asymptomatic *Giardia* infections in dogs. Parasitol Res 97:445–451

Choquette LPE, Gelinas LG (1950) The incidence of intestinal nematodes and protozoa in dogs in the Montreal district. Can J Comp Med 14(2):33–38

Conboy G (1996) Diagnostic parasitology. Can Vet J 37:181–182

Craige JE (1948) Differential diagnosis and specific therapy of dystenteries in dogs. J Am Vet Med Assoc 113:343–347

Craig CF, Faust EC (1945) Clinical parasitology, 4th edn. Lea & Febiger, Philadelphia

Croquette LPE (1950) Canine giardiasis and its treatment with atebrin. Can J Comp Med 14:230–235

Davies NJ, Harrison GC, Goldsmith JM (1993) A case of giardiasis in a dog in Tasmania – a cause of disease or incidental finding. Aust Vet J 70:32–33

De Lucio A, Bailo B, Aguilera M et al (2017) No molecular epidemiological evidence supporting household transmission of zoonotic *Giardia duodenalis* and *Cryptosporidium* spp. from pet dogs and cats in the province of Álava, Northern Spain. Acta Trop 170:48–56

De Souza CP, Verocai GG, Correia TB et al (2010) Eficácia do febantel, pamoato de pirantel e praziquantel no tratamento de *Giardia* spp. em cães naturalmente infectados. Rev Bras Med Vet 32(Suppl 1):21–24

Decock C, Cadiergues M-C, Roques M et al (2003) Evaluation de quatre traitements de la giardiose canine. Rev Med Vet 154(12):763–766

Faure L, Fournel S, Nicolas C et al (2018) A field clinical study to confirm the efficacy and safety of a metronidazole-based oral suspension in dogs naturally infested by giardiasis: comparison to fenbendazole. Int J Appl Res Vet Med 16(2):110–116

Fiechter R, Deplazes P, Schnyder M (2012) Control of *Giardia* infections with ronidazole and intensive and intensive hygiene management in a dog kennel. Vet Parasitol 187(1):93–98

Giangaspero A, Traldi G, Paoletti B et al (2002) Efficacy of pyrantel embonate, febantel and praziquantel against *Giardia* species in naturally infected adult dogs. Vet Rec 150:184–186

Gleason NN, Horwitz MS, Newton LH et al (1970) A stool survey for enteric organisms in Aspen, Colorado. Am J Trop Med Hyg 19(3):480–484

Gonzalez AE, Codd EE, Horton J et al (2019) Oxfendazole a promising agent for the treatment and control of helminth infections in humans. Expert Rev Anti-Infect Ther 17(1):51–56

Hegner RW, Chu HY (1930) Aa survey of protozoa parasitic in plants and animals of the Philippine Islands. Philipp J Sci 43(3):451–482

Hernández O, Gómez E, Blandiono T (1984) Effectividad del metronidazol contra la giardiasis del perro. [Spanish]. Rev Salud Anim 6:143–145

Hořejš R, Koudela B (1994) Giardióza psů v chovné stanici (In Polish). Vet Med 39(2/3):93–101

Itoh N, Muraoka N (2001) Giardiosis in dogs. Jpn J Vet Med 54(9):713–716

Itoh N, Muraoka N, Aoki M et al (2002) Treatment of canine giardiasis with benzimidazoles. [Japanese with English summary] Jpn J Vet Med Assoc 55(11):739–743

Jungmann R, Hiepe T, Scheffler (1986) Zur parasitären Intestinalfauna be Hund und Katze mit einem speziellen Beitrag sur *Giardia*-Infektion [German]. Mh Vet Med 41:309–311

Karahallı C, Ural K (2017) Single dose secnidazol treatment efficacy against naturally occurring *Giardia duodenalis* in dogs. Magyar Allat Lapja 139(Okt):621–630

Kotloff KL, Nataro JP, Blackwelder C et al (2013) Burden and aetiology of diarrhoeal disease in infants and young children in developing countries (the Global Enteric Multicenter Study, GEMS): a prospective, case-control study. Lancet 382:209–222

Lappin MR, Spindel M, Riggenbach L (2008) Infectious agent prevalence rates in dogs with diarrhea and response to administration of fenbendazole or nitazoxanide. Annual meeting ACVIM 2008, small animal infectious disease section

Leitsch D (2015) Drug resistance in the microaerophilic parasite *Giardia lamblia*. Curr Trop Med Rep 2:128–135

Leitsch D, Williams CF, Hrdy I (2019) Redox pathways as drug targets in microaerophilic parasites. Trends Parasitol 34(7):576–592

Levine ND (1973) Protozoan parasites of domestic animals and man. Burges, Minneapolis

Levine ND, Ivens V (1965) Prevalence of nematodes, Giardia and Demodex in Illinois dogs. Illinois Vet 8(23):19–23

Ma D-W, Lee M-R, Hong SH et al (2019) Molecular prevalence and genotypes of *Cryptosporidium parvum* and *Giardia duodenalis* in patients with acute diarrhea in Korea, 2013-2016. Korean J Parasitol 57(5):531–536

Matsubayashi M, Abe N, Kimata I et al (2004) Mixed infection of *Cryptosporidium* and *Giardia* in a pet dog. [Japanese]. J Jap Vet Med Assoc 57:317–320

Medema GJ, Schijven JF (2001) Modelling the sewage discharge and dispersion of *Cryptosporidium* and *Giardia* in surface water. Water Res Int 35(18):4307–4316

Miro G, Mateo M, Montoya A et al (2007) Survey of intestinal parasites in stray dogs in the Madrid area and comparison of the efficacy of three anthelmintics in naturally infected dogs. Parasitol Res 100:317–320

Montoya A, Dado D, Mateo M et al (2008) Efficacy of Drontal Flavour Plus (50 mg praziquantel, 144 mg pyrantel embonate, 150 mg febantel per tablet) against *Giardia sp* in naturally infected dogs. Parasitol Res 103:1141–1144

Moore GT, Cross WM, McGuire D et al (1969) Epidemic giardiasis at a ski resort. New Eng J Med 281:402–407

Mørch K, Hanevik K, Robertson LJ et al (2008) Treatment ladder and genetic characterisation of parasites in refractory giardiasis after an outbreak in Norway. J Infect 56(4):268–273

Morgan UM, Reynoldson JA, Thompson RCA (1993) Activities of several benzimidazoles and tubulin inhibitors against *Giardia* spp *in vitro*. Antimicrob Agent Chemother 37(2):328–331

Moron-Soto M, Gutierrez L, Sumano H et al (2017) Efficacy of nitazoxanide to treat natural *Giardia* infections in dogs. Parasit Vectors 10:52

Nesvadba J, Nesvadba J Jr, Nesvabda Z (1980) Beitrag zur klinischen Manifestation und Therapie der Giardiasis 26. Jahrestagung, Deutsche Veterinarmedizinische Gesellschaft Fachgruppe Kleintierkrankheiten in Verbindung mit der deutschsprachigen Gruppe der WSAVA, 13-- 15 November 1980 in Karlsruhe, pp 231–232

Painter JE, Gargano JW, Collier SA et al (2015) *Giardia* surveillance United States 2011–2012. MMWR 64(3):15–15

Reddy NRJ, Thimmappa M, Ranganath L et al (1992) Treatment of giardiasis with metronidazole in dogs. Indian Vet J 69:163–164

Rehbein S, Klotz C, Ignatius R et al (2018) *Giardia duodenalis* in small animals and their owners in Germany: a pilot study. Zoonoses Public Health 66(1):117–124

Rijpstra AC (1967) Proc Koninkl Nederl Akad Wet Ser C 70:395–401

Robertson LJ, Hermansen L, Gjerde BK et al (2006) Application of genotyping during an extensive outbreak of waterborne Giardiasis in Bergen, Norway, during Autumn and Winter 2004. Appl Environ Microbiol 72(3):2212–2217

Robertson LJ, Hanevik K, Escobedo AA et al (2010) Giardiasis: why do the symptoms sometimes never stop? Trends Parasitol 26(2):75–82

Rohde E (1983) Therapieversuche der Giardiasis des Hundes. Inaugural-Dissertationsur Erlangung der Doktorwürde der Veterinär-Medixinishcen Fakultät der Universität Zürich. 29 pp

Roudebush P (1978) Clinical giardiasis in small animals. Mo Vet 28:5–7

Saghaug CS, Klotz C, Kallio JP et al (2019) Genetic variation in metronidazole metabolism and oxidative stress pathways in clinical *Giardia lamblia* assemblage A and B isolates. Infect Drug Resist 12:1221–1235

Saleh MN, Gilley AD, Byrnes MK et al (2016) Development and evaluation of a protocol for control of *Giardia duodenalis* in a colony of group-housed dogs at a veterinary medical college. J Am Vet Med Assoc 249(6):644–649

Schlüshe A, Grewing M, Mehlhorn H et al (2002) Efficacy of febantel and pyrantel embonate (Drontal® Puppy) on *Giardia* infections in dogs and mice: a therapeutic and cytologic study. In: Olson BE, Olson ME, Wallis PM (eds) *Giardia* the cosmopolitan parasite. CABI, Wallingford, pp 201–208

Serradell MC, Saura A, Rupil L et al (2016) Vaccination of domestic animals with a novel oral vaccine prevents *Giardia* infections, alleviates signs of giardiasis and reduces transmission to humans. NPJ Vaccines 2016:16018

Serradell MC, Gargantini PR, Saura A et al (2018) Cytokines, antibodies, and histopathological profiles during *Giardia* infection and variant-specific surface protein-based vaccination. Infect Immun 86(6):e00773–e00717

Shields JM, Gleim ER, Beach MJ (2008) Prevalence of *Cryptosporidium* spp. and *Giardia intestinalis* in swimming pools, Atlanta, Georgia. Emerg Infect Dis 146:948–950

Simmons J, Passon TJ (1981) Diagnosis of *Giardia canis*, the elusive parasite. Vet Med Small Anim Clin 76(1):55–56

Sugano H, Ando K (1978) Giardiasis in dogs. [Japanese]. J Jpn Vet Med Assoc 31:635–638

Sugano H, Fukase T, Chinone S et al (1989) Giardiasis of puppies from a breeder's kennel. J Jpn Vet Med Assoc 42:68–71

Tantrongsup S, Scorza V (2010) Update on the diagnosis and management of *Giardia* spp infections in dogs and cats. Top Comp Anim Med 25(3):155–161

Thomas RE (1960) Incidence of intestinal parasites in German Shepherd dogs. J Am Vet Med Assoc 136(1):25–26

Tysnes KR, Skancke E, Robertson LJ (2014) Suclinica Giarida in dogs: a veterinary conundrum relevant to human infection. Trends Parasitol 39(11):521–527

Ural K, Gultekin M, Pasa S et al (2017) Chloroquine treatment against naturally occurring *Giardia duodenalis* infection in dogs. Med Wet 73(5):280–283

Villeneuve V, Beugnet Fm Bourdoiseau G et al (2000) Efficacy of oxfendazole for the treatment of giaridiosis in dogs. Experiments in dog breeding kennels. Parasite 7:221–226
Watson ADJ (1980) Giardiosis and colitis in a dog. Aust Vet J 56:444–447
Zajac AM, Leib MS, Burkholder WJ (1992) *Giardia* infection in a group of experimental dogs. J Small Anim Pract 33:257–260
Zimmer JF, Burrington DB (1986) Comparison of four protocols for the treatment of canine giardiasis. JAAHA 22:168–171

Chapter 5
Dogs and Their Role in the Eco-epidemiology of Chagas Disease

Ricardo E. Gürtler and Marta V. Cardinal

Abstract Chagas disease, caused by infection with *Trypanosoma cruzi*, is an important disease of humans and dogs in the Americas. Here, we review aspects of canine Chagas disease with emphasis on its diagnosis, eco-epidemiology, transmission and control, and the role of dogs as a risk factor for human infection and sentinel animals. Dogs may become infected through multiple routes mediated by triatomine bugs, including consumption of mammal preys; maintain *T. cruzi* in the absence of any other host species (a primary reservoir host); provide a bridge between sylvatic, peridomestic and domestic transmission cycles, and increase bug population size. Dog blood meals were reported in $\geq$31 triatomine species collected in domestic and/or peridomestic habitats. A systematic search including 309 published reports showed that the prevalence of *T. cruzi* infection in dogs typically varied from 10 to 30% across the Americas, and sometimes exceeded 50%. Serological cross-reactivity with other potentially co-endemic trypanosomatids supports the use of more specific methods for confirmation. Xenodiagnosis-based assessments showed that the mean infectiousness of *T. cruzi*-seropositive dogs to seropositive humans differed by 4–10×. Mathematical modeling supports that dogs amplify domestic parasite transmission. House residual spraying with pyrethroid insecticides provided high levels of protection to both humans and dogs in endemic areas.

Keywords *Trypanosoma cruzi* · Triatominae · Reservoir host · Dog · Population dynamics · Vector control · Infectiousness · Treatment · Molecular epidemiology · Host-feeding patterns

---

R. E. Gürtler (✉) · M. V. Cardinal
Facultad de Ciencias Exactas y Naturales, Laboratorio de Eco-Epidemiología, Universidad de Buenos Aires, Ciudad Universitaria, Buenos Aires, Argentina

Instituto de Ecología, Genética y Evolución de Buenos Aires, Consejo Nacional de Investigaciones Científicas y Técnicas, Ciudad Universitaria, Buenos Aires, Argentina
e-mail: gurtler@ege.fcen.uba.ar

C. Strube, H. Mehlhorn (eds.), *Dog Parasites Endangering Human Health*, Parasitology Research Monographs 13,
https://doi.org/10.1007/978-3-030-53230-7_5

5.1 Background

Dogs play a central role in current human societies across the world. Their numbers have been increasing approximately at the same pace as the human population and may number one billion worldwide (Gompper 2014). While the main role of dogs may be as pets and companion animals, they are used for guarding property against human or animal intruders, hunting, herding, law enforcement, assistance, research, transport, and as food (Macpherson et al. 2013). The close association of dogs with humans implies a direct dependence on humans for feeding and multiple favorable and unfavorable effects, such as dog bites, noise, fowling of pavements, and predation of threatened or non-threatened wildlife species (Gompper 2014).

Dogs are also involved in the persistence and transmission of ~65 zoonotic pathogens that circulate between animals and humans, including bacteria, virus, protozoa, helminths, and ectoparasites (Macpherson et al. 2013). Children and people with compromised immune systems or nutritional and metabolic disorders are most vulnerable to zoonotic infection and disease. Ranking high in the list since biblical times is rabies, which still causes the annual death of 65,000 people, mostly in Africa. Among vector-mediated zoonoses, dogs play a key role as reservoir hosts of *Trypanosoma cruzi*, a hemoflagellate protozoan that causes human Chagas disease (American Trypanosomiasis), transmitted by triatomine bugs in the Americas.

This chapter focuses on canine Chagas disease and reviews the key roles that dogs play as maintenance and amplifying hosts of *T. cruzi* in many transmission cycles covering diverse ecoregions and triatomine species. We update our previous review (Gürtler and Cardinal 2015) and expand into some contemporary issues (e.g., diagnosis and molecular epidemiology). After summarizing the rich diversity of *T. cruzi* and its life cycle in mammalian hosts and vectors, we describe the multiple routes of transmission in dogs, the course of infection, clinical signs and pathology, and current diagnostic methods. In the section on eco-epidemiology, we review the fundamental pieces of evidence that show that dogs are able to maintain *T. cruzi* in the absence of any other host species; increase pathogen and vector abundance, and maybe a bridge between sylvatic, peridomestic and domestic transmission cycles. We provide an overview of methods designed for prevention and treatment of dog infections.

5.2 Parasite Diversity

Trypanosoma cruzi (Kinetoplastida: Trypanosomatidae) is a multihost pathogen of mammals and a serious cause of morbidity and mortality affecting six to nine million people (WHO 2002, 2015). Human infection with *T. cruzi* follows a short acute phase and a lifelong chronic phase with an indeterminate (preclinical) form and a determinate form; the latter has cardiac, cardiodigestive, or digestive (megaesophagous and megacolon) manifestations with large geographic variations (Rassi et al. 2017). Approximately 200 species of nonhuman mammalian species have been found naturally infected with *T. cruzi*; the main hosts are opossums, armadillos, rodents, humans, dogs, and cats (Jansen et al. 2017).

Trypanosoma cruzi has a genetically diverse clonal structure classified into six genotypes (TcI-TcVI) denominated Discrete Typing Units or DTUs (Zingales 2018). Another genotype genetically close to TcI is mostly restricted to bats (TcBat) in sylvatic habitats and may enlarge the list of DTUs (Marcili et al. 2009; Zingales 2018). All DTUs can cause human or mammal infection and disease, and as yet there is no clear association between DTU and clinical presentation (Zingales et al. 2012).

The geographical distribution and frequency of occurrence of DTUs vary widely across the Americas (Brenière et al. 2016). TcI is closely associated with *Didelphis* opossums and predominates in domestic transmission cycles occurring north of the Amazon basin. TcIII and TcIV mainly circulate in sylvatic transmission cycles. TcII, TcV, and TVI predominate in domestic habitats of the Southern Cone countries. The six DTUs are currently associated with distinct ecological niches. This has implications for understanding the eco-epidemiology of *T. cruzi* and developing improved disease control and surveillance strategies (Miles et al. 2009).

5.2.1 *Life Cycle*

Trypanosoma cruzi reproduces both in the insect vector and in the mammalian host. An uninfected triatomine bug acquires the infection while blood-feeding on an infected mammal that carries bloodstream (non-replicating) trypomastigotes. The ingested parasites undergo several morphological changes before attaching to the walls of the hindgut as epimastigotes; they reproduce by binary fission, and ultimately develop as metacyclic trypomastigotes (i.e., the infective stage). When these trypomastigotes are defecated on the skin or mucosae of a susceptible host and succeed in traversing the skin layer, they penetrate the white cells (mainly macrophages) where they evade the immune system, lose the flagellum, and reproduce as intracellular amastigotes. The latter develops into trypomastigotes, which are released when the cell is ruptured, spread throughout and penetrate other cells, or are killed by the immune system or ingested by blood-feeding triatomines or other arthropods.

5.2.2 *Routes of Transmission*

Foundational work established the multiple routes of transmission of *T. cruzi* (references in Hoare 1972; Zeledón 1974; Minter 1976a). Host infection with *T. cruzi* is typically acquired through skin contamination with feces from infected triatomine bugs (stercorarian transmission). Infection may have an oral point of entry: it may occur through ingestion of infected triatomines, perhaps the main route in mammals that prey on insects; ingestion of food and beverages contaminated with feces of infected triatomines or with urine or feces of infected opossums or other reservoir hosts, as in foodborne outbreaks of human Chagas disease; licking furs contaminated with infected bug feces or infected blood; ingestion of domestic flies or cockroaches that had fed fresh feces from infected triatomines; and ingestion of *T. cruzi*-infected mammals or infected raw meat. Other routes include vertical transmission, either transplacental or through maternal milk; unprotected contact with infected body secretions or blood; blood transfusion; and organ transplantation. *Didelphis* opossums may maintain a vector-independent cycle through secretions of anal glands (Jansen et al. 2017). In triatomine bugs, *T. cruzi* is rarely transmitted horizontally (Schaub 1988), and there is no vertical transmission.

Several transmission routes involve the participation of triatomines directly or indirectly, yet others are restricted to clinical settings. *Trypanosoma cruzi* was isolated from saliva and urine of pups experimentally infected with massive infectious doses (Marsden and Hagstrom 1968). Although all transmission routes are relevant to dogs, vector-mediated and oral infections with *T. cruzi* appear to be more likely than other routes. In general, the relative contribution of each route to overall transmission in nonhuman mammals is hard to determine as several of them operate in conjunction (e.g., Cardinal et al. 2014).

Oral infections may be especially frequent in settings where intrusive sylvatic bugs infected with *T. cruzi* invade house premises and dogs prey upon them, likely leading to virulent infections similar to those recorded in foodborne outbreaks of human Chagas disease. Using raccoons as surrogates of dogs, the probability of infection of a host following consumption of a *T. cruzi*-infected vector was estimated to range between 0.177 and 1 (Kribs-Zaleta 2010, p. 4). This range is orders of magnitude greater than the range of estimates for typical stercorarian transmission to humans (Nouvellet et al. 2013). The other components involved in the oral transmission is the rate of host–vector contact followed by ingestion of the insect. This can be estimated from data collected in two insecticide trials in which untreated (control) dogs were housed overnight in closed experimental huts, and the numbers of triatomines (fourth- or fifth-instar nymphs and adults) initially released and lost (presumably consumed) were recorded over defined exposure periods. The daily rate of bug consumption per dog averaged 0.214 (81 bugs lost to 3 dogs over 126 days) (Reithinger et al. 2006) and 0.200 (27.5% of 240 initially released bugs lost to 3 dogs over 110 days) (Gürtler et al. 2009b). Thus, each dog consumed nearly 1 bug every 5 days of overnight exposure in a confined space, suggesting a large potential for repeated oral infections.

5.2.3 Other Trypanosomatids

Dogs may harbor other Trypanosomatids in the Americas. *Trypanosoma rangeli* infects humans, dogs, and other mammals through the bite (salivaria transmission) of several species of the genus *Rhodnius* (widespread from 30°S in the Southern Amazon Basin up to Central America) principally; *T. rangeli* is considered nonpathogenic for mammals and pathogenic for triatomines (Vallejo et al. 2009). Dogs may be also infected with *Trypanosoma evansi* (the etiologic agent of "surra" or "mal das cadeiras," a severe canine disease transmitted by hematophagous Diptera); *Leishmania infantum* (agent of canine visceral leishmaniasis, Quinnell and Courtenay 2009), *Leishmania braziliensis* and other *Leishmania* species that cause American cutaneous leishmaniasis (Reithinger and Davies 1999), all transmitted by sand fly bites. *Trypanosoma vivax* infects multiple ungulate species, not carnivores. These Trypanosomatids are widely distributed in Latin America; circulate co-endemically in some geographic areas; cross-react serologically with *T. cruzi* (Desquesnes 2017); and hinder diagnosis (e.g., Lucheis et al. 2005). For example, in the Pantanal of Mato Grosso do Sul, *T. evansi* infected 30% of dogs (Franke et al. 1994). *Trypanosoma caninum*, apparently a nonpathogenic trypanosome recently isolated from the intact skin of dogs in various parts of Brazil, may cross-react serologically with *Leishmania infantum* and *T. cruzi* (Pinto et al. 2010; Madeira et al. 2014). However, the antibody response of dogs to *T. caninum* appears to be poor and time-limited.

5.3 Course of Infection, Clinical Signs, and Pathogenesis

Primary infection with *T. cruzi* in dogs includes an acute phase with a short latent period of <1 and 2 weeks for parasitological and serological conversion, respectively, followed by a lifelong chronic phase (Lauricella et al. 1986; Machado et al. 2001; Barr 2009). Patient parasitemia appears as early as 3 days post-infection (DPI) and wanes approximately 1 month later, whereas specific antibodies to *T. cruzi* appear 2–3 weeks post-infection and persist lifelong in the absence of etiological treatment or immunosuppression. Parasitemia peaks at about 17 DPI, in coincidence with signs of generalized lymphadenopathy and acute myocarditis.

During the acute phase, lethargy (sometimes accompanied by splenomegaly and hepatomegaly) is common in young puppies, and much less severe or inapparent in dogs aged more than 6 or 12 months, in which the infection runs a milder, subclinical course (Barr 2009; Snowden and Kjos 2012). Clinical signs of *T. cruzi* infection vary depending on the phase of infection, inoculum size and route, breed and age of the dog, and virulence of parasite strain (Lana 2017). Young dogs aged less than 6 months that also present acute myocarditis and high parasitemia have a poor prognosis and may suffer large mortality (Barr 2009; Zeledón et al. 2012; Snowden and Kjos 2012). In Texas, 94% of 335 cases brought to veterinary clinics involved purebred dogs, and especially affected sporting and working dogs (Kjos et al. 2008). In endemic rural areas, however, early observations documented a large number of either fatal or uncomplicated acute infections with *T. cruzi* in well-nourished mongrel dogs aged less than 3 months, often with no pathognomonic signs of clinical disease (Mazza 1934; Mazza and Lobos 1937; Gürtler et al. 2007a).

Dogs that survive the acute phase enter a prolonged chronic phase showing no clinical signs and no ECG abnormalities (indeterminate form). Likewise infected humans, a fraction of the infected dogs develop chronic myocarditis with cardiac dilation and concomitant ECG abnormalities within a few years, which may result in sudden death (Barr 2009). A detailed account of the pathological findings recorded in experimentally and naturally infected dogs may be found elsewhere (Snowden and Kjos 2012; Lana 2017). Histopathology of the myocardium is characterized by multifocal interstitial mononuclear cellular infiltrates, perivasculitis, marked fibrosis, and rare parasitic pseudocysts. Meningoencephalitis is less common. Megacolon and megaesophagus occurred in a few dogs experimentally infected with *T. cruzi* (Lana 2017). In Texas (USA), dogs naturally infected with *T. cruzi* that survived the acute phase displayed enlarged heart, lethargy, anorexia, ascites, and cardiac conduction disturbances (Kjos et al. 2008); they were more likely to have ventricular arrhythmias (Meyers et al. 2019), and represented a significant veterinary problem. Electrocardiographic surveys of an unselected, large dog population in northwest Argentina documented a very low frequency of abnormalities (or unspecific ones) in dogs seropositive for *T. cruzi* (Lauricella et al. 1989), whereas significant cardiac alterations attributable to the parasite were revealed elsewhere (Meurs et al. 1998; Cruz-Chan et al. 2009; González-Vieyra et al. 2011). These alterations included first-degree heart block, right bundle-branch block, decreased-amplitude QRS complexes, ventricular premature complexes, atrial fibrillation, ventricular tachycardia, and third-degree atrioventricular block.

5.4 Diagnosis

Canine infections with *T. cruzi* are detected using the same methods as for human infections (Luquetti and Schmunis 2017). Chagas disease may be suspected in any dog with signs of myocarditis or cardiomyopathy or electrical conduction disturbance in the heart, and history of exposure to triatomine bugs (Barr 2009; Snowden and Kjos 2012).

Canine infections with *T. cruzi* may be detected using parasitological, serological, and molecular methods depending on whether the infection is acute or chronic. Giemsa- or Wright-stained thick or wet blood smears, the Strout method, and the microhematocrit centrifugation method (with diligent microscopical examination at 400×) are indicated for the acute phase but their sensitivity is limited (Barr 2009). Direct parasite detection by any of these methods, or pre- or postmortem identification of *T. cruzi* in histologic examinations of myocardium or other tissues, provides a definitive diagnosis of canine Chagas disease (Snowden and Kjos 2012).

Xenodiagnosis and hemoculture are more sensitive, costly, and laborious than the above-mentioned direct methods, and results usually take several weeks; their sensitivity is affected by the total amount of blood tested (i.e., number of laboratory-reared, uninfected triatomines, or hemoculture tubes) and other technical details (Luquetti and Schmunis 2017). Both methods are frequently used for parasite isolation and other research purposes rather than for clinical practice. Xenodiagnosis additionally measures infectiousness to the vector, and is more suitable for field surveys than hemoculture. In a large dog survey, xenodiagnosis detected a much greater number of *T. cruzi* infections than the Strout method (Lauricella et al. 1989). In dogs seropositive for *T. cruzi*, a fraction of the xenodiagnostic bugs that were microscope-negative (13%) subsequently tested positive by conventional kDNA-PCR (Enriquez et al. 2014). Although xenodiagnosis is highly specific, *T. rangeli* and *Blastochritidia triatomae* (Trypanosomatidae) may infect the triatomines. *Trypanosoma rangeli* frequently infects dogs (Pifano 1973; Pineda et al. 2011; Ramírez et al. 2013); it develops in the salivary glands of *Rhodnius* sp. and may also appear in its rectal contents (Vallejo et al. 2009). *Trypanosoma rangeli* may also develop in hemoculture media under the conditions used for *T. cruzi*; in co-endemic regions, trypanosome-positive cultures should be confirmed via microscopic examination of stained thin blood films. *Trypanosoma caninum* was unable to infect xenodiagnostic bugs (Madeira et al. 2009).

In clinical settings, thoracic radiography and echocardiography may provide evidence of cardiac enlargement, while ECG may reveal conduction abnormalities compatible with chronic Chagas disease (Snowden and Kjos 2012). Detection of chronic infections with *T. cruzi* is best accomplished by using serological methods targeting specific antibodies. An in-house ELISA (using the flagellar fraction of *T. cruzi* as antigen) and indirect immunofluorescence antibody test (using whole epimastigotes) displayed high (94%) sensitivity and specificity (96.2%) in population-based serosurveys of dogs residing in endemic rural areas of the dry Argentine Chaco that were apparently free of *T. rangeli* and *Leishmania* sp. (Lauricella et al. 1998). Indirect hemagglutination assays were consistently less sensitive than those tests. The large diversity of antigens in use (e.g., whole trypanosomes, semi-purified preparations, recombinant proteins, and synthetic peptides) affects the validity of tests. Some dipstick tests were very helpful for screening serosurveys of dog populations (Cardinal et al. 2006b; Nieto et al. 2009; Rosypal et al. 2011; Lizundia et al. 2014). In a large sample of naturally infected dogs, the sensitivity of the dipstick test was >96% and specificity >94% (Cardinal et al. 2006a). A subsequent dog survey in a different location corroborated the large sensitivity of the dipstick test (Enriquez et al. 2013b).

Despite its simplicity and low cost, a major caveat of serodiagnosis in some areas is cross-reactivity with *Leishmania* sp., *T. rangeli*, and *T. evansi* (Troncarelli et al. 2009; Desquesnes 2017). In areas where these trypanosomatids are co-endemic with *T. cruzi*, serosurveys of dog populations should rule out cross-reactivity between trypanosomatids, especially if the prevalence of seroreactivity to *T. cruzi* is moderate to substantial and parasitological evidence (by hemoculture or xenodiagnosis) is nil or marginal. A trans-sialidase inhibition assay (Sartor et al. 2011) and TESA (trypanosome excreted-secreted antigen) blot, but not ELISA using TESA or epimastigotes as antigens (Umezawa et al. 2009), were able to rule out canine infections with *Leishmania* sp. and *T. rangeli*.

Molecular methods based on the amplification of highly repetitive sequences of kinetoplast DNA were slightly more sensitive than xenodiagnosis (91% vs. 86%, respectively) for identifying naturally infected dogs seropositive for *T. cruzi*, but both methods failed to detect all seropositive individuals (Enriquez et al. 2013b). Quantitative real-time PCR detected all *T. cruzi*-seropositive dogs, including a fraction that tested xenodiagnosis negative (Enriquez et al. 2014). Molecular methods can distinguish between canine infections with *T. cruzi* and *T. rangeli* (Vallejo et al. 2009; Ramírez et al. 2013), *T. evansi* and *T. caninum*, and the latter with *Leishmania infantum* (Barros et al. 2012; Pinto et al. 2014; Oliveira-Porfirio et al. 2018).

5.5 Transmission Cycles

Three types of transmission cycles are usually recognized: domestic, peridomestic, and sylvatic, in consonance with the habitat where they occur (Fig. 5.1).

Domestic transmission cycles occur in human sleeping accommodations and typically include <20 domiciliated triatomine species that have adapted to colonize such structures, such as *T. infestans*, *Rhodnius prolixus*, *Panstrongylus megistus*, *Triatoma dimidiata*, and *Triatoma brasiliensis*, and humans and domestic or synanthropic mammals: dogs, cats, domesticated guinea pigs, rats, and mice (Gürtler et al. 2020). All these mammals are major domestic nonhuman reservoir hosts of *T. cruzi*: they are able to maintain *T. cruzi* in the absence of any other host species, and play key roles as amplifying hosts and parasite sources across many ecoregions and triatomine species (Gürtler and Cardinal 2015). House mice and rats contributed to domestic bug infection with *T. cruzi* in many settings (Barrett et al. 1980; Bustamante et al. 2014; Rosal et al. 2018). Humans constitute a lifelong reservoir of *T. cruzi* across many decades unless given an effective course of treatment. The prevalence of human infection with *T. cruzi* attests to the size of the human reservoir and to the potential magnitude of Chagas disease.

Peridomestic transmission cycles usually occur in anthropic habitats located around domestic habitats, such as storerooms and animal outhouses, and usually include dogs, cats, and other domestic or synanthropic mammals, all of which may freely move between domestic and sylvatic habitats.

Sylvatic transmission cycles classically involve sylvatic mammals and sylvatic triatomine species. Two archetypical cycles were identified across the Americas: an arboreal cycle involving didelphid marsupials and TcI, and a terrestrial cycle involving armadillos and TcIII (Yeo et al. 2005; Miles et al. 2009). Transmission cycles may be connected through a shared triatomine species that occupies all three main habitats (e.g., *R. prolixus* in Venezuela and *T. brasiliensis* in northeast Brazil) or by a mammalian host that ranges across habitats and is a source of parasites to sympatric triatomine species (see next section).

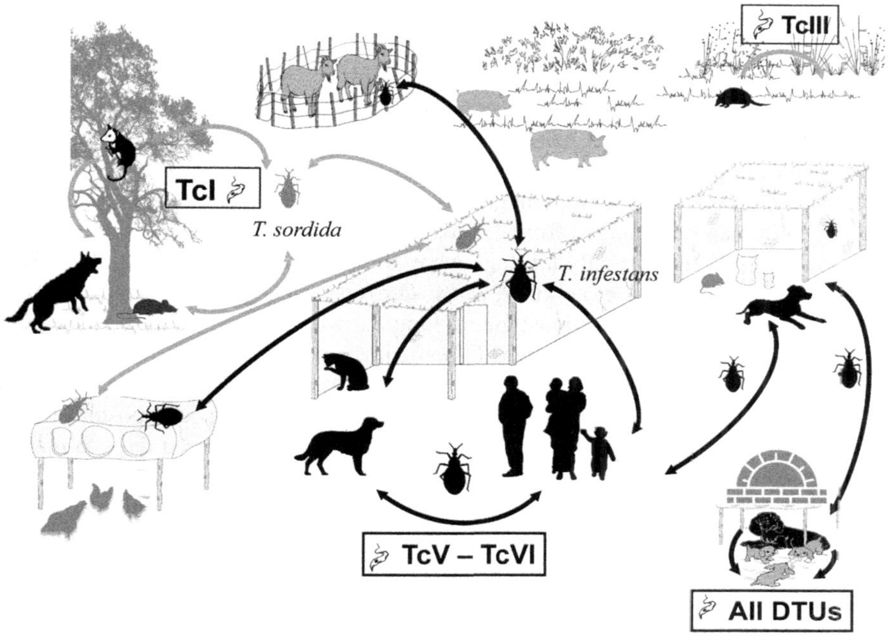

Fig. 5.1 Putative routes of transmission of *T. cruzi* related to dogs in domestic, peridomestic, and adjacent sylvatic habitats in northern Argentina. Dogs may become infected with the same DTUs as humans, cats, *Triatoma infestans* and other dogs in human habitations and auxiliary peridomestic structures, and by vertical transmission. Other DTUs from sylvatic enzootic cycles having armadillos and opossums (possibly including rodents) as primary reservoir hosts may be acquired by oral transmission or through sylvatic or peridomestic populations of a secondary triatomine species, *Triatoma sordida*. House invasion by these vector species may introduce DTUs from other houses or sylvatic habitats. Chickens, goats, and pigs are sources of blood meals, not parasites, for triatomines. Unbroken black lines indicate definite routes of transmission of TcV and TcVI in domestic and peridomestic habitats and dispersal of *T. infestans*. Unbroken gray lines represent likely transmission routes of TcI and TcIII and dispersal of *T. sordida* (Modified from Gürtler and Cardinal 2015)

5.6 Eco-epidemiology of Transmission

Several pieces of evidence combined demonstrate that dogs are key domestic reservoir hosts of *T. cruzi* and a risk factor for human infection in many locations: dog abundance; nighttime resting behavior in proximity to humans; frequent blood meal source of several triatomine species; high susceptibility to *T. cruzi*; high prevalence rates of infection; persistent, high infectiousness to the vector; co-occurrence of the same parasite genotypes detected in humans; and spatial association between canine, human and vector infections. The following sections review the evidence on these aspects.

Dog Demography The role of dogs in the epidemiology of Chagas disease in vast rural and urban areas of Latin America can be traced back to their main functions and dog husbandry practices deeply rooted in local culture (Matter and Daniels 2000). These aspects jointly determine a large number of dogs per household, directly correlated with the number of human residents; and high rates of recruitment of new (susceptible) dogs (by birth or in-migration) related to lack of fertility control. In multiple rural areas endemic for Chagas disease across the Argentine Chaco region, the dog-to-human ratio ranged from 1:2 to 1:3; each household owned nearly three dogs on average, and virtually all dogs were mongrel (Gürtler et al. 1990, 2007a; Castañera 1999; Cardinal et al. 2007, 2014; Orozco et al. 2013a). In general, dogs complied with four main functions: guard; hunting; working with livestock; and pet or companion. All dogs were owned (sometimes were shared between nearby households), and their movement usually was unrestrained, thus qualifying as free-ranging or—roaming dogs with eventually shifting resting sites and exposures over space and time. Undefined property lines and lack of fences allowed dogs to make frequent incursions into the adjacent forest. Both the median age and the mean life expectancy of dogs was approximately 3 years. The surplus production of dogs derived from lack of fertility control, combined with a low age at infection in infested houses, favored the putative dispersal of *T. cruzi*-infected dogs within and between communities and between rural and urban areas.

Demographic surveys of dog populations in resource-constrained areas of the Gran Chaco region invariably showed that a large fraction of dogs was in a poor health and nutritional status (e.g., Petersen et al. 2001; Cardinal et al. 2014; Enriquez et al. 2014) and rested in human sleeping accommodations and other peridomestic structures during the nighttime (Fig. 5.2). In general, dogs lacked both veterinary care and vaccination against virulent canine diseases; suffered from chronic malnutrition and anemia; and frequently had coinfections with up to six parasite species (Enriquez et al. 2019) and protein-deficient diets. Host conditions varied widely within the same village, and even more within the same region, depending on local levels of socioeconomic deprivation. The socioeconomic position of individuals and households, rooted in structural social determinants of health (e.g., income and educational level), determine housing conditions and husbandry practices relevant to domestic and peridomestic infestation with triatomine bugs and parasite transmission (Fernández et al. 2019). Thus, dogs owned by poor households were concomitantly in poor body condition elsewhere (Fung et al. 2014). These aspects most likely influenced survival, population turnover, and infectiousness (see below), and the risks of human and dog infection with *T. cruzi*.

Despite the apparent impact of *T. cruzi* infection on dog survival (see Sect. 5.3), its overall effects on dog demography appeared to be marginal in the presence of competing risks of death (i.e., unvaccinated dogs, untreated helminth infections, evident malnutrition, and venomous snake bites) in resource-constrained, endemic rural areas of the Argentine Chaco (Fig. 5.2). When house infestation with triatomines and transmission was eliminated community-wide in a sustained fashion, the prevalence of dog infection with *T. cruzi* declined exponentially, as predicted by a simple demographic model assuming non-differential mortality attributable to the

Fig. 5.2 Domestic dogs resting under a bed in a typical rural house (**a**) and beside a storeroom in peridomestic habitats (**b**) in northern Argentina

parasite (Gürtler and Cardinal 2015, Fig. 5.2). The dog population retained a stable age structure skewed toward young ages despite the fast removal of *T. cruzi*-infected dogs, which were almost immediately replaced with noninfected young dogs (Gürtler et al. 1990; Castañera 1999). Young dogs apparently are more susceptible to *T. cruzi*; have greater infectiousness, and remain more attached to domestic premises during the initial months of life. The combined effects of large mortality and large fertility generated fast annual turnover rates (20–30%) and a nearly stationary dog population size. Both the age distribution and rate of decay in the

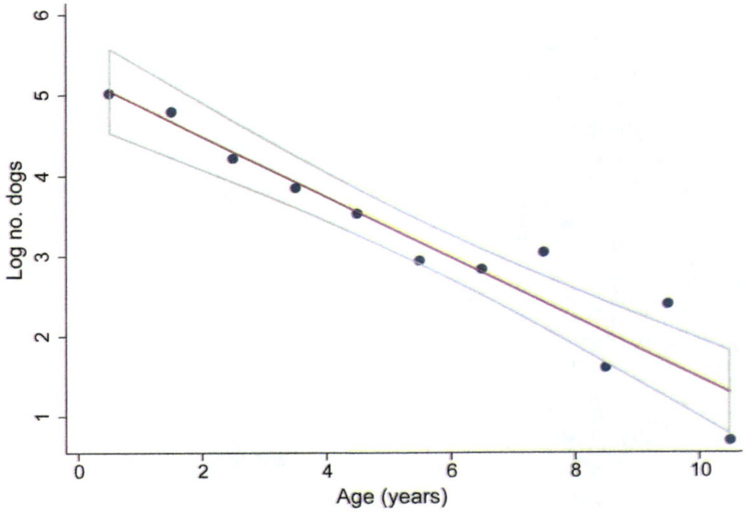

Fig. 5.3 Linear regression of log-transformed numbers of dogs (y) by mid-point age class (x) in Pampa del Indio, 2008. Data taken from Cardinal et al. (2014). Linear regression equation of log-transformed age frequencies ($y = -0.318x + 4.951$, $n = 16$, Adj $R^2 = 0.907$, $P < 0.001$). The belt above and below the regression line represents a 95% confidence interval for individual residuals

frequency of dogs between successive age groups were similar in another rural location of the Argentine Chaco (Fig. 5.3).

Host–Vector Contact Triatomine host-feeding patterns tend to reflect the relative abundance and proximity of local hosts combined with host attractiveness and defensive behavior (Gürtler et al. 2009a, 2014). Thus, host availability and host resting behavior play major roles in the eco-epidemiology of Chagas disease, and vary widely within and between settings. In host choice experiments conducted in small huts housing two individuals of different host species, *T. infestans* bugs preferred dogs to chickens or cats, and engorged more on dogs than on the other hosts (Gürtler et al. 2009a).

The main blood meal hosts of Triatominae in domestic or peridomestic habitats were humans, chickens ("avian hosts"), dogs, cats, and rodents, as determined from a large database including 159 reports of host-feeding patterns identified through immunological methods (Rabinovich et al. 2011). We used this database to search for trends and patterns across triatomine species collected at least in domestic habitats (numbering 75 reports, including domestic only, domestic/peridomestic, and domestic/peridomestic/sylvatic bug collections). Dog blood meals averaged 8% (range, 0–50%) of the tested bugs and human blood meals averaged 23% (range, 0–96%). Human and dog blood meals jointly occurred in 54 (72%) reports, both events being highly significantly and positively associated (Fisher's test, $P < 0.001$). Overall, there was evidence that 17 triatomine species collected in those habitats had

Table 5.1 Distribution of triatomine species with dog blood meals detected by immunological and molecular methods

Genus	Species	Method	References
Triatoma	*T. barberi*	Immunological	Rabinovich et al. (2011)
	T. brasiliensis	Immunological	Rabinovich et al. (2011)
	T. dimidiata	Immunological	Rabinovich et al. (2011)
	T. eratyrusiformis	Immunological	Cecere et al. (2016)
	T. gerstaeckeri	Molecular	Curtis-Robles et al. (2017b)
	T. indictiva	Molecular	Kjos et al. (2013)
	T. infestans	Immunological	Rabinovich et al. (2011)
	T. maculata	Molecular	Cantillo-Barraza et al. (2015)
	T. longipennis	Molecular	Brenière et al. (2004)
	T. pallidipennis	Immunological	Rabinovich et al. (2011)
	T. petrochiae	Immunological	Silva et al. (2017)
	T. phyllosoma	Immunological	Villalobos et al. (2011)
	T. pseudomaculata	Immunological	Rabinovich et al. (2011)
	T. protracta	Molecular	Stevens et al. (2012)
	T. rubida	Molecular	Stevens et al. (2012)
	T. rubrofasciata	Immunological	Rabinovich et al. (2011)
	T. sanguisuga	Molecular	Waleckx et al. (2014)
	T. sordida	Immunological	Rabinovich et al. (2011)
	T. vitticeps	Immunological	Rabinovich et al. (2011)
Rhodnius	*R. nasutus*	Immunological	Rabinovich et al. (2011)
	R. neglectus	Immunological	Rabinovich et al. (2011)
	R. pallescens	Immunological	Rabinovich et al. (2011)
	R. pictipes	Immunological	Rabinovich et al. (2011)
	R. prolixus	Immunological	Rabinovich et al. (2011)
Panstrongylus	*P. herreri*	Immunological	Rabinovich et al. (2011)
	P. geniculatus	Immunological	Carrasco et al. (2005)[a]
	P. lutzi	Immunological	Silva et al. (2017)
	P. megistus	Immunological	Rabinovich et al. (2011)
	P. tupynambai	Immunological	Rabinovich et al. (2011)
Belminus	*B. ferroae*	Immunological	Rabinovich et al. (2011)
Psammolestes	*P. tertius*	Immunological	Silva et al. (2017)

[a]Undefined collection sites: triatomines were collected by householders
Includes triatomines collected in domestic habitats only, domestic/peridomestic or domestic/peridomestic/sylvatic

fed on dogs (Table 5.1). Using the same selection criteria for a systematic search of host-feeding studies using molecular methods, several *Triatoma* species (mainly from North America) prolong the list (Table 5.1). In summary, dogs are a blood meal source of $\geq$31 triatomine species almost exclusively collected in domestic and/or peridomestic habitats. While a meta-analysis of published reports does not provide a bias-free representation of eco-epidemiological scenarios for any triatomine species, it does provide a crude assessment of the relative generality of dog-triatomine

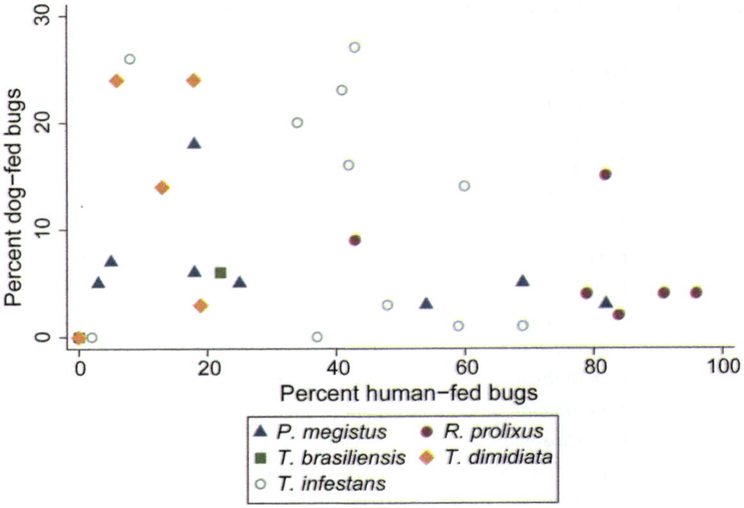

Fig. 5.4 Relationship between the percentage of triatomines collected in domestic and/or peridomestic habitats fed on dogs and humans regardless of other blood meal sources. Includes *T. infestans*, *T. brasiliensis*, *T. dimidiata*, *P. megistus*, and *R. prolixus* (Data compiled by Rabinovich et al. 2011)

contact across study areas. This inventory may increase as more population-based studies with larger sample sizes and better definition of bug collection sites are conducted.

For the five major domestic vector species (including *T. infestans*, *R. prolixus*, *P. megistus*, *T. dimidiata*, and *T. brasiliensis*) collected in habitats as defined above, the fraction of bugs that fed on dogs tended to be negatively related to the fraction fed on humans (Fig. 5.4). Dog blood meals ranked high in domestic *T. infestans* in the Argentine Chaco (Gürtler et al. 1997, 2014). For any triatomine species, the host-feeding patterns displayed large variability between locations (e.g., Bustamante et al. 2014; Lima-Cordón et al. 2018). At least five species strictly collected in sylvatic areas (*Mepraia spinolai*, *Triatoma rubrovaria*, *T. brasiliensis*, *R. neglectus*, and *R. pallescens*) had blood meals attributed to canid hosts, which may correspond to domestic dogs or other wild canids because of serological cross-reactivity.

Using molecular methods, dogs were second to humans as blood meal sources of *T. dimidiata* in Yucatan, Mexico (Dumonteil et al. 2018; Moo-Millan et al. 2019) and Guatemala (Pellecer et al. 2013), and for *T. maculata* in Colombia (Cantillo-Barraza et al. 2015). Dogs and humans were important blood meal and parasite sources for several triatomine species in the USA (Stevens et al. 2012; Kjos et al. 2013; Gorchakov et al. 2016; Curtis-Robles et al. 2017b) and Mexico (Brenière et al. 2004; Mota et al. 2007; Villalobos et al. 2011; Ibáñez-Cervantes et al. 2013). Humans, raccoons, dogs, and cats were frequent blood meal sources of *T. sanguisuga* in a site in Louisiana (USA) where an acute case of human Chagas disease had been detected (Waleckx et al. 2014). PCR-based blood meal identifications using primers for the mitochondrial 12S ribosomal gene also have the limitation of not distinguishing between dogs, coyotes, and wolves (Gorchakov et al. 2016).

Mixed blood meals including dog and human blood occurred frequently in domestic *T. infestans* (e.g., Gürtler et al. 1996, 2007a; Pizarro and Stevens 2008)—related to both host availability and shifting host resting sites—suggesting the putative flow of *T. cruzi* between these hosts. The association between blood meal source and bug infection with *T. cruzi* (i.e., the infective blood meal index: Zárate et al. 1980) suggested that dogs, cats, rodents, opossums, and humans played significant roles as parasite sources in several transmission cycles (Minter 1976b). Dog-fed bugs had higher infective indices than those fed on cats, humans, and chickens in northern Argentina (Wisnivesky-Colli et al. 1982; Gürtler et al. 2007a).

Prevalence of Infection A systematic review of the literature including 309 reports shows that dogs were naturally infected with *T. cruzi* from as far north as Ohio and Virginia (USA) to southern Argentina and Chile (Gürtler and Cardinal 2015, Appendix A). Dog infection was frequent even in areas of the USA where sylvatic transmission cycles are dominant (reviewed in Zeledón et al. 2012). All six DTUs were identified in dogs (Cardinal et al. 2008; Roellig et al. 2008; Enriquez et al. 2013b, 2014; Ramírez et al. 2013), albeit at different frequencies (Brenière et al. 2016). Coinfections with different DTUs are frequent (e.g., Ramírez et al. 2013; Monje-Rumi et al. 2015; Ortiz et al. 2016; Curtis-Robles et al. 2017b). Barring classical domestic or peridomestic vector-borne transmission cycles, there is sparse evidence on which species of triatomines and routes were implicated in transmission to dogs.

In general, the prevalence of *T. cruzi* infection in dogs varied from 10 to 30%, but in several areas infested with *T. infestans* and *R. prolixus*, dog infection exceeded 50% and was more prevalent than in the local human populations (Gürtler and Cardinal 2015, Appendix A). The majority of surveys assessed dog infection status via serodiagnosis, but parasitological and molecular methods predominated in different periods. Large variations in diagnostic methods (including performance) and in the local history of vector control actions hinder direct comparisons between reported prevalence rates. Sustained vector control and surveillance combined with the fast turnover rates of dog populations determine that the prevalence of dog infection declines rapidly after implementing effective interventions. Notwithstanding these caveats, there is an increasing time trend in the prevalence of dog infection (e.g., Brazil, Costa Rica, Mexico, and Peru), in part related to the use of more sensitive methods and to increased awareness of the role of dogs.

In the absence of vector or disease control actions, the age-specific seroprevalence rate of infection with *T. cruzi* increased monotonically, reflecting the cumulative risk of infection over the life course (Fig. 5.5). The age-prevalence data may be used to estimate the instantaneous rate of per capita incidence (force of infection) using a susceptible-infected (irreversible catalytic) mathematical model of transmission (Anderson and May 1991; Cucunubá et al. 2017). The force of *T. cruzi* infection in dogs varied from 43.2 to 72.7 per 100 dog-years in two highly infested, endemic

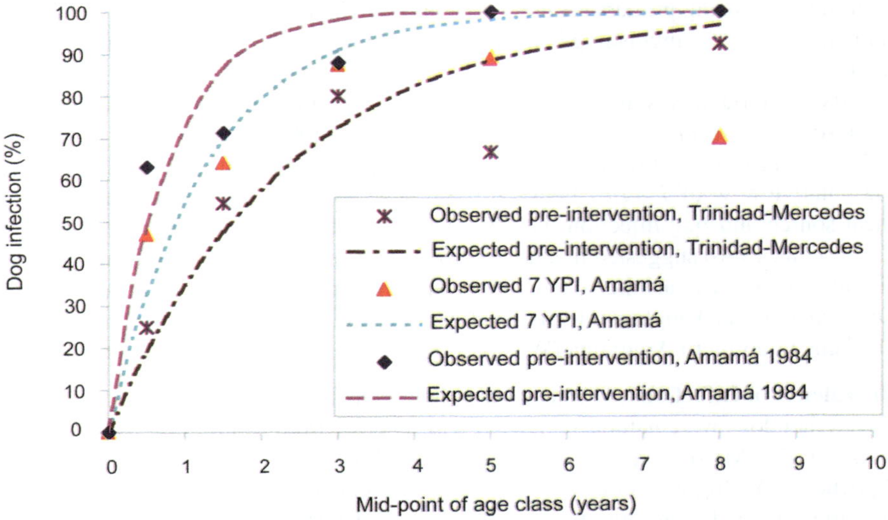

Fig. 5.5 Observed (dots) and expected (line) age-specific seroprevalence of *T. cruzi* infection in dogs from two highly infested villages (Amamá and Trinidad-Mercedes) in northern Argentina. The line is the fit of the catalytic model with constant force of infection over time and age (Reproduced from Gürtler et al. 2007b, SI Fig. 5.4)

rural locations in northern Argentina, implying an average age at primary infection of 5–6 months (Gürtler et al. 2005, 2007b). Using these estimates and mean life expectancy, the basic reproduction number (R_0) of *T. cruzi* infection in dogs before control interventions was 8.2. The prevalence of dog infection also increased with infected-bug density, an index of transmission risk (Cardinal et al. 2007, 2014).

Infectiousness The intensity of host infectiousness to the vector, combined with host infection prevalence and vector–host contact rates, may be used to assess the relative contribution of a host species to onward parasite transmission. Host infectiousness has usually been measured by the proportion of uninfected vectors that become infected after a full blood meal on an infected host using xenodiagnosis with laboratory-reared, uninfected triatomines. This metric correlated closely with the concentration of *T. cruzi* DNA measured by quantitative real-time PCR (qPCR) in dogs, which provides an alternative way to identify infectious dogs (Enriquez et al. 2014).

From the published reports of dog infectiousness compiled elsewhere (Gürtler and Cardinal 2015, Table 5.1), we estimated that the median percentage of xenodiagnostic bugs that became infected after feeding on a *T. cruzi*-seropositive dog was 53% (range, 29–56%). This summary estimate largely exceeds the infectiousness of *T. cruzi*-seropositive humans, who infected 1.9–3.1% to 13.6–27.6% of third- or fourth- and fifth-instar nymphs, respectively (Gürtler et al. 1996). The mean infectiousness of *T. cruzi*-seropositive dogs to seropositive humans differed by a factor of 10 by xenodiagnosis, and by 4 using real-time PCR (see below). In a population-based survey of *T. cruzi*-seropositive humans, infectiousness to fourth instars of *T. infestans* averaged 5.2% and conformed to the 80–20 rule (Macchiaverna et al. 2020). Xenodiagnosis-based measures of host infectiousness are modified by vector competence, blood meal size, duration of exposure to the host, and other relevant factors beyond control in the field (e.g., parasite strain, multiclonal infection, health condition of dog, and coinfections).

The lifetime course of infectiousness is important from a transmission standpoint. In experimentally infected dogs, infectiousness decreased with time or remained negative (by xenodiagnosis) over a prolonged period of 5–12 years post-infection (Machado et al. 2001; Araujo et al. 2002), or persisted at similar levels over a 2-year follow-up (Lauricella et al. 1986). Successive reinoculation of dogs with two genotypes of *T. cruzi* generated few symptoms during the acute phase; modified the intensity of infectiousness transiently or not at all; allowed both genotypes to persist albeit at different levels; and boosted the antibody response of the infected dogs (Machado et al. 2001).

In surveys of naturally infected dog populations, the prevalence of xenodiagnosis-positive individuals and their intensity of infectiousness to the vector were either age-independent or slightly declined with age (reviewed in Gürtler and Cardinal 2015), and were not significantly related to specific antibody titers (Gürtler et al. 2007a). In general, the potential contribution of pups to bug infection was 50% greater than that of older age groups combined (Gürtler et al. 1996), in part due to their slightly higher infectiousness and greater relative abundance; consideration of age-related exposure patterns may increase the gap. Dog infectiousness correlated negatively with body condition (reflecting impaired health and nutritional status) and positively with helminth coinfections, and was not significantly modified by parasite DTU and age of dog (Enriquez et al. 2014). Nor was dog infectiousness modified after suppression of domestic bug infestations in a 3-year follow-up (Gürtler et al. 1992). Individual dog infectiousness varied little over time and was moderately aggregated at the population level, implying that some dogs contributed disproportionally to transmission and acted as "superspreaders" (Gürtler et al. 2007a; Enriquez et al. 2014).

Another related metric of host infectiousness, the prevalence of xenodiagnosis-positive individuals in the host population, corroborated the large magnitude of the dog reservoir, which frequently exceeded the estimates for the local human population in other areas infested with *R. prolixus*, *T. dimidiata*, and *T. infestans* (e.g., Freitas 1950; Pifano 1973; Zeledón et al. 1975; Lauricella et al. 1989). Further studies of dog infectiousness in Brazil, the USA, and other endemic areas are needed to gauge the external validity of the patterns recorded in the Chaco and other areas, and the extent to which dog infectiousness is modified by host species-level factors and ecological-level factors (i.e., socioeconomic, exposure patterns, and parasite strains).

Dogs as a Risk Factor and Sentinel Hosts Several studies have shown a direct association between the household presence or number of dogs and domestic infestation with triatomines (Cecere et al. 1998; Gurevitz et al. 2011; Dumonteil et al. 2013; Bustamante et al. 2014). Moreover, the presence or number of *T. cruzi*-infected dogs significantly increased the relative odds of domestic bug or human infection with *T. cruzi* in rural villages of the Argentine Chaco (Gürtler and Cardinal 2015).

These close associations suggested that domestic dogs may function as natural sentinels of exposure to *T. cruzi*. This type of application for disease surveillance, in general, has been underused (Halliday et al. 2007). Several features of dogs make them appropriate candidates for a sentinel host: they are highly susceptible to *T. cruzi* and develop a measurable, consistent response to it; share habitats with humans; are accessible, easy to enumerate and handle, and abundant (Gürtler and Cardinal 2015). Having a simple, inexpensive, sensitive, and specific detection method is essential.

The strong links between the household presence of infected dogs, infected children, and *T. infestans* in northern Argentina supported the use of dogs as sentinels of the domestic and peridomestic transmission of *T. cruzi* (Gürtler et al. 1990; Castañera et al. 1998; Cardinal et al. 2006a). The very few dogs infected with *T. cruzi* detected during sustained vector surveillance and control were linked to vertical transmission, in-migration from other infested villages, and transient domestic infestations (Castañera et al. 1998; Cardinal et al. 2006a, 2007). The co-occurrence of infected dogs and infected children was also documented in Brazil, Mexico, Peru, and Venezuela (Mott et al. 1978; Estrada-Franco et al. 2006; Crisante et al. 2006; Castillo-Neyra et al. 2015).

The information derived from surveys of dog infection may be used for other purposes. The spatial distribution of *T. cruzi* infection in dogs may be used for risk stratification at district- (Cardinal et al. 2014) or state-wide levels (Tenney et al. 2014), and for gauging the intensity of parasite transmission (Cardinal et al. 2006a, 2007). In a novel application, a trained dog was able to detect hitherto unknown sylvatic foci of *T. infestans* in the Paraguayan Chaco (Rolón et al. 2011).

Domestic dogs may also act as sentinels of sylvatic sources of *T. cruzi* infection in connection to their roles as predators or scavengers. For example, TcIII is usually found in armadillos and skunks, and more rarely in dogs and humans (Chapman et al. 1984; Marcili et al. 2009; Monje-Rumi et al. 2015; Brenière et al. 2016). In population-based surveys conducted in two well-defined rural districts of northern Argentina, TcIII was frequently detected in nine-banded (*Dasypus novemcinctus*) or three-banded armadillos (*Tolypeutes matacus*) and hog-nosed skunks (*Conepatus chinga*), and rarely in dogs and domestic or peridomestic *T. infestans* (Ceballos et al. 2006; Cardinal et al. 2008; Enriquez et al. 2013a; Orozco et al. 2016). These patterns suggested the introduction of a sylvatic DTU into domestic hosts and vectors probably through dogs preying on armadillos or other sylvatic mammals or when feeding dogs with these hosts' raw viscera or fresh blood. The latter is a widespread practice in the region, and implies the human-mediated introduction of a sylvatic DTU in the household environment (Cardinal et al. 2014). In this location (Pampa del Indio), the most frequently reported prey of dogs with a hunting habit were armadillos (by 59% of hunting dogs), gray-brown brocket deer (*Mazama gouazoubira*, 26%), and white-eared opossums (*Didelphis albiventris*, 21%). Both armadillos and opossums had large infectiousness to the vector (Orozco et al. 2013b), which would increase the chances of transmission via fomites or ingestion. Also in Pampa del Indio, *Didelphis* opossums had a high prevalence of infection with TcI, as presumably had wild rodent species (Orozco et al. 2014). TcI rarely occurred in dogs, cats, *T. infestans*, and peridomestic *T. sordida* (Maffey et al. 2012; Enriquez et al. 2013a; Macchiaverna et al. 2015), and was not detected in the local human population (Macchiaverna et al. 2018). These patterns suggest that dog infections with TcI were a spillover from non-domestic transmission routes.

In rural areas of Brazil affected by foodborne outbreaks of human Chagas disease linked to putative sylvatic sources, dogs had high infection prevalence with *T. cruzi* associated with the presence of transmission-competent small wild mammals (Xavier et al. 2012). Elsewhere in Brazil, triatomines, dogs and various species of small wild mammals shared the same DTU and most likely were part of the same transmission cycle (Rocha et al. 2013; Xavier et al. 2014).

Mathematical Models of Transmission Theory predicts that parasite circulation among multiple host species differing in reservoir competence (i.e., humans, dogs, and chickens) might favor pathogen persistence (maintenance), increase pathogen abundance (amplification), or reduce both of them (the dilution effect) (Begon 2008). An empirically based mathematical model of the domestic transmission of *T. cruzi* in northern Argentina predicted that having two or more infected dogs disproportionally increased the prevalence of infection in humans and domestic vectors, and increased human incidence (Cohen and Gürtler 2001). Both the model and the empirical data showed that dogs acquired the infection within a few months of exposure (before their human counterparts), and 90–100% of dogs were infected by 3 years of age. Other model specifications also supported that dogs amplified parasite transmission in domestic habitats under specific circumstances (Spagnuolo et al. 2012; Fabrizio et al. 2014; Nouvellet et al. 2015; Peterson et al. 2015; Flores-Ferrer et al. 2019). Moreover, the presence or number of dogs was associated with increased triatomine abundance in domestic or peridomestic habitats despite exerting some predatory effects on triatomines (as documented above), and ultimately increased human-vector contact rates and transmission risks.

The role of chickens in the domestic transmission of *T. cruzi* has been more controversial, in part because all birds are insusceptible to the parasite and all chicken-vector contacts represent wasted opportunities for parasite transmission. However, it is well-known that indoor-resting chickens boost domestic or peridomestic triatomine population size across species and settings (e.g., Cecere et al. 1998; Gurevitz et al. 2011; Dumonteil et al. 2013; Bustamante et al. 2014). Because of the frequent host shifts of *T. infestans* and other domestic triatomines to transmission-competent hosts such as dogs and humans, the net effects of chickens were to increase the equilibrium abundance of *T. cruzi*-infected *T. infestans* and human prevalence of infection in northern Argentina (Cohen and Gürtler 2001). Other models for *T. dimidiata* in Yucatan, fitted to a vector with intrusive behavior and distinctive habitat use, predicted that the addition of a large number of chickens would exert zooprophylactic effects in the long term (Flores-Ferrer et al. 2019).

5.7 Molecular Epidemiology of Transmission Cycles

The issue of whether parasite transmission cycles overlap has important repercussions as it may threaten control efforts directed at curtailing domestic transmission (Miles et al. 2003). Overlapping transmission cycles were recorded in multiple locations: between TcI and TcVI in Yucatan (López-Cancino et al. 2015); TcI circulating both in domestic and in sylvatic habitats in Guatemala (Pennington et al. 2015); and in Venezuela, where *Rhodnius* bugs infested both houses and palm trees (Miles et al. 2003, 2009). A classic example of separate transmission cycles occurred in eastern Bahia (northeast Brazil), where TcII circulated in houses infested with *P. megistus* while TcI circulated between *Triatoma tibiamaculata* and white-eared opossums in bromeliad epiphytes; in contrast, both TcI and TcII infected humans in western Bahia (Barrett et al. 1980; Miles et al. 2003).

Molecular methods support that dogs play a key role in diverse eco-epidemiological settings. In several locations of the Argentine Chaco, TcV and TcVI circulated in domestic or peridomestic habitats among dogs, cats, humans, and *T. infestans* (Cardinal et al. 2008; Maffey et al. 2012; Enriquez et al. 2013a, b; Fernández et al. 2014; Macchiaverna et al. 2018). In contrast, TcI and TcIII were mainly restricted to sylvatic hosts (see Sect. 5.6, Dogs as a Risk Factor and Sentinel Hosts; Diosque et al. 2003; Cardinal et al. 2008; Lucero et al. 2016). Dogs frequently harbored the same DTUs as did domestic or peridomestic *T. infestans* in the same house compound or in nearby habitats (i.e., co-occurrence of parasite genotypes) (Cardinal et al. 2014; Macchiaverna et al. 2018). The apparent segregation of TcVI from TcV (the former isolated mainly from dogs or cats and the latter from humans) was inconsistent with the evidence reviewed above. Subsequent direct typing of blood samples revealed a large frequency of mixed infections with TcI and TcVI in dogs, and TcV and TcVI in humans (Monje-Rumi et al. 2015), suggesting that parasite isolation and amplification procedures may have artificially selected some DTUs (Gürtler and Cardinal 2015). A lineage-specific rapid test of 480 serum samples from humans, dogs, and cats further corroborated the predominance of TcV and TcVI in local domestic transmission cycles (Murphy et al. 2019), and subsequent genotyping of human isolates revealed a substantial fraction of TcVI (Macchiaverna et al. 2018). Nonoverlapping transmission cycles were also inferred in the Paraguayan Chaco (Acosta et al. 2017; Yeo et al. 2005), though identification of parasite DTUs from sympatric domestic hosts remains sparse.

Dogs also play a key role in other scenarios where a non-domiciliated vector mediates transmission. For example, in Yucatan, *T. dimidiata* invades house compounds from sylvatic habitats and blood-feeds on humans, dogs, cats, synanthropic rodents, and birds (Dumonteil et al. 2018). TcI was identified as the predominant DTU (Monteón et al. 2016). An overlapping transmission cycle (including dogs, sylvatic mammals, TcI, and *T. pallidipennis*) was also recorded elsewhere in Mexico (Ramsey et al. 2012). In Brazil, where *T. infestans* were nearly eliminated countrywide and its direct contribution to transmission stopped (Belisário et al. 2017), the role of other triatomine species became more patent. For example, *T. brasiliensis* infected with *T. cruzi* co-occurred with infected dogs in rural houses from four municipalities of Rio Grande do Norte (Freitas et al. 2018; Araújo-Neto et al. 2019), and harbored TcII and TcIII, suggesting putative links between sylvatic and peridomestic transmission cycles (Barbosa-Silva et al. 2016).

Dogs may participate in sympatric transmission cycles through different routes, including the oral route. They may acquire *T. cruzi* when exposed to triatomines in established domestic or peridomestic infestations of kennels or other outdoor structures (Curtis-Robles et al. 2017b), and during transient exposure to intrusive sylvatic triatomines. This may be the case in Texas and several other states in the USA, with extensive sylvatic transmission cycles including *Neotoma* woodrats, *Peromyscus* rodents, *Didelphis* opossums, armadillos, and raccoons (Zeledón et al. 2012). Examination of a recent inventory of 6343 DTUs (Brenière et al. 2016, S1 Table) shows that dogs in the USA were infected with TcI and TcIV; both DTUs were prevalent in wood rats, opossums, raccoons, and several triatomine species. At least six species of triatomines occasionally established colonies in peridomestic habitats (more rarely in domestic premises) and transmitted *T. cruzi* (Zeledón et al. 2012). Infected triatomines frequently invaded houses during late spring, suggesting seasonal disease threats to dogs in Arizona (Reisenman et al. 2012).

In Texas, dogs had a high prevalence of infection and were mainly infected with TcI, as were several local triatomine species, which harbored blood meals on dogs, raccoons, and humans (Kjos et al. 2013; Curtis-Robles et al. 2016, 2017a, b; Gorchakov et al. 2016; Meyers et al. 2017). In a local national park, TcI was frequently detected in dogs and triatomines, and three species of sylvatic triatomines had dog and human blood meals (Curtis-Robles et al. 2018). The magnitude of sylvatic transmission cycles in the USA combined with the intrusive behavior of triatomines suggests the potential for human and dog exposure to sylvatic triatomines infected with enzootic DTUs over an extended region.

5.8 Prevention and Therapy

The menu of control methods for dogs potentially exposed to or infected with *T. cruzi* includes housing improvement; residual insecticide spraying to suppress or prevent domestic and peridomestic infestations with triatomine bugs; insecticide-impregnated collars, sprays, pour-ons and ivermectin; and etiologic treatment. For primary prevention, dogs should not be fed with blood or uncooked meat of mammal prey. As a public health measure, culling of *T. cruzi*-infected dogs finds no justification: keeping the vector out of domestic and peridomestic premises will interrupt the transmission of *T. cruzi* from an infected dog to triatomines and other domestic hosts, and will afford protection to the new dogs recruited. Direct transmission of *T. cruzi* from infected dogs to humans is possible only through unprotected contact with blood and rarely with other secretions. Elimination of synanthropic rodent reservoirs may also afford indirect protection to dogs and other domestic hosts by suppressing an immediate source of *T. cruzi* to peridomestic and domestic triatomines (de Urioste-Stone et al. 2015).

Housing Improvement Likewise for human dwellings, doghouses and other structures used by dogs as resting sites should have as few refuges for triatomines as possible. Plastering walls and thatched roofs with appropriate mortar that does not crack easily or use of materials that provide a continuous surface (e.g., sheets of corrugated metal or wood) may render the structure virtually resistant to bug colonization (Gürtler and Cecere 2020). Other prevention measures include installing mechanical barriers, using outside lighting that does not attract flying triatomines, and housing dogs to reduce exposure to triatomines when these are active.

Residual Insecticide Spraying Residual house spraying with pyrethroid insecticides remains the mainstay of triatomine control over the last 40 years (Gürtler and Cecere 2020). Pyrethroids exert excito-repellents on the insects, cause hyperactivity, incoordination, paralysis, and death. When applied indoors at recommended doses, pyrethroids in a suspension concentrate formulation retain residual effects for 3–-12 months post-application. Residual effects quickly diminish from exposure to sunlight, rain, and dust in outdoor locations such as corrals, chicken coops, and open storage areas. Evidence on whether targeted residual spraying with pyrethroids of kennels or other habitual dog resting sites may afford enhanced prevention of dog infections with *T. cruzi*, especially in areas with intrusive triatomines, is needed. Reduced susceptibility of Triatominae to pyrethroids has been reported in some settings (Gürtler and Cecere 2020).

Community-wide insecticide spraying campaigns followed by sustained vector surveillance and control have provided strong evidence in support of triatomine-mediated transmission being the main route in endemic rural areas of the Argentine Chaco (Cardinal et al. 2006a, 2014; Gürtler and Cardinal 2015, Fig. 5.2). Putative cases of oral transmission of TcIII were rare, and only a few cases of *T. cruzi* infection in young dogs were compatible with vertical transmission. However, in women with a chronic *T. cruzi* infection, the probability of transmission to offspring averaged 5% (it may reach 10% using molecular methods), and increased with parasitemia levels (Carlier and Truyens 2017).

Xenointoxication Xenointoxication is a targeted vector control strategy that involves the application of pesticides on nonhuman hosts to kill the bugs that contact or blood-feed on them. Deltamethrin-impregnated collars applied to dogs reduced the blood-feeding success and engorgement rates of *T. infestans* (Reithinger et al. 2005) and suppressed triatomine populations kept in closed experimental huts (Reithinger et al. 2006). In contrast, fipronil, either impregnated in collars or applied as a spray or pour-on formulation on dogs, exerted limited effects on bug populations (Gürtler et al. 2009b; Amelotti et al. 2012). Subcutaneous injection of dogs with ivermectin increased the mortality of *T. infestans* and *R. neglectus* nymphs relative to control dogs up to 7 days post-treatment (Dias et al. 2005).

Some novel compounds hold the promise of longer-lasting effects. Oral administration of fluralaner (a novel ectoparasiticide) to kenneled dogs killed first- to fourth-instar nymphs of *T. infestans* up to 51 days post-treatment (Loza et al. 2017). Moreover, a single oral dose of fluralaner administered to dogs in an endemic rural area of the Argentine Chaco killed both pyrethroid-susceptible and pyrethroid-resistant *T. infestans* up to 120 days post-treatment, and exerted no anti-feeding effects (Laiño et al. 2019). Whether treatment with fluralaner would provide individual protection against contact with triatomines remains to be determined. A small-scale controlled trial of fluralaner administered to dogs in this pyrethroid-resistant rural area substantially reduced the local abundance of *T. infestans* over a 5-month follow-up (Laiño et al. unpublished data).

Drug Treatment Treatment of infected dogs with the two available parasiticidal drugs used for human Chagas disease (benznidazole and nifurtimox) is considered to be more effective during the acute phase (Haberkorn and Gönnert 1972; Guedes et al. 2002). Nifurtimox administered to experimentally infected beagle dogs at 16–30 mg/kg during 3–4 months suppressed parasitemia; exerted no adverse side effects during treatment; and did not lead to serorecovery in chronic infections followed up over a four-month period (Haberkorn and Gönnert 1972). Barr (2009) considered that nifurtimox caused serious adverse reactions and recommended a dose of 2–7 mg/kg every 6 h over 4 months, usually associated with corticosteroids. Benznidazole should be administered at a daily dose of 5–7 mg/kg over 2 months (Snowden and Kjos 2012), and may cause vomiting. In chemotherapy trials of dogs experimentally infected with *T. cruzi*, benznidazole combined with itraconazole or not was apparently effective in treating acute infections (Cunha et al. 2019). In another drug trial of chronic dog infections, benznidazole suppressed parasitemia during the first year post-treatment and reduced systolic cardiac alterations, but did not prevent the development of cardiomyopathy (Santos et al. 2012). Some strains of *T. cruzi* have shown natural resistance to benzanidazole and nifurtimox. For dogs with chronic Chagas disease, supportive treatment should be directed toward myocardial failure and ventricular arrhythmias (Barr 2009).

Other newer drugs tested in humans were also tested in dogs. Ravuconazole (Figueiredo Diniz et al. 2010) and albaconazole (Guedes et al. 2004) exerted trypanocidal activity in vivo and little or no adverse effects attributable to the drug, although some parasite strains appeared to be naturally resistant to albaconozale. Other compounds (ketoconazole, gossypol, allopurinol, imidazole, and verapamil) have been ineffective in the treatment of canine Chagas disease, and itraconazole in combination with amiodarone showed promising results for treatment of natural infections in dogs (Madigan et al. 2019).

Vaccines As yet there is no effective vaccine against *T. cruzi* (Padilla et al. 2017; Travi 2019). A field trial of a vaccine using live-attenuated antigens of *T. cruzi* reduced the incidence of dog infection and infectiousness in northwest Argentina (Basombrío et al. 1993). Similar reduction in transmission rates was obtained in dogs experimentally vaccinated with *T. rangeli* antigens (Basso et al. 2007). Experimentally infected mongrel dogs had reduced parasitemia, cardiac inflammation, and parasite burden after receiving a preventive and therapeutic DNA vaccine (Quijano-Hernández et al. 2013).

5.9 Conclusions

Our current understanding of the composition, structure, and degree of connectivity of transmission cycles is severely curtailed by the lack of integrated eco-epidemiological and genetic marker information at relevant, well-defined spatiotemporal scales. Despite the rich literature on the eco-epidemiology of Chagas disease, most research efforts have focused on selected (one or two) components (vector- or host-parasite) in one type of habitat (i.e., domestic versus sylvatic) at a defined time point (cross-sectional surveys). More systematic studies, covering a wide range of sylvatic and domestic host species and their associated triatomine species in relevant locations over at least a few years, are needed to advance our current knowledge on complex system dynamics in a scenario of global change.

Our review highlights the distinct roles that dogs may play in the eco-epidemiology and control of Chagas disease. The rich diversity of ecological situations, biological factors, and sociocultural patterns across the Americas does not warrant broad generalizations. For example, in some circumstances, dogs maybe dead-end hosts of *T. cruzi* in developed rural areas with modern housing kept free of house infestations with triatomines, and where dog contact with intrusive triatomines is minimized; Mennonite populations in the Central Paraguayan Chaco would be a relevant example. In highly endemic rural areas such as those in the Argentine or Bolivian Chaco, our empirical data showed that two alternative transmission routes of *T. cruzi* (vertical and oral) were not able to maintain dog infections in the absence of domestic or peridomestic infestations with *T. infestans*. In contrast, the available data for Texas (Kjos et al. 2008) show that dogs may maintain the infection across a decade in the near absence of established house infestations with triatomines, systematic triatomine control actions, and human infection with *T. cruzi*. In this type of scenario, the hypothetical administration of a fully protective vaccine of *T. cruzi* to humans only would not affect parasite persistence or disease incidence in the dog population, a question that mathematical modeling efforts may address. In most circumstances, however, dogs and humans are part of the same reservoir of infection (sensu Haydon et al. 2002).

Acknowledgments We are grateful to M Carla Cecere, M Pilar Fernández, M Sol Gaspe, and Natalia Macchiaverna for helpful discussions. Gustavo Enriquez provided helpful comments on an earlier version of the manuscript. This work was supported by the University of Buenos Aires (UBACYT 20020130100843BA), PICT 2014-2661, and PICT 2015-2921.

References

Acosta N, López E, Lewis MD et al (2017) Hosts and vectors of *Trypanosoma cruzi* discrete typing units in the Chagas disease endemic region of the Paraguayan Chaco. Parasitology 144:884–898

Amelotti I, Catalá SS, Gorla DE (2012) Effects of fipronil on dogs over *Triatoma infestans*, the main vector of *Trypanosoma cruzi*, causative agent of Chagas disease. Parasitol Res 111:1457–1462. https://doi.org/10.1007/s00436-012-2979-6

Anderson RM, May R (1991) Infectious diseases of humans: dynamics and control. Oxford University Press, Oxford, p 757

Araujo FMG, Bahia MT, Magalhaes NM et al (2002) Follow-up of experimental chronic Chagas disease in dogs: use of polymerase chain reaction (PCR) compared with parasitological and serological methods. Acta Trop 81:21–31

Araújo-Neto VT, Medeiros Honorato NR, Oliveira Santana R et al (2019) *Trypanosoma cruzi* circulating among dogs and triatomines in the endemic countryside of the State of Rio Grande do Norte, Brazil. Acta Trop 200:105067

Barbosa-Silva AN, da Câmara ACJ, Martins K, Nunes DF, de Oliveira PIC, de Azevedo PRM, Chiari E, da Cunha Galvão LM (2016) Characteristics of Triatomine infestation and natural Trypanosoma cruzi infection in the State of Rio Grande do Norte, Brazil. Rev Soc Bras Med Trop 49(1):57–67

Barr SC (2009) Canine Chagas' disease (American trypanosomiasis) in North America. Vet Clin North Am Small Anim Pract 39:1055–1064, v–vi. https://doi.org/10.1016/j.cvsm.2009.06.004

Barrett TV, Hoff RH, Mott KE et al (1980) Epidemiological aspects of three *Trypanosoma cruzi* zymodemes in Bahia State, Brazil. Trans R Soc Trop Med Hyg 74:84–90. https://doi.org/10.1016/0035-9203(80)90016-4

Barros JHS, Almeida ABPF, Figueiredo FB et al (2012) Occurrence of *Trypanosoma caninum* in areas overlapping with leishmaniasis in Brazil: what is the real impact of canine leishmaniasis control? Trans R SocTrop Med Hyg 106:419–423

Basombrío MA, Segura MA, Mora MC et al (1993) Field trial of vaccination against American Trypanosomiasis (Chagas' disease) in dogs. Am J Trop Med Hyg 49:143–151

Basso B, Castro I, Introini V et al (2007) Vaccination with *Trypanosoma rangeli* reduces the infectiousness of dogs experimentally infected with *Trypanosoma cruzi*. Vaccine 25:3855–3858. https://doi.org/10.1016/j.vaccine.2007.01.114

Begon M (2008) Effects of host diversity on disease dynamics. In: Ostfeld RS, Keesing F, Eviner ET (eds) Infectious disease ecology. Effects of ecosystems on disease and of disease on ecosystems. Princeton University Press, Princeton, pp 12–29

Belisário CJ, Pessoa GC, Silva EM et al (2017) Genetic characterization of residual *Triatoma infestans* populations from Brazil by microsatellites. Genetica 145:105–114. https://doi.org/10.1007/s10709-017-9949-y

Brenière SF, Pietrokovsky S, Gastélum EM et al (2004) Feeding patterns of *Triatoma longipennis* Usinger (Hemiptera, Reduviidae) in peridomestic habitats of a rural community in Jalisco state, Mexico. J Med Entomol 41:1015–1020

Brenière SF, Waleckx E, Barnabé C (2016) Over six thousand *Trypanosoma cruzi* strains classified into discrete typing units (DTUs): attempt at an inventory. PLoS Negl Trop Dis 10:e0004792

Bustamante DM, de Urioste-Stone SM, Cruz JG et al (2014) Ecological, social and biological risk factors for continued *Trypanosoma cruzi* transmission by *Triatoma dimidiata* in Guatemala. PLoS One 9:e104599

Cantillo-Barraza O, Garcés E, Gómez-Palacio A et al (2015) Eco-epidemiological study of an endemic Chagas disease region in northern Colombia reveals the importance of *Triatoma maculata* (Hemiptera: Reduviidae), dogs and *Didelphis marsupialis* in *Trypanosoma cruzi* maintenance. Parasit Vectors 8:482

Cardinal MV, Castañera MB, Lauricella MA et al (2006a) A prospective study of the effects of sustained vector surveillance following community-wide insecticide application on *Trypanosoma cruzi* infection of dogs and cats in rural northwestern Argentina. Am J Trop Med Hyg 75:753–761

Cardinal MV, Reithinger R, Gürtler RE (2006b) Use of an immunochromatographic dipstick test for rapid detection of *Trypanosoma cruzi* in sera from animal reservoir hosts. J Clin Microbiol 44:3005–3007

Cardinal MV, Lauricella MA, Marcet PL et al (2007) Impact of community-based vector control on house infestation and *Trypanosoma cruzi* infection in *Triatoma infestans*, dogs and cats in the Argentine Chaco. Acta Trop 103:201–211

Cardinal MV, Lauricella MA, Ceballos LA et al (2008) Molecular epidemiology of domestic and sylvatic *Trypanosoma cruzi* infection in rural northwestern Argentina. Int J Parasitol 38:1533–1543. https://doi.org/10.1016/j.ijpara.2008.04.010

Cardinal MV, Orozco MM, Enriquez GF et al (2014) Heterogeneities in the eco-epidemiology of *Trypanosoma cruzi* infection in rural communities of the Argentinean Chaco. Am J Trop Med Hyg 90:1063–1073

Carlier Y, Truyens C (2017) Maternal-fetal transmission of *Trypanosoma cruzi*. In: Telleria J, Tibayrenc M (eds) American Trypanosomiasis Chagas disease: one hundred years of research, 2nd edn. Elsevier, London, pp 517–559

Carrasco HJ, Torrellas A, García C et al (2005) Risk of *Trypanosoma cruzi* I (Kinetoplastida: Trypanosomatidae) transmission by *Panstrongylus geniculatus* (Hemiptera: Reduviidae) in Caracas (metropolitan district) and neighboring states, Venezuela. Int J Parasitol 35:1379–1384

Castañera MB (1999) Contribución de los perros a la transmisión de *Trypanosoma cruzi* en un área rural de Argentina: demografía, rol centinela y modelado matemático. Tesis doctoral. Universidad de Buenos Aires. Buenos Aires, Argentina, 215 p, vol, 2, pp 33–53

Castañera MB, Lauricella MA, Chuit R et al (1998) Evaluation of dogs as sentinels of the transmission of *Trypanosoma cruzi* in a rural area of north-western Argentina. Ann Trop Med Parasitol 92:671–683

Castillo-Neyra R, Chu LC, Quispe-Machaca V et al (2015) The potential of canine sentinels for reemerging *Trypanosoma cruzi* transmission. Prev Vet Med 120:349–356

Ceballos LA, Cardinal MV, Vazquez-Prokopec GM et al (2006) Long-term reduction of *Trypanosoma cruzi* infection in sylvatic mammals following deforestation and sustained vector surveillance in northwestern Argentina. Acta Trop 98:286–296

Cecere MC, Gürtler RE, Canale DM et al (1998) Factors limiting the domiciliary density of *Triatoma infestans*, vector of Chagas' disease, in north-west Argentina: a longitudinal study. Bull World Health Organ 76:373–384

Cecere MC, Leporace M, Fernández MP et al (2016) Host-feeding sources and infection with *Trypanosoma cruzi* of *Triatoma infestans* and *Triatoma eratyrusiformis* (Hemiptera, Reduviidae) from the Calchaqui Valleys in Northwestern Argentina. J Med Entomol 53:666–673

Chapman MD, Baggaley RC, Godfrey-Fausset PF et al (1984) *Trypanosoma cruzi* from the Paraguayan Chaco: isoenzyme profiles of strains isolated at Makthlawaiya. J Protozool 31:482–486

Cohen JE, Gürtler RE (2001) Modeling household transmission of American trypanosomiasis. Science 293:694–698

Crisante G, Rojas A, Teixeira MMG et al (2006) Infected dogs as a risk factor in the transmission of human *Trypanosoma cruzi* infection in western Venezuela. Acta Trop 98:247–254. https://doi.org/10.1016/j.actatropica.2006.05.006

Cruz-Chan JV, Bolio-González M, Colín-Flores R (2009) Immunopathology of natural infection with *Trypanosoma cruzi* in dogs. Vet Parasitol 162:151–155. https://doi.org/10.1016/j.vetpar.2009.02.024

Cucunubá ZM, Nouvellet P, Conteh L et al (2017) Modelling historical changes in the force-of-infection of Chagas disease to inform control and elimination programmes: application in Colombia. BMJ Glob Health 2:e000345

Cunha ELA, Vieira da Silva Torchelsen FK, Cunha LM et al (2019) Benznidazole, itraconazole and their combination in the treatment of acute experimental Chagas disease in dogs. Exp Parasitol 204:107711

Curtis-Robles R, Lewis BC, Hamer SA (2016) High *Trypanosoma cruzi* infection prevalence associated with minimal cardiac pathology among wild carnivores in Central Texas. Int J Parasitol Parasit Wildl 5:117–123. https://doi.org/10.1016/j.ijppaw.2016.04.001

Curtis-Robles R, Zecca IB, Roman-Cruz V et al (2017a) *Trypanosoma cruzi* (agent of Chagas disease) in sympatric human and dog populations in "Colonias" of the lower Rio Grande Valley of Texas. Am J Trop Med Hyg 96:805–814

Curtis-Robles R, Snowden KF, Dominguez B et al (2017b) Epidemiology and molecular typing of *Trypanosoma cruzi* in naturally-infected hound dogs and associated triatomine vectors in Texas, USA. PLoS Negl Trop Dis 11:e0005298. https://doi.org/10.1371/journal.pntd.0005298

Curtis-Robles R, Meyers AC, Auckland LA et al (2018) Parasitic interactions among *Trypanosoma cruzi*, triatomine vectors, domestic animals, and wildlife in Big Bend National Park along the Texas-Mexico border. Acta Trop 188:225–233. https://doi.org/10.1016/j.actatropica.2018.09.002

de Urioste-Stone SM, Pennington PM, Pellecer E et al (2015) Development of a community-based intervention for the control of Chagas disease based on peridomestic animal management: an eco-bio-social perspective. Trans R Soc Trop Med Hyg 109:159–167

Desquesnes M (2017) Veterinary aspects. In: Telleria J, Tibayrenc M (eds) American Trypanosomiasis Chagas disease: one hundred years of research, 2nd edn. Elsevier, London, pp 283–298

Dias JCP, Schofield CJ, Machado EM et al (2005) Ticks, ivermectin, and experimental Chagas disease. Mem Inst Oswaldo Cruz 100:829–832

Diosque P, Barnabé C, Padilla AM et al (2003) Multilocus enzyme electrophoresis analysis of *Trypanosoma cruzi* isolates from a geographically restricted endemic area for Chagas' disease in Argentina. Int J Parasitol 33:997–1003. https://doi.org/10.1016/S0020-7519(03)00139-5

Dumonteil E, Nouvellet P, Rosecrans K et al (2013) Eco-bio-social determinants for house infestation by nondomiciliated *Triatoma dimidiata* in the Yucatan peninsula, Mexico. PLoS Negl Trop Dis 7:e2466

Dumonteil E, Ramirez-Sierra MJ, Pérez-Carrillo S et al (2018) Detailed ecological associations of triatomines revealed by metabarcoding and next-generation sequencing: implications for triatomine behavior and *Trypanosoma cruzi* transmission cycles. Sci Rep 8:4140

Enriquez GF, Cardinal MV, Orozco MM et al (2013a) Discrete typing units of *Trypanosoma cruzi* identified in rural dogs and cats in the humid Argentinean Chaco. Parasitology 140:303–308. https://doi.org/10.1017/S003118201200159X

Enriquez GF, Cardinal MV, Orozco MM et al (2013b) Detection of *Trypanosoma cruzi* infection in naturally infected dogs and cats using serological, parasitological and molecular methods. Acta Trop 126:211–217

Enriquez GF, Bua J, Orozco MM et al (2014) High levels of *Trypanosoma cruzi* DNA determined by qPCR and infectiousness to *Triatoma infestans* support dogs and cats are major sources of parasites for domestic transmission. Infect Genet Evol 25:36–43. https://doi.org/10.1016/j.meegid.2014.04.002

Enriquez GF, Macchiaverna NP, Argibay HD et al (2019) Zoonotic parasites and polyparasitism in dogs from a rural area of the Argentine Chaco. Vet Parasitol Reg Studies Rep 16:100287

Estrada-Franco JG, Bhatia V, Diaz-Albiter H et al (2006) Human *Trypanosoma cruzi* infection and seropositivity in dogs, Mexico. Emerg Infect Dis 12:624–630

Fabrizio M, Schweigmann NJ, Bartoloni N (2014) Modeling American trypanosomiasis in an endemic zone: application to the initial spread of household infection in the Argentine Chaco. Zoonoses Public Health 61:545–559

Fernández MP, Cecere MC, Lanati LA et al (2014) Geographic variation of *Trypanosoma cruzi* discrete typing units from *Triatoma infestans* at different spatial scales. Acta Trop 140:10–18

Fernández MP, Gaspe MS, Sartor P et al (2019) Human *Trypanosoma cruzi* infection risk is driven by eco-social interactions in rural communities of the Argentine Chaco. PLoS Negl Trop Dis 13:e0007430

Figueiredo Diniz L, Caldas IS, Guedes PMM et al (2010) Effects of ravuconazole treatment on parasite load and immune response in dogs experimentally infected with *Trypanosoma cruzi*. Antimicrob Ag Chemother 54:2979–2986

Flores-Ferrer A, Waleckx E, Rascalou G et al (2019) *Trypanosoma cruzi* transmission dynamics in a synanthropic and domesticated host community. PLoS Negl Trop Dis 13:e0007902. https://doi.org/10.1371/journal.pntd.0007902

Franke CR, Greiner M, Mehlitz D (1994) Investigation on naturally occurring *Trypanosoma evansi* infections in horses, cattle, dogs and capybaras (*Hydrochaeris hydrochaeris*) in Pantanal de Poconé (Mato Grosso, Brazil). Acta Trop 58:159–169

Freitas JLP (1950) Observações sôbre xenodiagnósticos practicados em reservatórios domésticos e silvestres do *Trypanosoma cruzi* em uma localidade endémica da moléstia de Chagas no estado de São Paulo. Hospital 38:63–71

Freitas YBN, Souza CDSF, Magalhães JME et al (2018) Natural infection by *Trypanosoma cruzi* in triatomines and seropositivity for Chagas disease of dogs in rural areas of Rio Grande do Norte, Brazil. Rev Soc Bras Med Trop 51:190–197. https://doi.org/10.1590/0037-8682-0088-2017

Fung HL, Calzada J, Saldaña A et al (2014) Domestic dog health worsens with socio-economic deprivation of their home communities. Acta Trop 135:67–74. https://doi.org/10.1016/j.actatropica.2014.03.010

Gompper ME (2014) Free-ranging dogs and wildlife conservation. Oxford University Press, Oxford, p 314

González-Vieyra SD, Ramírez-Durán N, Sandoval-Trujillo AH et al (2011) *Trypanosoma cruzi* in dogs: electrocardiographic and echocardiographic evaluation, in Malinalco, State of Mexico. Res Rep Trop Med 2:155–161

Gorchakov R, Trosclair LP, Wozniak EP et al (2016) *Trypanosoma cruzi* infection prevalence and bloodmeal analysis in triatomine vectors of Chagas disease from rural peridomestic locations in Texas, 2013-2014. J Med Entomol 53:911–918

Guedes PMDM, Veloso VM, Tafuri WL (2002) The dog as model for chemotherapy of the Chagas' disease. Acta Trop 84:9–17. https://doi.org/10.1016/S0001-706X(02)139-0

Guedes PMDM, Urbina JA, Lana M et al (2004) Activity of the new triazole derivative albaconazole against *Trypanosoma (Schizotrypanum) cruzi* in dog hosts. Antimicrob Ag Chemother 48:4286–4292

Gurevitz JM, Ceballos LA, Gaspe MS et al (2011) Factors affecting infestation by *Triatoma infestans* in a rural area of the humid Chaco in Argentina: a multi-model inference approach. PLoS Negl Trop Dis 5:e1365

Gürtler RE, Cardinal MV (2015) Reservoir host competence and the epidemiologic role of domestic and commensal hosts in the transmission of *Trypanosoma cruzi*. Acta Trop 151:32–50. https://doi.org/10.1016/j.actatropica.2015.05.029

Gürtler RE, Cecere MC (2020) Chagas disease vector control. In: Guarneri AA, Lorenzo MG (eds) Triatominae: the biology of Chagas disease vectors. Springer

Gürtler RE, Kravetz FO, Petersen RM et al (1990) The prevalence of *Trypanosoma cruzi* and the demography of dog populations after insecticidal spraying of houses: a predictive model. Ann Trop Med Parasitol 84:313–323

Gürtler RE, Petersen RM, Lauricella MA et al (1992) Infectivity to the vector *Triatoma infestans* of dogs infected with *Trypanosoma cruzi* in north-west Argentina. Ann Trop Med Parasitol 86:111–119

Gürtler RE, Cecere MC, Castañera MB et al (1996) Probability of infection with *Trypanosoma cruzi* of the vector *Triatoma infestans* fed on infected humans and dogs in northwest Argentina. Am J Trop Med Hyg 55:24–31

Gürtler RE, Cohen JE, Cecere MC et al (1997) Shifting host choices of the vector of Chagas disease *Triatoma infestans* in relation to the availability of hosts in houses in north-west Argentina. J Appl Ecol 34:699–715

Gürtler RE, Cecere MC, Lauricella MA et al (2005) Incidence of *Trypanosoma cruzi* infection among children following domestic reinfestation after insecticide spraying in rural northwestern Argentina. Am J Trop Med Hyg 73:95–103

Gürtler RE, Cecere MC, Lauricella MA et al (2007a) Domestic dogs and cats as sources of *Trypanosoma cruzi* infection in rural northwestern Argentina. Parasitology 134:69–82

Gürtler RE, Kitron U, Cecere MC et al (2007b) Sustainable vector control and management of Chagas disease in the Gran Chaco, Argentina. Proc Natl Acad Sci USA 104:16194–16199

Gürtler RE, Ceballos LA, Ordóñez-Krasnowski P et al (2009a) Strong host-feeding preferences of the Chagas disease vector *Triatoma infestans* modified by vector density: implications for the epidemiology of Chagas disease. PLoS Negl Trop Dis 3:e447. https://doi.org/10.1371/journal.pntd.0000447

Gürtler RE, Ceballos LA, Stariolo R et al (2009b) Effects of topical application of fipronil spot-on on dogs against the Chagas disease vector *Triatoma infestans*. Trans R Soc Trop Med Hyg 103:298–304. https://doi.org/10.1016/j.trstmh.2008.09.018

Gürtler RE, Cecere MC, Vázquez-Prokopec GM et al (2014) Domestic animal hosts strongly influence human-feeding rates of the Chagas disease vector *Triatoma infestans* in Argentina. PLoS Negl Trop Dis 8:e2894

Gürtler RE, Fernández MP, Cardinal MV (2020) Eco-epidemiology of the domestic vector-borne transmission of *Trypanosoma cruzi*. In: Guarneri AA, Lorenzo MG (eds) Triatominae: the biology of Chagas disease vectors. Springer

Haberkorn A, Gönnert R (1972) Animal experimental investigation into the activity of nifurtimox against *Trypanosoma cruzi*. Arzneim Forsch (Drur Res) 22:1570–1582

Halliday JEB, Meredith AL, Knobel DL et al (2007) A framework for evaluating animals as sentinels for infectious disease surveillance. J R Soc Interface 4:973–984. https://doi.org/10.1098/rsif.2007.0237

Haydon DT, Cleaveland S, Taylor LH et al (2002) Identifying reservoirs of infection: a conceptual and practical challenge. Emerg Infect Dis 8:1468–1473. https://doi.org/10.3201/eid0812.010317

Hoare C (1972) The trypanosomes of mammals. A zoological monograph. Blackwell, Oxford

Ibáñez-Cervantes G, Martínez-Ibarra A, Nogueda-Torres B et al (2013) Identification by Q-PCR of *Trypanosoma cruzi* lineage and determination of blood meal sources in triatomine gut samples in Mexico. Parasitol Int 62:36–43

Jansen AM, Xavier SC, Roque AL (2017) Ecological aspects of *Trypanosoma cruzi*: wild hosts and reservoirs. In: Telleria J, Tibayrenc M (eds) American Trypanosomiasis Chagas disease: one hundred years of research, 2nd edn. Elsevier, London, pp 243–264

Kjos S, Snowden K, Craig T et al (2008) Distribution and characterization of canine Chagas disease in Texas. Vet Parasitol 152:249–256

Kjos S, Marcet PL, Yabsley MJ et al (2013) Identification of bloodmeal sources and *Trypanosoma cruzi* infection in triatomine bugs (Hemiptera: Reduviidae) from residential settings in Texas, the United States. J Med Entomol 50:1126–1139

Kribs-Zaleta C (2010) Estimating contact process saturation in sylvatic transmission of *Trypanosoma cruzi* in the United States. PLoS Negl Trop Dis 4:e656

Laiño M, Cardinal MV, Enriquez GF et al (2019) An oral dose of Fluralaner administered to dogs kills pyrethroid resistant and susceptible Chagas disease vectors for at least four months. Vet Parasitol 268:98–104

Lana M (2017) Experimental studies of Chagas disease in animal models. In: Telleria J, Tibayrenc M (eds) American Trypanosomiasis Chagas disease: one hundred years of research, 2nd edn. Elsevier, London, pp 299–320

Lauricella MA, Riarte A, Lazzari JO et al (1986) Enfermedad de Chagas en perros experimentalmente infectados con *Trypanosoma cruzi*. Medicina (B. Aires) 46:195–200

Lauricella MA, Sinagra AJ, Paulone I et al (1989) Natural *Trypanosoma cruzi* infection in dogs of endemic areas of the Argentine Republic. Rev Inst Med Trop São Paulo 31:63–70

Lauricella MA, Castañera MB, Gürtler RE et al (1998) Immunodiagnosis of *Trypanosoma cruzi* (Chagas' disease) infection in naturally infected dogs. Mem Inst Oswaldo Cruz 93:501–507

Lima-Cordón RA, Stevens L, Solórzano Ortíz E et al (2018) Implementation science: epidemiology and feeding profiles of the Chagas vector *Triatoma dimidiata* prior to Ecohealth intervention for three locations in Central America. PLoS Negl Trop Dis 12:e0006952. https://doi.org/10.1371/journal.pntd.0006952

Lizundia R, Picado A, Cordero M et al (2014) Molecular and serological rapid tests as markers of *Trypanosoma cruzi* infection in dogs in Costa Rica. Trop Parasitol 4:111–114

López-Cancino SA, Tun-Ku E, Keynes H et al (2015) Landscape ecology of *Trypanosoma cruzi* in the southern Yucatan Peninsula. Acta Trop 151:58–72. https://doi.org/10.1016/j.actatropica.2015.07.021

Loza A, Talaga A, Herbas G et al (2017) Systemic insecticide treatment of the canine reservoir of *Trypanosoma cruzi* induces high levels of lethality in *Triatoma infestans,* a principal vector of Chagas disease. Parasit Vectors 10:344

Lucero RH, Brusés BL, Cura CI et al (2016) Chagas' disease in aboriginal and creole communities from the Gran Chaco region of Argentina: seroprevalence and molecular parasitological characterization. Infect Genet Evol 41:84–92. https://doi.org/10.1016/j.meegid.2016.03.028

Lucheis SB, Da Silva AV, Araújo JP Jr et al (2005) Trypanosomatids in dogs belonging to individuals with chronic Chagas disease living in Botucatu town and surrounding region. São Paulo State, Brazil. J Venom Anim Toxins Incl Trop Dis 11:492–509

Luquetti AO, Schmunis GA (2017) Diagnosis of *Trypanosoma cruzi* infection. In: Telleria J, Tibayrenc M (eds) American Trypanosomiasis Chagas disease: one hundred years of research, 2nd edn. Elsevier, London, pp 687–730

Macchiaverna NP, Gaspe MS, Enriquez GF et al (2015) *Trypanosoma cruzi* infection in *Triatoma sordida* before and after community-wide residual insecticide spraying in the Argentinean Chaco. Acta Trop 143:97–102. https://doi.org/10.1016/j.actatropica.2014.12.010

Macchiaverna NP, Enriquez GF, Buscaglia CA et al (2018) New human isolates confirm the predominance of hybrid lineages in the Argentine domestic cycle of *Trypanosoma cruzi*. Infect Gen Evol 66:229–235

Macchiaverna NP, Enriquez GF, Bua J et al (2020) Human infectiousness and parasite load in chronic patients seropositive for *Trypanosoma cruzi* in a rural area of the Argentine Chaco. Infect Gen Evol 78:104062

Machado EMM, Fernandes AJ, Murta SMF et al (2001) A study of experimental reinfection by *Trypanosoma cruzi* in dogs. Am J Trop Med Hyg 65:959–965

Macpherson CLN, Meslin FX, Wandeler AI (eds) (2013) Dogs, zoonoses and public health, 2nd edn. CABI, New York, p 288

Madeira MF, Sousa MA, Barros JH et al (2009) *Trypanosoma caninum* sp. (Protozoa: Kinetoplastida) isolated from intact skin of a domestic dog (*Canis familiaris*) captured in Rio de Janeiro, Brazil. Parasitology 136:411–423

Madeira MF, Almeida AB, Barros JH et al (2014) *Trypanosoma caninum*, a new parasite described in dogs in Brazil: aspects of natural infection. J Parasitol 100:231–234

Madigan R, Majoy S, Ritter K et al (2019) Investigation of a combination of amiodarone and itraconazole for treatment of American trypanosomiasis (Chagas disease) in dogs. J Am Vet Med Assoc 255:317–329. https://doi.org/10.2460/javma.255.3.317

Maffey L, Cardinal MV, Ordóñez-Krasnowski PC et al (2012) Direct molecular identification of *Trypanosoma cruzi* discrete typing units in domestic and peridomestic *Triatoma infestans* and *Triatoma sordida* from the Argentine Chaco. Parasitology 139:1570–1579

Marcili A, Lima L, Cavazzana M et al (2009) A new genotype of *Trypanosoma cruzi* associated with bats evidenced by phylogenetic analyses using SSU rDNA, cytochrome b and histone H2B genes and genotyping based on ITS1 rDNA. Parasitology 136:641–655. https://doi.org/10.1017/S0031182009005861

Marsden PD, Hagstrom JWC (1968) Experimental *Trypanosoma cruzi* infection in beagle puppies. The effect of variations in dose and source of infecting trypanosomes and the route of inoculation on the course of the infection. Trans R Soc Trop Med Hyg 62:816–824

Matter HC, Daniels TJ (2000) Dog ecology and population biology. In: Macpherson CNL, Meslin FX, Wandeler AI (eds) Dogs, zoonoses and public health. CABI, New York, pp 17–50

Mazza S (1934) Difusión de la infección natural por *Schyzotripanum cruzi* en perros de la Provincia de Jujuy. Mis Est Patol Reg Arg 17:23–28

Mazza S, Lobos MM (1937) Casos de enfermedad de Chagas y animales domésticos infectados naturalmente con *S. cruzi* comprobados en el Departamento de Trancas provincia de Tucumán. Mis Est Patol Reg Arg 32:18–33

Meurs KM, Anthony MA, Slater M et al (1998) Chronic *Trypanosoma cruzi* infection in dogs: 11 cases (1987-1996). J Am Vet Med Assoc 213:497–500

Meyers AC, Meinders M, Hamer SA (2017) Widespread *Trypanosoma cruzi* infection in government working dogs along the Texas-Mexico border: discordant serology, parasite genotyping and associated vectors. PLoS Negl Trop Dis 11:e0005819. https://doi.org/10.1371/journal pntd.0005819

Meyers AC, Hamer SA, Matthews D et al (2019) Risk factors and select cardiac characteristics in dogs naturally infected with *Trypanosoma cruzi* presenting to a teaching hospital in Texas. J Vet Intern Med 33:1695–1706

Miles MA, Feliciangeli MD, Rojas de Arias A (2003) American trypanosomiasis (Chagas' disease) and the role of molecular epidemiology in guiding control strategies. BMJ 326:1444–1448

Miles MA, Llewellyn MS, Lewis MD et al (2009) The molecular epidemiology and phylogeography of *Trypanosoma cruzi* and parallel research on *Leishmania*: looking back and to the future. Parasitology 136:1509–1528. https://doi.org/10.1017/S0031182009990977

Minter DM (1976a) Effects on transmission to man of the presence of domestic animals in infested households. In: New approaches in American Trypanosomiasis research. Pan American Health Organization No. 318, Washington, pp 330–337

Minter DM (1976b) Feeding patterns of some Triatominae vectors. In: New approaches in American Trypanosomiasis research, Pan American Health Organization Scientific Publication No. 318, Washington, pp 33–47

Monje-Rumi MM, Brandán CP, Ragone PG et al (2015) *Trypanosoma cruzi* diversity in the Gran Chaco: mixed infections and differential host distribution of TcV and TcVI. Infect Genet Evol 29:53–59. https://doi.org/10.1016/j.meegid.2014.11.001

Monteón V, Triana-Chávez O, Mejía-Jaramillo A et al (2016) Circulation of Tc Ia discrete type unit *Trypanosoma cruzi* in Yucatan Mexico. J Parasit Dis 40:550–554

Moo-Millan JI, Arnal A, Pérez-Carrillo S et al (2019) Disentangling *Trypanosoma cruzi* transmission cycle dynamics through the identification of blood meal sources of natural populations of *Triatoma dimidiata* in Yucatán, Mexico. Parasit Vectors 12:572

Mota J, Chacon JC, Gutierrez-Cabrera AE et al (2007) Identification of blood meal source and infection with *Trypanosoma cruzi* of Chagas disease vectors using a multiplex cytochrome b polymerase chain reaction assay. Vector-Borne Zoonotic Dis 7:617–628

Mott KE, Mota EA, Sherlock I et al (1978) *Trypanosoma cruzi* infection in dogs and cats and household seroreactivity to *T. cruzi* in a rural community in Northeast Brazil. Am J Trop Med Hyg 27:1123–1127

Murphy N, Macchiaverna NP, Cardinal MV et al (2019) Lineage-specific rapid diagnostic tests can resolve *Trypanosoma cruzi* TcII/V/VI ecological and epidemiological associations in the Argentine Chaco. Parasit Vectors 12:424. https://doi.org/10.1186/s13071-019-3681-7

Nieto PD, Boughton R, Dorn PL (2009) Comparison of two immunochromatographic assays and the indirect immunofluorescence antibody test for diagnosis of *Trypanosoma cruzi* infection in dogs in South Central Louisiana. Vet Parasitol 165:241–247

Nouvellet P, Dumonteil E, Gourbière S (2013) The improbable transmission of *Trypanosoma cruzi* to human: the missing link in the dynamics and control of Chagas disease. PLoS Negl Trop Dis 7:e2505

Nouvellet P, Cucunubá ZM, Gourbière S (2015) Ecology, evolution and control of Chagas disease: a century of neglected modelling and a promising future. Adv Parasitol 87:135–191. https://doi.org/10.1016/bs.apar.2014.12.004

Oliveira-Porfirio GE, Martins Santos F, Carvalho de Macedo G et al (2018) Maintenance of *Trypanosoma cruzi*, *T. evansi* and *Leishmania* spp. by domestic dogs and wild mammals in a rural settlement in Brazil-Bolivian border. Int J Parasitol: Parasit Wildlife 7:398–404

Orozco MM, Ceballos LA, Pino MC et al (2013a) Local threats and potential infectious hazards to maned wolves (*Chrysocyon brachyurus*) in the southeastern argentine Chaco. Mammalia 78:339–349. https://doi.org/10.1515/mammalia-2013-0067

Orozco MM, Enriquez GF, Alvarado-Otegui JA et al (2013b) New sylvatic hosts of *Trypanosoma cruzi* and their reservoir competence in the humid Chaco of Argentina: a longitudinal study. Am J Trop Med Hyg 88:872–882

Orozco MM, Piccinali RV, Mora MS et al (2014) The role of sigmodontine rodents as sylvatic hosts of *Trypanosoma cruzi* in the Argentinean Chaco. Infect Gen Evol 22:12–22

Orozco MM, Enriquez GF, Cardinal MV et al (2016) A comparative study of *Trypanosoma cruzi* infection in sylvatic mammals from a protected and a disturbed area in the Argentine Chaco. Acta Trop 155:34–42

Ortiz S, Ceballos MJ, González CR et al (2016) *Trypanosoma cruzi* diversity in infected dogs from areas of the north coast of Chile. Vet Parasitol Reg Studies Rep 5:42–47

Padilla AM, Brandan CP, Basombrío MA (2017) Vaccine development for Chagas disease. In: Telleria J, Tibayrenc M (eds) American Trypanosomiasis Chagas disease: one hundred years of research, 2nd edn. Elsevier, London, pp 773–796

Pellecer MJ, Dorn PL, Bustamante DM et al (2013) Vector blood meals are an early indicator of the effectiveness of the Ecohealth approach in halting Chagas transmission in Guatemala. Am J Trop Med Hyg 88:638–644

Pennington PM, Messenger LA, Reina J et al (2015) The Chagas disease domestic transmission cycle in Guatemala: parasite-vector switches and lack of mitochondrial co-diversification between *Triatoma dimidiata* and *Trypanosoma cruzi* subpopulations suggest non-vectorial parasite dispersal across the Motagua valley. Acta Trop 151:80–87. https://doi.org/10.1016/j.actatropica.2015.07.014

Petersen RM, Gürtler RE, Cecere MC et al (2001) Association between nutritional indicators and infectivity of dogs seroreactive for *Trypanosoma cruzi* in a rural area of northwestern Argentina. Parasitol Res 87:208–214

Peterson JK, Bartsch SM, Lee BY et al (2015) Broad patterns in domestic vector-borne *Trypanosoma cruzi* transmission dynamics: synanthropic animals and vector control. Parasit Vectors 8:537

Pifano CF (1973) La dinámica epidemiológica de la enfermedad de Chagas en el Valle de los Naranjos, Estado Carabobo, Venezuela. II. La infección chagásica en la población rural del área. Arch Venez Med Trop Parasitol Med 5:31–45

Pineda V, Saldaña A, Monfante I et al (2011) Prevalence of trypanosome infections in dogs from Chagas disease endemic regions in Panama, Central America. Vet Parasitol 178:360–363. https://doi.org/10.1016/j.vetpar.2010.12.043

Pinto AGS, Schubach TM, Figueiredo FB et al (2010) Isolation of *Trypanosoma caninum* in domestic dogs in Rio de Janeiro. Rev Bras Parasitol Vet 137:1653–1660

Pinto AG, Toma HK, Figueiredo FB et al (2014) Evaluation of 18S rDNA PCR assay using skin fragments as a diagnostic test for *Trypanosoma caninum*. Vet Parasitol 205:343–346

Pizarro JC, Stevens (2008) A new method for forensic DNA analysis of the blood meal in Chagas disease vectors demonstrated using *Triatoma infestans* from Chuquisaca, Bolivia. PLoS One 3:3585

Quijano-Hernández IA, Castro-Barcena A, Vázquez-Chagoyán JC et al (2013) Preventive and therapeutic DNA vaccination partially protect dogs against an infectious challenge with *Trypanosoma cruzi*. Vaccine 31:2246–2252. https://doi.org/10.1016/j.vaccine.2013.03.005

Quinnell RJ, Courtenay O (2009) Transmission, reservoir hosts and control of zoonotic visceral leishmaniasis. Parasitology 136:1915–1934

Rabinovich JE, Kitron UD, Obed Y et al (2011) Ecological patterns of blood-feeding by kissing-bugs (Hemiptera: Reduviidae: Triatominae). Mem Inst Oswaldo Cruz 106:479–494

Ramírez JD, Turriago B, Tapia-Calle G et al (2013) Understanding the role of dogs (*Canis lupus familiaris*) in the transmission dynamics of *Trypanosoma cruzi* genotypes in Colombia. Vet Parasitol 196:216–219. https://doi.org/10.1016/j.vetpar.2012.12.054

Ramsey JM, Gutiérrez-Cabrera AE, Salgado-Ramírez L et al (2012) Ecological connectivity of *Trypanosoma cruzi* reservoirs and *Triatoma pallidipennis* hosts in an anthropogenic landscape with endemic Chagas disease. PLoS One 7:e46013. https://doi.org/10.1371/journal.pone.0046013

Rassi A, Rezende JM, Luquetti AO et al (2017) Clinical phases and forms of Chagas disease. In: Telleria J, Tibayrenc M (eds) American Trypanosomiasis Chagas disease: one hundred years of research, 2nd edn. Elsevier, London, pp 653–686

Reisenman CE, Savary W, Cowles J et al (2012) The distribution and abundance of triatomine insects, potential vectors of Chagas disease, in a metropolitan area in southern Arizona, United States. J Med Entomol 49:1254–1261

Reithinger R, Davies CR (1999) Is the domestic dog (*Canis familiaris*) a reservoir host of American cutaneous leishmaniasis? A critical review of the current evidence. Am J Trop Med Hyg 61:530–541

Reithinger R, Ceballos LA, Stariolo R et al (2005) Chagas disease control: deltamethrin-treated collars reduce *Triatoma infestans* feeding success on dogs. Trans R Soc Trop Med Hyg 99:502–508

Reithinger R, Ceballos LA, Stariolo R et al (2006) Extinction of experimental *Triatoma infestans* populations following continuous exposure to dogs wearing deltamethrin-treated collars. Am J Trop Med Hyg 74:766–771

Rocha FL, Roque ALR, Arrais RC et al (2013) *Trypanosoma cruzi* TcI and TcII transmission among wild carnivores, small mammals and dogs in a conservation unit and surrounding areas, Brazil. Parasitology 140:160–170. https://doi.org/10.1017/S0031182012001539

Roellig DM, Brown EL, Barnabé C et al (2008) Molecular typing of *Trypanosoma cruzi* isolates, United States. Emerg Infect Dis 14:1123–1125. https://doi.org/10.3201/eid1407.080175

Rolón M, Vega MC, Román F et al (2011) First report of colonies of sylvatic *Triatoma infestans* (Hemiptera: Reduviidae) in the Paraguayan Chaco, using a trained dog. PLoS Negl Trop Dis 5: e1026. https://doi.org/10.1371/journal.pntd.0001026

Rosal GG, Nogueda-Torres B, Villagrán ME et al (2018) Chagas disease: importance of rats as reservoir hosts of *Trypanosoma cruzi* (Chagas, 1909) in western Mexico. J Infect Public Health 11:230–233

Rosypal AC, Hill R, Lewis S (2011) Evaluation of a rapid immunochromatographic dipstick test for detection of antibodies to *Trypanosoma cruzi* in dogs experimentally infected with isolates obtained from opossums (*Didelphis virginiana*), armadillos (*Dasypus novemcinctus*), and dogs (*Canis familiaris*) from the United States. J Parasitol 97:140–143

Santos FM, Lima WG, Gravel AS et al (2012) Cardiomyopathy prognosis after benznidazole treatment in chronic canine Chagas' disease. J Antimicrob Chemother 67:1987–1995. https://doi.org/10.1093/jac/dks135

Sartor PA, Cardinal MV, Orozco MM et al (2011) Trans-Sialidase neutralizing antibody detection in *Trypanosoma cruzi*-infected domestic reservoirs. Clin Vaccine Immunol 18:984–989

Schaub GA (1988) Direct transmission of *Trypanosoma cruzi* between vectors of Chagas disease. Acta Trop 45:11–19

Silva MB, Menezes KR, Farias MC et al (2017) Description of the feeding preferences of triatominae in the Chagas disease surveillance study for the State of Pernambuco, Brazil (Hemiptera: Reduviidae). Rev Soc Brasil Med Trop 50:543–536

Snowden KF, Kjos SA (2012) American trypanosomiasis. In: Greene CE (ed) Infectious diseases of the dog and cat, 4th edn. Elsevier/Saunders, St. Louis, pp 722–730

Spagnuolo AM, Shillor M, Kingsland L et al (2012) A logistic delay differential equation model for Chagas disease with interrupted spraying schedules. J Biol Dyn 6:377–394. https://doi.org/10.1080/17513758.2011.587896

Stevens L, Dorn PL, Hobson P et al (2012) Vector blood meals and Chagas disease transmission potential, United States. Emerg Infect Dis 18:646–649

Tenney TD, Curtis-Robles R, Snowden KF et al (2014) Shelter dogs as sentinels for *Trypanosoma cruzi* transmission across Texas. Emerg Infect Dis 20:1323–1326. https://doi.org/10.3201/eid2008.131843

Travi BL (2019) Considering dogs as complementary targets of Chagas disease control. Vector-Borne Zoonotic Dis 19:90–94

Troncarelli MZ, Camargo JB, Machado JG et al (2009) *Leishmania* spp. and/or *Trypanosoma cruzi* diagnosis in dogs from endemic and nonendemic areas for canine visceral leishmaniasis. Vet Parasitol 164:118–123

Umezawa ES, Souza AI, Pinedo-Cancino V et al (2009) TESA-blot for the diagnosis of Chagas disease in dogs from co-endemic regions for *Trypanosoma cruzi*, *Trypanosoma evansi* and *Leishmania chagasi*. Acta Trop 111:15–20

Vallejo GA, Guhl F, Schaub GA (2009) Triatominae-*Trypanosoma cruzi/T.rangeli*: vector-parasite interactions. Acta Trop 110:137–147. https://doi.org/10.1016/j.actatropica.2008.10.001

Villalobos G, Martínez-Hernández F, de la Torre P et al (2011) Entomological indices, feeding sources, and molecular identification of *Triatoma phyllosoma* (Hemiptera: Reduviidae) one of the main vectors of Chagas disease in the Istmo de Tehuantepec, Oaxaca, Mexico. Am J Trop Med Hyg 85:490–497

Waleckx E, Suarez J, Richards B et al (2014) *Triatoma sanguisuga* blood meals and potential for Chagas disease, Louisiana, USA. Emerg Infect Dis 20:2141–2143

Wisnivesky-Colli C, Gürtler RE, Solarz ND et al (1982) Feeding patterns of *Triatoma infestans* (Hemiptera, Reduviidae) in relation to transmission of American Trypanosomiasis in Argentina. J Med Entomol 19:645–654

World Health Organization (2002) Control of Chagas disease. Second report of the WHO Expert Committee. WHO Technical Report Series 905, Geneva

World Health Organization (2015) Chagas disease in Latin America: an epidemiological update based on 2010 estimates. Wkly Epidemiol Rec 90:33–44

Xavier SCC, Roque ALR, Lima VS et al (2012) Lower richness of small wild mammal species and Chagas disease risk. PLoS Negl Trop Dis 6:e1647. https://doi.org/10.1371/journal.pntd.0001647

Xavier SCC, Roque ALR, Bilac D et al (2014) Distantiae transmission of *Trypanosoma cruzi*: a new epidemiological feature of acute Chagas disease in Brazil. PLoS Negl Trop Dis 8:e2878. https://doi.org/10.1371/journal.pntd.0002878

Yeo M, Acosta N, Llewellyn MS et al (2005) Origins of Chagas disease: *Didelphis* species are natural hosts of *Trypanosoma cruzi* I and armadillos hosts of *Trypanosoma cruzi* II, including hybrids. Int J Parasitol 35:225–233. https://doi.org/10.1016/j.ijpara.2004.10.024

Zárate LG, Zárate RJ, Tempelis CH et al (1980) The biology and behavior of *Triatoma barberi* (Hemiptera: Reduviidae) in Mexico. I. Blood meal sources and infection with *Trypanosoma cruzi*. J Med Entomol 17:103–116

Zeledón R (1974) Epidemiology, modes of transmission and reservoir hosts of Chagas' disease. In: Trypanosomiasis and leishmaniasis with special reference to Chagas' disease. Ciba foundation symposium 20 (new series). Amsterdam, pp 51–85

Zeledón R, Solano G, Burstin L et al (1975) Epidemiological pattern of Chagas' disease in an endemic area of Costa Rica. Am J Trop Med Hyg 24:214–225

Zeledón R, Beard CB, Dias JCP et al (2012) An appraisal of the status of Chagas disease in the United States. Elsevier, London, p 112

Zingales B (2018) *Trypanosoma cruzi* genetic diversity: something new for something known about Chagas disease manifestations, serodiagnosis and drug sensitivity. Acta Trop 184:38–52. https://doi.org/10.1016/j.actatropica.2017.09.017

Zingales B, Miles MA, Campbell DA et al (2012) The revised *Trypanosoma cruzi* subspecific nomenclature: rationale, epidemiological relevance and research applications. Infect Genet Evol 12:240–253. https://doi.org/10.1016/J.MEEGID.2011.12.009

Chapter 6
Echinococcus Species: Tiny Tapeworms

Heinz Mehlhorn

Abstract The adult stages of the very tiny tapeworms of *Echinococcus* live in the intestines of dogs and foxes. Their larvae, however, parasitize not only in many animals, but unfortunately also in humans, where they may cause deadly effects. This chapter shows how to diagnose infections, how to avoid them and how to treat foxes, dogs, and humans with antiparasitic drugs.

Keywords Life cycle · Transmission · Tumor-like disease

Some stages of these species had already been described by ancient scientists like Hippocrates (460–370 BC), who had studied the cyst-like, fluid-filled larvae (hydatids). However, intensive scientific research started only before about 200 years ago, when German scientists described the genus *Echinococcus* (Rudolphi 1801) and several species such as *E. granulosus* (Bartsch 1786), *E. multilocularis* (Leuckart 1863), *E. oligarthrus* (Diesing 1863), *E. vogeli* (Rausch and Bernstein 1972). Then it took until recent times, when more and more molecular biological details had been discovered, which among others allow a better and quicker diagnosis.

1. *Name*: Greek: *echinos* = hedgehog; *kokkos* = nucleus. Latin: *granulus* = with grains; *multilocularis* = with many tiny chambers/hollows.
2. *Geographic distribution/epidemiology*: See Table 6.1.
3. *Biology, morphology*: As it can be seen in Tables 6.2 and 6.3, more and more *Echinococcus* species are described in recent times. However, these species induce mostly very similar clinical symptoms, so that it remains very often very difficult to diagnose the exact species. This, however, does not cause real problems in treatment, since the common drugs act significantly well on the most important ones of them.

H. Mehlhorn (✉)
Department of Parasitology, Heinrich Heine University, Düsseldorf, Nordrhein-Westfalen, Germany
e-mail: mehlhorn@uni-duesseldorf.de

C. Strube, H. Mehlhorn (eds.), *Dog Parasites Endangering Human Health*, Parasitology Research Monographs 13,
https://doi.org/10.1007/978-3-030-53230-7_6

Table 6.1 Diagrammatic representation of the geographic distribution of the *Echinococcus* species and the induced types of echinococcosis

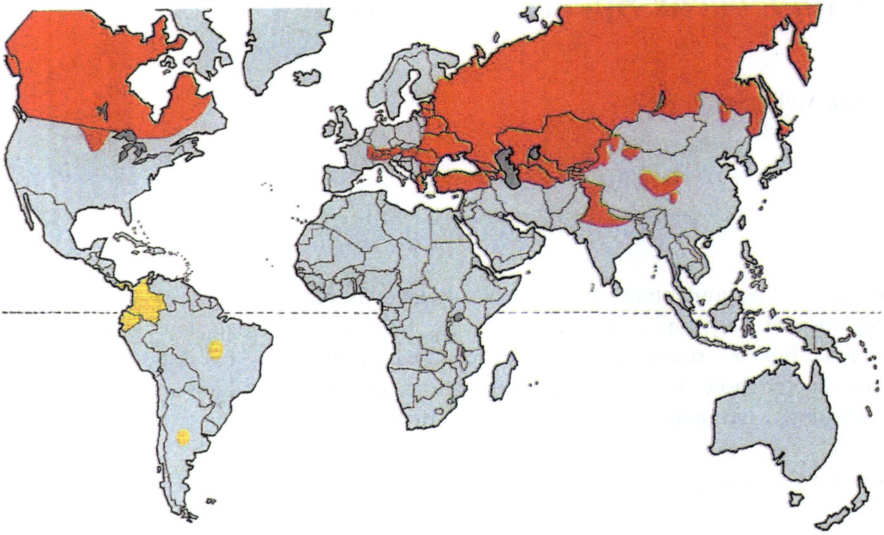

Red stained regions: alveolar echinococcosis
Yellow stained regions: polycystic echinococcosis

Table 6.2 Differentiation between the gravid specimens of *E. multilocularis* and *E. granulosus* (after Piekarski 1962)

Characteristics	*E. multilocularis*	*E. granulosus*
Length of the body	1.11–2.71 mm, average: 2.13 mm	2.10–5.02 mm, average: 3.36 mm
Length of the terminal segment	Smaller than half the whole length of the body: 0.44–1.11 mm, average: 0.85 mm	Usually longer than half the length of the whole body: 1.02–3.2 mm, average: 1.95 mm
Numbers of segments	3–5 (in dogs usually 4)	3
Sexually mature segment	The last but two	The penultimate one
Number of testes	14–31, average: 22	38–52, average: 44
Number of testes in front of the cirrus sac	0–5, average: 2	9–23, average: 15
Uterus of the gravid segment	Without lateral branches	Lateral branches usually distinct

Table 6.3 Species distribution (*D*), genotypes (*G*), and hosts of tapeworms of the genus *Echinococcus*

Species, strains, genotypes (*G*) Distribution (*D*)	Final hosts (Adult worms in the intestine)	Intermediate hosts (Seat of metacestodes)	Amounts of proglottids of adult worms	Disease (Infected organs)
Echinococcus granulosus (*G*1, *G*2, *G*3) *D*: worldwide	Canids (dogs, dingos, shakal, etc.)	Sheep, cattle, goats, yaks, camels, pigs, *humans*	3	*Cystic* echinococcosis (CE) in liver, lung and—less common—on other organs
Echinococcus equinus (*G*4) *D*: Rare in Europe, North and South Africa, Middle East	Dogs	Horses, donkeys, zebras, *unknown: humans*	3	*Cystic* liver echinococcosis
Echinococcus ortleppi (*G*5) *D*: Europe, Africa, Central/ South America	Dogs	*Rarely humans*; cattle, sheep, goats	3	*Cystic* echinococcosis in lung, liver
Echinococcus intermedius Camel *G*6, other *G*9 *D*: East Europe, Middle East, Asia, Argentina	Dogs	Camels (G6), pigs (G9), goats, rarely *humans* (G9)	3	*Cystic* echinococcosis in lung, liver
Echinococcus canadensis *G*8, *G*10 *D*: North Eurasia, North America	Wolves, dogs	Cervids, rarely *humans*	3	*Cystic* echinococcosis in lung, liver
Echinococcus felidis *D*: Africa	Lions	Warthog, other wildlife animals	3	*Cystic* echinococcosis, human cases unknown
Echinococcus multilocularis *D*: Northern hemisphere (1.2–4.5 mm)	Foxes, wolves, racoon dogs, dogs, rarely cats	Rodents, *humans*, monkeys, horses, pigs, dogs	Up to 5	*Alveolar* echinococcosis, secondary organs
Echinococcus vogeli *D*: Central, South America (3.9–5.6 mm)	Bush dogs (*Speothos venaticus*)	Rodents (paca, agoutis), potentially *humans*	3	*Polycystic* echinococcosis in liver, lung, diaphragm

(continued)

Table 6.3 (continued)

Species, strains, genotypes (*G*) Distribution (*D*)	Final hosts (Adult worms in the intestine)	Intermediate hosts (Seat of metacestodes)	Amounts of proglottids of adult worms	Disease (Infected organs)
Echinococcus oligarthrus *D*: Central, South America (2.2–2.9 mm)	Wild cats (puma, jaguar)	Rodents (agoutis, rats), pakas, occasionally *humans*	3	*Unicystic* echinococcosis in skin, muscles, internal organs, orbita
Echinococcus shiquicus *D*: China, Tibet (1.3–1.7 mm)	Tibetan fox (*Vulpes ferrilata*)	Pica; *Ochrotona curzoniae*	3	Human cases not documented yet

3.1. *Echinococcus granulosus*:

This species prefers as *final host* apparently the dog, while cats and foxes are less often infected (Fig. 6.1). The adult worms, which live in the intestine of their final hosts, reach a length of 2.5–6 mm, whereby the last (=gravid), 3rd proglottis is considerably larger (~3 mm) than all others before (Figs. 6.1, 6.2, 6.3, 6.4, and 6.5). The pores of the genital open just before or just behind the middle of the proglottis. The uterus of the excreted terminal proglottis shows lateral protrusions and fills in contrast to that of *E. multilocularis* the whole proglottis (see in Fig. 6.1 drawing 3, 3.1). The scolex of this rather tiny tapeworm is attached with the help of four suckers and a crown of hooks at the intestinal wall and enters deeply into the intestinal villi (Fig. 6.2). The hooks of the scolex are of different size: there are small ones measuring 19–35 μm and large ones 25–40 μm. All together 30–42 hooks may occur. In most cases, large crowds of worms are present being anchored very close to each other, so that fertilizations may occur by neighboring worms. The eggs of *E. granulosus* belong to the *Taenia* type (Fig. 6.1) and contain already prior to their discharge the oncosphaera larva, which is provided with six characteristic hooks. The strains of *E. granulosus* use a very broad spectrum of intermediate hosts; e.g., pigs, ruminants, and horses are common hosts, but humans may also become infected. There is a broad spectrum of animals, which are believed to harbor separate strains or even separate species (see Table 6.3). Also, the so-called hydatid larval stages have slightly different shapes in these various intermediate hosts. Inside these hydatids—the name was given since they are filled with a fluid—other bladder-like compartments of which are formed along the inner layer, wherein the protoscolices (=heads) of the later tapeworms are produced asexually (Fig. 6.1). Each of these protoscolices grows up to a fertile worm, if a final host (dogs, cats, and foxes) ingests such pieces of the hydatid. Thus, this life cycle contains a

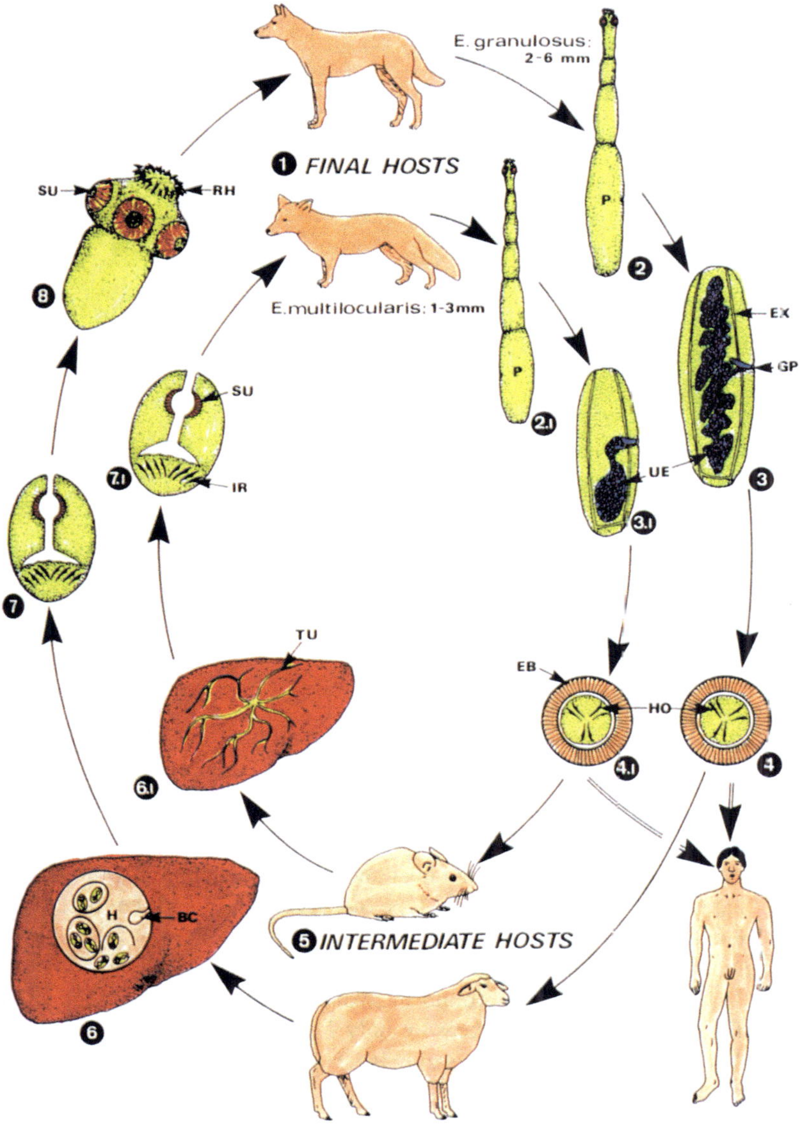

Fig. 6.1 Diagrammatic representation of the life cycles of *Echinococcus granulosus* (1–8) and *E. multilocularis* (1.1–7.1). (**1, 1.1**) Final hosts may be dog, cat, or fox with clear, species-specific preference. (**2–2.1**) Adult worms, which live in the small intestine of the final host, may be differentiated according to the size of the terminal proglottids (P), shape of uterus (UE), and size of rostellar hooks. (**4, 4.1**) Eggs containing an infectious oncosphaera larva are released from the detached drying proglottid in the feces of the host; eggs are indistinguishable from those of *Taenia* spp. (**5, 5.1**) Eggs are orally ingested by intermediate hosts or man with contaminated food. (**6, 6.1**) Inside the intestine of the intermediate hosts (including man), the oncosphaera hatches, enters the wall, and may migrate (via blood) to many organs. Cysts are formed mostly in the liver and lung; in *E. granulosus* large unilocular hydatids occur, which are filled with fluid (containing thousands of protoscolices), whereas in *E. multilocularis* a tubular system infiltrates the whole organ (giving rise

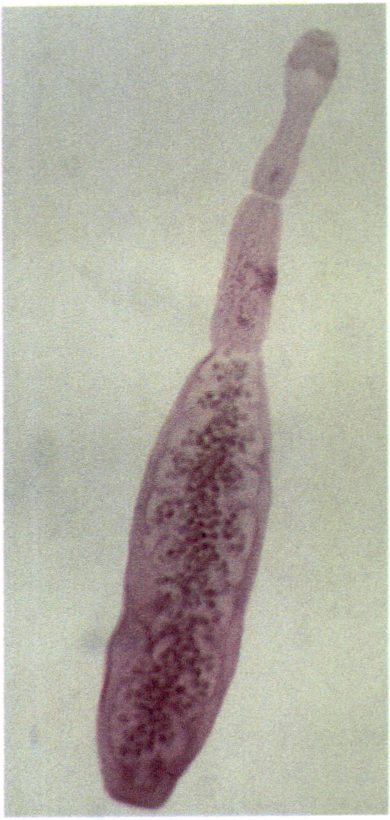

Fig. 6.2 Lightmicrograph of an adult worm (stained) of *Echinococcus granulosus* showing the large, egg-filled last proglottis. It has its place inside the dog intestine. Reprinted by permission from Springer Nature: Springer International Publishing "Animal Parasites: Diagnosis, Treatment, Prevention" by Heinz Mehlhorn © 2016

sexual generation in the final host and asexual generations in the intermediate hosts (Fig. 6.1). This type of lifecycle would be described as *metagenesis*.

3.2. *Echinococcus multilocularis*:

This worm is mainly found in foxes and only rarely in dogs and cats (Fig. 6.1). Adult worms measure only 1.5–4.5 mm and thus are considerably smaller than *E. granulosus* (Figs. 6.1, 6.6, 6.7, 6.8 and 6.9). The adult specimens—although shorter—have more proglottids than *E. granulosus*. The last proglottis is often larger than all others together. The genital pores

Fig. 6.1 (continued) to alveolar aspects in sections). (**7–8.1**) In brood capsules of both cyst types, protoscolices are formed, which may become evaginated (**8**) even inside their cysts. Evaginated or not, protoscolices are fully capable to infect final hosts when these animals ingest infected organs of intermediate hosts. *BC* brood capsule, *EB* embryophore of the egg, *EX* excretory vessels, *GP* genital pore, *H* hydatid, *HO* hooks of oncosphaera, *IR* invaginated rostellar hooks, *P* proglottid, *RH* rostellar hooks, *SU* sucker, *TU* tubular system, *UE* uterus containing eggs. Reprinted by permission from Springer Nature: Springer International Publishing "Animal Parasites: Diagnosis, Treatment, Prevention" by Heinz Mehlhorn © 2016

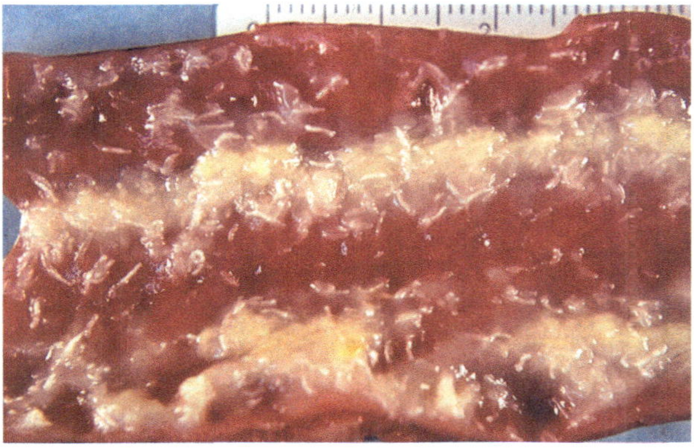

Fig. 6.3 Section of a dog intestine filled with many whitish appearing adults of *E. granulosus*. Reprinted by permission from Springer Nature: Springer International Publishing "Animal Parasites: Diagnosis, Treatment, Prevention" by Heinz Mehlhorn © 2016

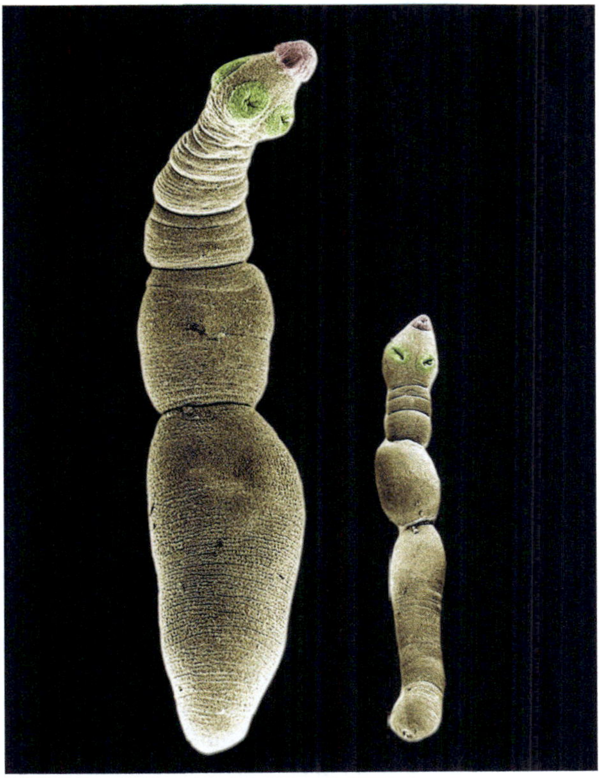

Fig. 6.4 Scanning electron micrograph of adult worms of *Echinococcus granulosus* (left) and *Echinococcus multilocularis*. Reprinted by permission from Springer Nature: Springer International Publishing "Animal Parasites: Diagnosis, Treatment, Prevention" by Heinz Mehlhorn © 2016

of the proglottids are situated clearly before the midst of the proglottis. The uterus of the gravid proglottid mostly does not show lateral protrusions, but appears sack like (Fig. 6.1). The scolex (Figs. 6.5 and 6.6) possesses a crown of 26–36 hooks, where the large ones (25–29 μm), as well as the

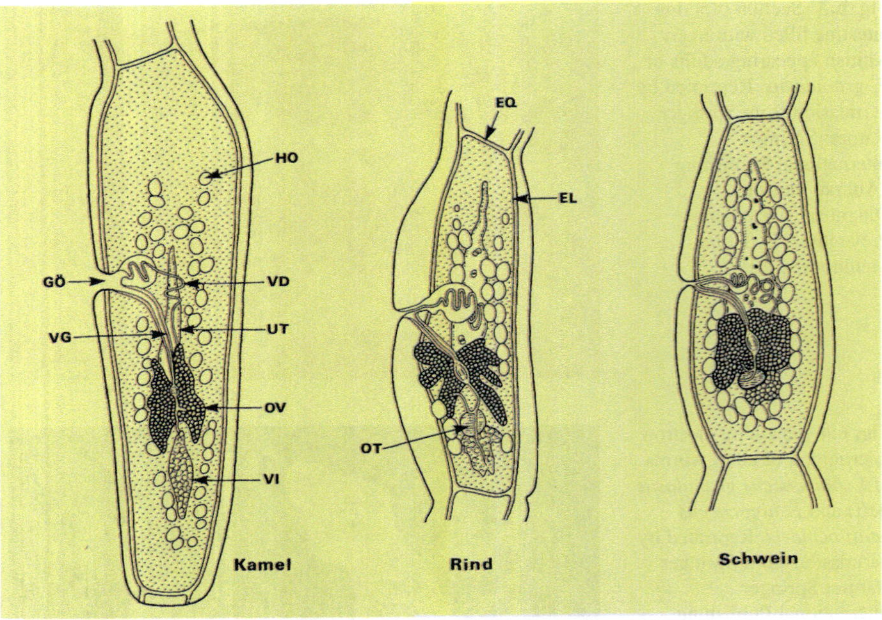

Fig. 6.5 *Echinococcus granulosus*: diagrammatic representation of the morphological investigation of defined strains of *Echinococcus granulosus* (Eckert et al. 1993) showing different proglottids. *EL* longitudinal excretion channel, *EQ* cross-running excretion channel, *GÖ* genital opening, *HO* testis, *OT* ootype with Mehlis' glands, *OV* ovary (= germarium), *UT* uterus, *VD* vas deferens, *VG* vagina, *VI* vitellarium; German: Kamel = camel; Rind = cattle; Schwein = pig. Reprinted by permission from Springer Nature: Springer International Publishing "Animal Parasites: Diagnosis, Treatment, Prevention" by Heinz Mehlhorn © 2016

small ones (19–24 μm), are smaller than those of *E. granulosus* worms. The eggs (Fig. 6.1) of this species can only hardly (if at all) be differentiated from those of *E. granulosus* or from those of *Taenia* species. *Intermediate hosts* are mainly rodents (field mice, rats). However, also humans may become infected as intermediate hosts and would develop the multilocular cyst inside its tissues, if they ingest an egg containing the typical *oncosphaera* larva (Fig. 6.1). These multilocular cysts consist of widespread tubules that do not contain fluid but grow by solid protrusions of 10–20 μm in diameter entering the tissues of many organs (liver etc.; Figs. 6.8, 6.9, and 6.10). Inside these tubular strands, protoscolices are formed, which infect final hosts, if these ingest such pieces of meat. During surgery of infected humans, such fine strands with multitasking cells are set free. They become distributed via lymph and bloodstream and start to produce new cysts in other organs.

Thus, surgery of humans has to be done very cautiously; otherwise, single cells being set free from these multilocular cysts will act like cancer cells and start permanent reproduction.

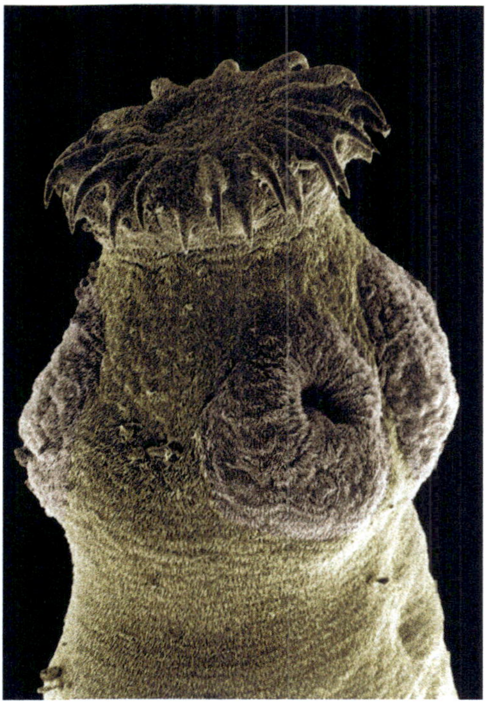

Fig. 6.6 Scanning electron micrograph of the scolex region of *Echinococcus multilocularis*. Reprinted by permission from Springer Nature: Springer International Publishing "Animal Parasites: Diagnosis, Treatment, Prevention" by Heinz Mehlhorn © 2016

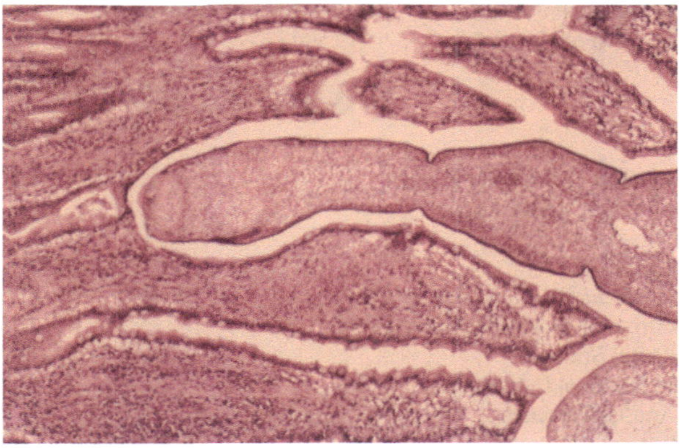

Fig. 6.7 Microscopical aspect of a section of the intestine of a fox showing the anterior end of *Echinococcus multilocularis* sticking between the intestinal villi. Reprinted by permission from Springer Nature: Springer International Publishing "Animal Parasites: Diagnosis, Treatment, Prevention" by Heinz Mehlhorn © 2016

4. *Genome*:

   4.1 *E. granulosus* worms contain nine pairs of chromosomes in their diploid body cells. Sequencing revealed 151.6 Mb and 11,325 genes. According to WHO data 10 genotypes (G1–G10) are distinguishable (Table 6.1).

   4.2 *E. multilocularis*. Here too the body cells contain nine pairs of chromosomes. Sequencing of the genome showed about 115 Mb coding for 10,345 genes and 2 noncoding regions.

5. *Symptoms of disease*: The *final hosts* (dogs, cats, and foxes) mainly do not show symptoms of disease when infected with the *Echinococcus* species, so that

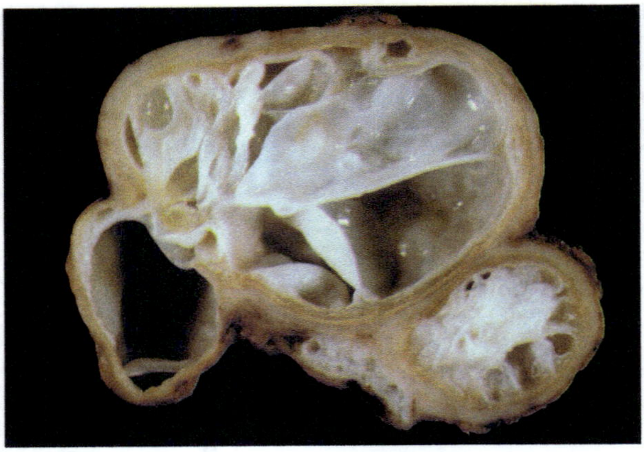

Fig. 6.8 Macrophoto of a human liver containing the strands of a so-called alveococcus cyst of the fox tapeworm *Echinococcus multilocularis*. Reprinted by permission from Springer Nature: Springer International Publishing "Animal Parasites: Diagnosis, Treatment, Prevention" by Heinz Mehlhorn © 2016

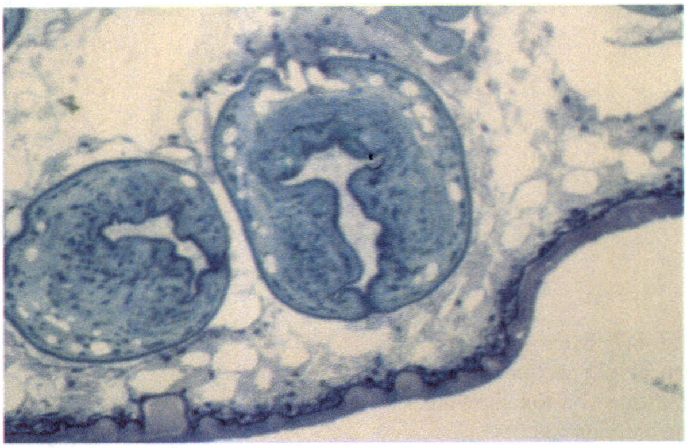

Fig. 6.9 Light micrograph of a semi-thin section through the periphery of a cyst of *Echinococcus multilocularis* showing two infectious protoscolices. Reprinted by permission from Springer Nature: Springer International Publishing "Animal Parasites: Diagnosis, Treatment, Prevention" by Heinz Mehlhorn © 2016

Fig. 6.10 Lightmicrograph of a young protruding protoscolex of *Echinococcus multilocularis*. Reprinted by permission from Springer Nature: Springer International Publishing "Animal Parasites: Diagnosis, Treatment, Prevention" by Heinz Mehlhorn © 2016

infections often remain undetected for long. This endangers humans as intermediate hosts, which live together in close neighborhood with these *final hosts*. Only rarely occur hemorrhagic or catarrhalic symptoms in final hosts, so that infections remain undetected. Humans as *intermediate hosts*, however, show organ-specific symptoms due to the presence of growing cysts, which are formed by both *Echinococcus* species. The clinical discourse of the alveolar echinococcosis is often very variable. Often occur for years unnoted infections of the liver, within which an alveolar tumor grows up very slowly, but often reaching a large size. However, the compression of important blood vessels may often induce cholestasis, portal hypertension, and secondary liver cirrhosis. By infiltrative growth and lympho- respectively hematogenous spreading into other tissues (brain, lung, peritoneum, etc.) become seriously infected, which leads—untreated—to death.

6. *Diagnosis*: Infected final hosts excrete in their feces the tiny whitish appearing, 1–2 mm long proglottids, which can easily be missed, if there are only worms present in the intestine. However, in cases of mass infections, the feces may appear like covered by white "dust" due to large amounts of proglottids, which, however, shrink quickly in drying feces. *Important*: Proglottids do not appear during each defecation! However, fresh proglottids show internal structures, which give hints to which strains of *Echinococcus granulosus* the excreted proglottids belong.
7. *Pathway of infection: Final hosts*: Oral uptake of protoscolices within raw or undercooked organs of infected intermediate hosts. *Intermediate hosts*: Humans may become infected by oral uptake of eggs excreted within the feces of final hosts (cats, dogs, and foxes). These worm eggs may also be transported by flies from fox or dog feces onto human food (Gestmann et al. 2012; Förster et al. 2012).
8. *Prophylaxis*: Do not feed raw meat to dogs and cats and avoid contact with their feces. Regular deworming of final hosts helps to keep away infections from humans and animals around humans. Keep farm animals away from fox feces.
9. *Incubation period: Final hosts*: 4–5 weeks; *intermediate hosts*: eventually years.
10. *Prepatent period: Final hosts*: Depending on the age and type of the host: in the case of *E. granulosus*: at least 35–42 days; in *E. multilocularis*: about 35 days.
11. *Patency*: *E. granulosus*: at least 6–7 months, rarely up to 2 years; *E. multilocularis*: rarely more than 5–6 months.
12. *Therapy: A:Final hosts (dogs and cats)*: Application of praziquantel (15 mg/kg bodyweight, orally, or subcutaneously). Importance: Hot cleaning of sleeping places of infected domestic dogs/cats, washing their fur after treatment, since eggs may stick herein. In regions with many infected mice, treatment should be repeated every 2 months. *B: Intermediate hosts: Humans*: Cysts due to *E. granulosus* can be surgically removed; however, cysts due to *E. multilocularis* must be treated practically lifelong with *albendazole* or *mebendazole*, since these products do not kill the growing cysts but only stop their enlargement. Note: Swiss veterinarians showed that dogs may also become intermediate hosts of *E. multilocularis* (www.escap.ch).
13. *Lethality*: Before introduction of drugs lethality was 90% within 10 years after infection. Torgerson et al. (2010, 2011) calculated that alveolar echinococcosis (AE) results in humans 666,434 DALYS per year (see also Tables 6.4 and 6.5).

Parasite details have been revised and are fully discussed in Mehlhorn (2016a).

Table 6.4 Distribution of confirmed echinococcosis cases and rates per 100,000 population by country, EU/EEA, 2013–2017

	2013		2014		2015		2016		2017			
Country	Number	Rate	Number	Rate	Number	Rate	Number	Rate	Confirmed cases	Rate	ASR	Reported cases
Austria	11	0.13	14	0.16	8	0.09	26	0.30	50	0.57	0.57	50
Bulgaria	278	3.82	302	4.17	313	4.35	269	3.76	218	3.07	3.19	218
France	34	0.05	32	0.05	48	0.07	38	0.06	48	0.07	0.07	48
Germany	132	0.16	131	0.16	157	0.19	177	0.22	123	0.15	0.16	123
Greece	10	0.09	13	0.12	13	0.12	18	0.17	15	0.14	0.13	15
Lithuania	23	0.77	22	0.75	33	1.13	26	0.90	53	1.86	1.71	53
Netherlands	33	0.20	37	0.22	64	0.38	33	0.19	38	0.22	0.23	38
Poland	39	0.10	48	0.13	47	0.12	64	0.17	75	0.20	0.19	75
Spain	94	0.20	70	0.15	83	0.18	87	0.19	83	0.18	0.17	83
Sweden	16	0.17	21	0.22	26	0.27	27	0.27	34	0.34	0.36	34

Table 6.5 Officially diagnosed cases of alveolar echinococcosis in Germany during the years 2001–2018

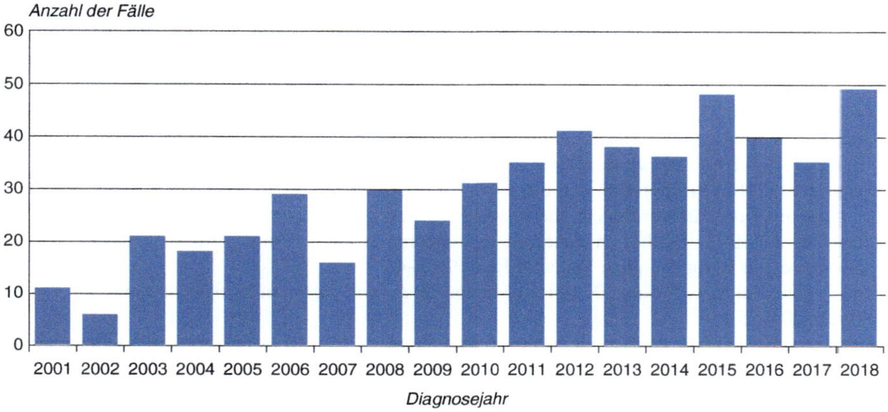

Further Reading

Ahmed ME et al (2013) First report on circulation of *Echinococcus ortleppi* in humped camel. BMC Vet Res 9:127

Alvarez Rojas CA, Romig T, Lightowlers MW (2014) *Echinococcus granulosus* sensu lato genotypes infecting humans—review of current knowledge. Int J Parasitol 44:9–18

Boulana B et al (2013) First report of *Echinococcus shiquicus* in dogs (China). Acta Trop 127:21–24

Bonfana B et al (2015) *Echinococcus equinus* and *E. granulosus* from the UK. Int J Parasitol 95:161–166

Brehm K (2017) Die Echinococcose– Eine Übersicht und neue Erkenntnisse der Diagnostik, Therapie und Parasitenbiologie. Epid Bull 15:127–132

Butt A, Khan JA (2019) Cystic echinococcosis: a 10 year experiment from a middle-income country. Trop Dr 5:49475519891338

Carmena D, Cardona GA (2014) *Echinococcus* in wild carnivorous species. Vet Parastiol 202:69–94

Chaignat V et al (2015) The brown hare as novel intermediate host for *E. multilocularis* in Europe. Parasitol Res 114:3167–3169

Chaligannis I et al (2015) *Echinococcus granulosus* infection dynamics in livestock in Greece. Parasitol Res 114:3167–3169

Craig PS, Hegglin D, Lightowler MW, Torgerson PR et al (2017) Echinococcosis: control and prevention. In: Thompson RCA, Deplazes P, Lymbery AJ (eds) *Echinococcus* and echinococcosis, Part A. Academic, London, pp 55–158

Comte S, Rabon V, Raoul F et al (2013) Fox baiting against *Echinococcus multilocularis*: contrasted achievements among two medium size cities. Prev Vet Med 111:147–155

Coucher MA et al (2016) Cystic echinococcosis in South America: systematic review. Trop Med Int Health 21:166–175

Cvejic D et al (2016) Efficacy of a single dose of milbemycin oxime/praziquantel combination tablets against adult *Echinococcus multilocularis*. Parasitol Res 115:1195–1202

Deplazes P, van Knapen F, Schweiger A, Overgaauw PA (2011) Role of pet dogs and cats in the transmission of helminthic zoonoses in Europe, with a focus on echinococcosis and toxocarosis. Vet Parasitol 182:41–53

Deplazes P, Rinaldi L, Alvarez Rojas CA, Torgerson PR, Harandi MF, Romig T, Antolova D, Schurer JM, Lahmar S, Cringoli G, Magambo J, Thompson RC, Jenkins EJ (2017a) Global distribution of alveolar and cystic echinococcosis. Adv Parasitol 95:315–493

Deplazes P, Rinaldi L, Alvarez Rojas CA, Torgerson PR, Harandi MF, Romig T et al (2017b) Global distribution of alveolar and cystic echinococcosis. In: Thompson RCA, Deplazes P, Lymbery AJ (eds) *Echinococcus* and echinococcosis, Part A. Academic, London, pp 315–493

Eckert J et al (1983) Proliferation and metastases formation of larval *Echinococcus multilocularis*. I. Animal model, macroscopical and histological findings. Parasitol Res 69:737–748

Eckert J et al (1993) Further evidence of the occurrence of a distinct strain of *Echinococcus granulosus* in European pigs. Parasitol Res 79:42–48

Eckert J, Gemmel MA, Meslin FX (2010) Manual on *Echinococcus* in humans and animals. CAB Direct 2010:286

Eckert J et al (2011) Alveolar echinococcosis. In: Palmer SR et al (eds) Textbook of zoonoses, 2nd edn. Oxford University Press, Oxford

European Centre for Disease Prevention and Control (2020) Echinococcosis. ECDC, Stockholm

European Centre for Disease Prevention and Control (2017) Introduction to the Annual Epidemiological Report. In: ECDC. Annual epidemiological report for 2017 [Internet]. ECDC, Stockholm [cited 11 December 2018]. Available from: http://ecdc.europa.eu/annual-epidemiological-reports/methods

European Food Safety Authority and European Centre for Disease Prevention and Control (2018) The European Union summary report on trends and sources of zoonoses, zoonotic agents and food-borne outbreaks in 2017. EFSA J 16(12):5500, 262 pp. Available from: https://www.ecdc.europa.eu/sites/default/files/documents/zoonoese-food-borne-outbreaks-surveillance-2017-updated.pdf

Grenouillet F et al (2014) *Echinococcus ortleppi* infections in humans and cattle, France. Emerg Infect Dis 20:2100–2112

Grubor NM, Jovanova-Nesic KD, Shoenfeld Y (2017) Liver cystic echinococcosis and human host immune and autoimmune follow-up: a review. World J Hepatol 9:1176–1189

Hegglin D, Deplazes P (2008) Control strategy for *Echinococcus multilocularis*. Emerg Infect Dis 14:1626–1628

Hüttner M, Romig T (2009) *Echinococcus* species in African wildlife. Parasitology 136:1089–1095

Hüttner M et al (2007) Genetic characterization and phylogenetic position of *Echinococcus felidis*. Int J Parasitol 38:861–868

Ikeda T, Yoshimura M, Onoyama K, Oku Y, Nonaka N, Katakura K (2014) Where to deliver baits for deworming urban red foxes for *Echinococcus multilocularis* control: new protocol for micro-habitat modeling of fox denning requirements. Parasit Vectors 7:357

Janko C, König A (2011) Disappearance rate of praziquantel-containing bait around villages and small towns in southern Bavaria, Germany. J Wildl Dis 47:373–380

Kern P, Menezes da Silva A, Akhan O, Müllhaupt B, Vizcaychipi KA, Budke C, Vuitton DA (2017) The echinococcoses: diagnosis, clinical management and burden of disease. Adv Parasitol 96:259–369

Knapp J et al (2015) Taxonomy and molecular epidemiology of *Echinococcus multilocularis*. Vet Parasitol 213:85–91

König A, Romig T, Janko C et al (2008) Integrating baiting concept against *Echinococcus multilocularis* is successful in Southern Bavaria, Germany. Eur J Wildl Res 54:439–447

Larrieu E et al (2015) Pilot field trial of the EG95 vaccine against ovine echinococcosis. PLoS Negl Dis e0004134

Larrieu E, Gavidia CM, Lightowlers MW (2019a) Control of cystic echinococcosis: background and prospects. Zoonoses Public Health 66:889–899

Larrieu E, Mujica G, Araya D, Labanchi JL, Arezo M, Herrero E, Santillán G, Vizcaychipi K, Uchiumi L, Salvitti JC, Grizmado C, Calabro A, Talmon G, Sepulveda L, Galvan JM, Cabrera M, Seleiman M, Crowley P, Cespedes G, GarcíaCachau M, Gino L, Molina L, Daffner J, Gauci CG, Donadeu M, Lightowlers MW (2019b) Pilot field trial of the EG95 vaccine against ovine cystic echinococcosis in Rio Negro, Argentina: 8 years of work. Acta Trop 191:1–7

Liu CN et al (2015) Discrimination between *E. granulosus* sensustricto, *E. multilocularis* and *E. shiquicus* a multiplex PCR assay. PLOS Negl Trop Dis e0004084

Mehlhorn H et al (1983) Proliferation and metastasis formation of larval *Echinococcus multilocularis*: II. Ultrastructural investigation. Parasitol Res 69:749–763

Mehlhorn H (2016a) Animal parasites. Diagnosis, treatment, prevention, 1st edn. Springer International, Cham

Mehlhorn H (2016b) Human parasites. Diagnosis, treatment, prevention, 1st edn. Springer International, Cham

Oksanen A, Lavikainen A (2015) *Echinococcus canadensis* transmission in the North. Vet Parasitol 213:182–186

Otero-Abad B, Torgerson PR (2013) A systematic review of the epidemiology of echinococcosis in domestic and wild animals. PLoS Negl Trop Dis 7:e2249

Pourseif MM, Moghaddam G, Saeedi N, Barzegari A, Dehghani J, Omidi Y (2018) Current status and future prospective of vaccine development against *Echinococcus granulosus*. Biologicals 51:1–11

Raoul F et al (2015) Trophic ecology, behaviour and host population dynamics of *E. multilocularis*. Vet Parasitol 213:162–171

Robert Koch Institut (2018) Infektionsbiologisches Jahrbuch meldepflichtiger Krankheiten für 2018. Echinococcose 6(13):77–80

Romig T, Bilger B, Dinkel A, Merli M et al (2007) Impact of praziquantel baiting on intestinal helminths of foxes in Southwestern Germany. Helminthologica 44:137–144

Saarma U, Jõgisalu I, Moks E, Varcasia A, Lavikainen A, Oksanen A, Simsek S, Andresiuk V, Denegri G, González LM, Ferrer E, Gárate T, Rinaldi L, Maravilla P (2009) A novel phylogeny for the genus *Echinococcus*, based on nuclear data, challenges relationships based on mitochondrial evidence. Parasitology 136:317–328

Schweiger A, Ammann RW, Candinas D, Clavien PA, Eckert J, Gottstein B, Halkic N, Muellhaupt B, Prinz BM, Reichen J, Tarr PE, Torgerson PR, Deplazes P (2007) Human alveolar echinococcosis after fox population increase, Switzerland. Emerg Infect Dis 13:878–882

Sikó SB, Deplazes P, Ceica C, Tivadar CS, Bogolin I, Popescu S, Cozma V (2011) *Echinococcus multilocularis* in south-eastern Europe (Romania). Parasitol Res 108:1093–1097

Simsek S et al (2015) First report of *Echinococcus equinus* in a donkey in Turkey. Korean J Parasitol 53:731–735

Soares MCP et al (2013) Anatomo-chemical and molecular description of liver neotropical echinococcosis caused by *E. oligarthrus* in human host. Actatropica 125:110–114

Soares MPC et al (2014) Neotropical echinococcosis: *E. vogeli* natural infection and its main definitive host: the bush dog (*Speathos venaticus*). Parasitol Int 63:485–487

Takahashi K, Uraguchi K, Romig T et al (2002) Preliminary report on *Echinococcus multilocularis* control by fox baiting with praziquantel. Rep Hokkaido Inst Public Health 52:61–63

Takahashi K, Uraguchi K, Hatakeyama H, Giraudoux P, Romig T (2013) Efficacy of anthelmintic baiting of foxes against *Echinococcus multilocularis* in northern Japan. Vet Parasitol 198:122–126

Tamarozzi F, Akhan O, Cretu CM, Vutova K, Akinci D, Chipeva R et al (2018) Prevalence of abdominal cystic echinococcosis in rural Bulgaria, Romania, and Turkey: a cross-sectional, ultrasound-based, population study from the HERACLES project. Lancet Infect Dis 8:769–778

Thompson RC (2008) The taxonomy phylogeny and transmission of *Echinococcus*. Exp Parasitol 119:439–446

Thompson RCA (2015) Neglected zoonotic helminth: *Hymenolepis nana*, *Echinococcus canadensis* and *Ancylostoma ceylanicum*. Clin Microbiol Infect 21:426–432

Thompson RCA et al (eds) (2017) *Echinococcus* and echinococcosis, Part B. Academic, London, pp 55–158

Torgerson PR, Keller K, Magnotta M, Ragland N (2010) The global burden of alveolar echinococcosis. PLoS Negl Trop Dis 4(6):e722

Torgerson PR, MacPherson CML, Vuitton D (2011) Cystic echinococcosis. In: Palmer SR et al (eds) Oxford textbook of zoonosis, 2nd edn. Oxford University Press, Oxford, pp 650–689

Vuitton DA, Azizi A, Richou C, Vuitton L, Blagosklonov O, Delabrousse E, Mantion GA, Bresson-Hadni S (2016) Current interventional strategy for the treatment of hepatic alveolar echinococcosis. Expert Rev Anti Infect Ther 14:1179–1194

Zait H, Kouidri M, Grenouillet FE, Umhang G, Millon L, Hamrioui B, Grenouillet F (2016) Molecular characterization of *Echinococcus granulosus* sensu stricto and *Echinococcus canadensis* in humans and livestock from Algeria. Parasitol Res 115:2423–2431

Zhang W, Zhang Z, Wu W, Shi B, Li J, Zhou X, Wen H, McManus DP (2015) Epidemiology and control of echinococcosis in central Asia, with particular reference to the People's Republic of China. Acta Trop 141(Pt B):235–243

Chapter 7
An Update on the Status of Hydatidosis/ Echinococcosis in Domestic Animals, Wildlife and Humans in Australia

David J. Jenkins

Abstract *Echinococcus granulosus* is widespread in wildlife over much of Australia, especially in the eastern side of the country in areas associated with the Great Dividing Range. The predominant transmission pattern is sylvatic, with wildlife acting as a major conduit for infection in domestic livestock, especially cattle, but also sheep and domestic dogs and more rarely humans. Hydatid disease in cattle is the cause of important financial loss to the Australian beef industry, losses to the sheep meat industry are low. Conventional meat inspection methods detect less than one-third of bovine hepatic hydatidosis infections. Cases of infection in humans and domestic dogs are becoming less common. Transmission of hydatid disease to humans in the island state of Tasmania has ceased. Occasional cases of hydatid infection in Tasmanian cattle are seen, mostly in animals imported from the Australian mainland.

Keywords Australia · *Echinocuccus granulosus* sensustricto · Livestock · Wildlife · Humans

7.1 Introduction

There have been several review articles published on *Echinococcus* in Australia since 2003 (Jenkins and Macpherson 2003; Jenkins et al. 2005a, b; Jenkins 2006; Thompson and Jenkins 2014; Mackenstedt et al. 2015; Romig et al. 2017). *Echinococcus granulosus* sensustricto, the causative agent of cystic hydatid disease, is not native to Australia, the parasite was introduced with imported sheep and domestic dogs during settlement in the late 1700s and early 1800s (Gemmell 1990). The life

---

D. J. Jenkins (✉)
School of Animal and Veterinary Sciences, Charles Sturt University, Wagga Wagga, NSW, Australia
e-mail: djjenkins@csu.edu.au

C. Strube, H. Mehlhorn (eds.), *Dog Parasites Endangering Human Health*, Parasitology Research Monographs 13,
https://doi.org/10.1007/978-3-030-53230-7_7

cycle of the parasite was elucidated in Germany by von Siebold in 1852 and the results of his experimental infections published in 1853 (von Siebold 1853), well after the arrival of the first Australian settlers and their animals in 1788. Due to total ignorance regarding the life cycle of *E. granulosus* in Australia at the time of settlement, and for several decades thereafter, the parasite spread rapidly after introduction and by the 1860s had become an important public health problem, causing the deaths of many colonists. The sheep industry in Australia grew steadily, initially based on wool and tallow, but after the successful introduction of refrigeration in 1882 (see Gemmell 1990), meat could also be exported. As a result, the size of the national sheep flock expanded into new areas of the country (reaching 20 million by 1860). At the same time the population of rural domestic dogs, mainly kept for herding the sheep and hunting, was also growing and spreading. Consequently, the extent of the public health problem due to hydatid disease also increased. During the 1860s, a number of reports appeared describing the frequency of infection in the colonists (see Gemmell 1990). Five hundred and eighty four people died from hydatid disease in Victoria between 1862 and 1881, a ratio of 3 deaths per 1000 deaths for all other reasons. Human hydatid disease was reported from all states with the highest prevalence in Victoria and South Australia (reviewed in Gemmell 1990).

An important factor that also enhanced the spread of hydatid disease was the native fauna of Australia had evolved without *E. granulosus* and were highly susceptible to infection. The top-order predator, the dingo (*Canis lupus dingo*) was a placental canid that preyed mainly on a range of herbivorous macropodid marsupials but was also large enough to prey on sheep. Predation on sheep by dingoes became, and remains, an important agricultural issue in the sheep rearing areas of Australia. Sheep and domestic dogs, during early settlement, and for about 150 years thereafter, were commonly infected with *E. granulosus* (Kumaratalake and Thompson 1982). Through contamination of the environment, particularly pasture, with *Echinococcus* egg-laden faeces by rural domestic dogs, it was not long before macropod populations living in the sheep-rearing areas of south-eastern Australia also became infected with cystic hydatid disease through accidental ingestion of eggs whilst grazing. Through predation on sheep and hydatid-infected macropods, the dingo population soon also became infected with *E. granulosus* tapeworms. In addition, an agricultural practice, transhumant grazing, undertaken annually in south-eastern Australia (King 1959) would have undoubtedly contributed to the transmission of *E. granulosus* to wildlife in many remote areas. Transhumance grazing is the practice of moving livestock from lowland pasture during the dryer summer months to alpine pasture, and after 2–3 months returning to the lowland pasture. This practice was undertaken mainly in Victoria and New South Wales for more than 100 years. Many hundreds of thousand sheep have moved annually to remote pasture in the Snowy Mountains, areas that now constitute the Kosciuszko National Park. The sheep were accompanied by numerous sheep-herding dogs and human drovers. Periodically, sheep were killed for human consumption and dog food, with the sheep offal commonly fed to the dogs. The sheep herding dogs introduced *E. granulosus* eggs onto the alpine pasture leading to eventual widespread infection in the local macropod populations. At the same time, some sheep were also predated on by dingoes. This interaction between the domestic and the wild animals soon led to sylvatic transmission *E. granulosus* becoming firmly established in native wildlife populations the length of eastern Australia, mainly in areas associated with the Great Dividing Range (Thompson and Jenkins 2014). Sylvatic transmission remains the most important cycle of transmission for *E. granulosus* in Australia maintaining the parasite over a wide area and acting as a potential source of infection to both humans, domestic dogs and livestock.

7.2 Wildlife Hosts and Aspects of Their Biology Contributing to Transmission and Perpetuation of *E. granulosus* in Australia

7.2.1 Dingoes and Wild Dogs

Dingoes were introduced into Australia by south-east Asian seafarers 4000 and 5000 years ago (Breckwoldt 1988; Corbett 1995) but in many parts of Australia the populations of dingoes are no longer pure-bred. Since the introduction of domestic dogs periodic incidences of cross-breeding of dingoes with domestic dogs have led to hybrid populations now present in many parts of Australia, particularly south-eastern Australia (Claridge et al. 2014). Within these populations, a few pure-bred dingoes may still occur but the hybrid animals appear to be similarly highly susceptible to infection with *E. granulosus* as pure-bred dingoes. These hybrids also have similar behaviour (Claridge et al. 2009) and physiology as pure-bred dingoes. They live in packs, have defined home ranges, hunt wildlife and importantly female hybrids breed only once per year (Cursino et al. 2017), compared to domestic dogs that can breed twice per year. These populations of dingoes and hybrids are collectively referred to as wild dogs. Wild dog packs commonly comprise family groups residing in defined home ranges that vary greatly in size depending on the availability of food and water. In a study of wild dogs living in forested alpine areas of south-eastern Australia, home range sizes varied between 1500 and 26,000 ha (Claridge et al. 2009), whilst wild dogs living in more arid areas where food and water resources were scarce have home ranges that are larger, 44.5–113.2 km^2 (Thomson 1992). Wild dogs mark their territorial boundaries with urine and faeces, therefore, wild dogs infected with *E. granulosus* have the capacity to spread eggs over wide areas. The prevalence *of E. granulosus*in wild dogs in south-eastern Australia (Victoria and New South Wales) and eastern Queensland is commonly high with worm burdens in individual wild dogs ranging from a few hundred tapeworms to several hundred thousand (Jenkins and Morris 1991, 2003; Grainger and Jenkins 1996; Jenkins et al. 2008; Harriott et al. 2019). Wild dogs commonly have worm burdens of *E. granulosus* that are far higher than those seen in domestic dogs. Wild dogs infected with more than 5000–10,000 *E. granulosus* is common. Particularly in eastern Australia, the heavy worm burdens in many wild dogs, the high prevalence of infection in wild dog populations and the highly mobile nature of wild dogs ensures that large numbers of *E. granulosus* eggs are distributed in the environment and available for ingestion by grazing macropod marsupials, over wide areas. These eggs are further distributed by agents such as wind and coprophagous flies (Gemmell and Lawson 1986). In the more temperate areas of Australia eggs of *E. granulosus* can remain viable in the environment for several months to one year (Gemmell and Lawson 1986) but may be longer (Thevenet et al. 2005).

7.2.2 Wild Dogs in Urban Environments

The presence of wild dogs in urban environments in Australia is becoming more common (Allen et al. 2013). It has been shown that wild dogs inhabiting urban environments may be infected with *E. granulosus* (Jenkins et al. 2008; Harriott et al. 2019). The prevalence and worm burdens of urban wild dogs are high, similar to those of wild dogs living in the bush, with some individuals carrying more than 100,000 *E. granulosus* (Jenkins et al. 2008). Hitherto, residents living in urban environments were considered not at risk of infection with hydatid disease but this encroachment of *E. granulosus*-infected wild dogs into urban centres in Queensland (Jenkins et al. 2008; Harriott et al. 2019) and New South Wales (Jenkins unpublished data) has changed the situation. It remains to be seen if in the next 10–20 years cases of human hydatidosis begin to occur in residents of urban areas into which wild dogs have encroached.

7.2.3 Foxes

E. granulosus infection in foxes was reviewed most recently in Thompson and Jenkins (2014) and there have been no new studies. Foxes occur widely in Australia but are absent in the tropical regions of northern Australia. Foxes are opportunistic feeders with a marked seasonal variation in the diet (Coman 1973). They commonly prey on small vertebrates, particularly rabbits and rodents (Coman 1973) but foxes have also been shown to feed on wallabies and less frequently kangaroos (Coman 1973; Roberts et al. 2006). The macropod remains present in scats and stomach contents of foxes were previously thought to have been the result of scavenging but there is an increasing body of data to indicate foxes are capable of catching and killing smaller macropodid species (Hornsby 1982; Meek and Wishart 2017). Should these small macropods have pulmonary hydatidosis they would be easier for the foxes to catch. It is also tempting to speculate that foxes predating on macropods will be exposed to *E granulosus* from time to time, it is therefore curious that the prevalence of infection and worm burden of *E. granulosus* in foxes is commonly low, usually fewer than 50 tapeworms (Gemmell 1959a; Thompson et al. 1985; Obendorf et al. 1989; Jenkins and Craig 1992; Jenkins and Morris 2003), however, Obendorf et al. (1989) did report a worm burden of 280 *E. granulosus* in one fox. The age at which foxes become exposed and the number of protoscoleces administered may be important. Young foxes with an immature immune system appear more susceptible to infection than adults with a fully developed immune system. Thompson (1983) infected three young foxes (6–8 weeks old) each with between 40,000 and 200,000 protoscoleces on one occasion and recovered between 283 and 26,100 *E. granulosus* when the foxes were euthanased 35 days later. In a study by Jenkins (unpublished data), two foxes 8–12 weeks old and two 14–16 weeks old were each fed 100,000 protoscoleces and between 646 and 13,830 *E. granulosus* were recovered 35 days later. In the same experiment, Jenkins fed 100,000 protoscoleces to each of 3 adult foxes (14–18 months old) and recovered 0, 12 and 16 *E. granulosus*. There has been one study (Reichel et al. 1994) where the burden of naturally acquired *E. granulosus* in one of two wild foxes was reported as being between 1000 and 10,000. No study before or since has reported naturally infected foxes with such a heavy worm burden. This report should be kept in mind as maybe some foxes, as with some wild dogs (Jenkins and Morris 1991), may contain unusually heavy infections of *E. granulosus.*

From the perspective of contaminating the environment with eggs of *E. granulosus*, foxes are of minor importance. However, where their importance may be greater is in urban environments, in and around places where people congregate, such as barbecue and picnic sites in parks and other public recreation spaces (Jenkins and Craig 1992).

7.2.4 Feral Cats

Australian domestic or feral cats infected with *E. granulosus* in Australia have never been reported. The contents from the intestines of 23 feral cats, collected in areas of high *E. granulosus* transmission were examined microscopically and did not contain any adult *E. granulosus* (Jenkins and Morris 2003). These cats would have had the opportunity, periodically, to scavenge macropods killed by wild dogs or killed in collisions with cars on bush roads. However, none was found infected with *E. granulosus* and it is generally accepted that feral cats in Australia play no part in the transmission of *E. granulosus*.

7.2.5 Macropodid Marsupials

There are numerous species of macropodid marsupial living in Australia, several of which have been reported susceptible to infection with *E. granulosus*, often with heavy infections (Jenkins and Morris 2003; Barnes et al. 2008). The macropodid species reported with natural infections of *E. granulosus* include swamp wallabies

(*Wallabia bicolor*), red-necked wallabies (*Macropus rufogriseus*) and eastern grey kangaroos (*Macropus giganteus*) in Victoria and New South Wales (Grainger and Jenkins 1996; Jenkins and Morris 2003) (Fig. 7.1); black-striped wallabies (*Macropus dorsalis*), swamp wallabies, eastern grey kangaroos, wallaroos (*M. robustus*), bridled nailtail wallabies (*Onychogalea fraenata*), brush-tailed rock wallabies (*Petrogale penicillata*), pademelons (*Thylogale stigmata*) and Lumholtz's tree kangaroos (*Dendrolagus lumholtzi*) in Queensland (Durie and Riek 1952; Johnson et al. 1998; Turni and Smales 2001; Banks et al. 2006a; Barnes et al. 2008; Shima et al. 2018); western grey kangaroos (*Macropus fuliginosus*) in Western Australia (Thompson et al. 1988); tammar wallabies (*Macropus eugenii*) (infected experimentally) in the ACT (Barnes et al. 2011). Hydatid disease has never been reported in macropodids in Tasmania (Howkins 1966). The top-order predator in Tasmania at the time of settlement was a large dasyurid, the thylacine (*Thylacinus cynocephalus*), and the apparent resistance of dasyurids to infection with *E. granulosus* (Jenkins et al. 2005a, b) has been suggested as a reason for the lack of *E. granulosus* transmission to Tasmanian wildlife (Jenkins 2006).

Infection with *E. granulosus* in macropodid marsupials almost exclusively occurs in the lungs commonly causing serious negative health impacts that may lead to the death of some individuals (Johnson et al. 1998; Barnes et al. 2007, 2011). Swamp wallabies appear to be particularly susceptible to infection (Jenkins and Morris 2003), commonly being infected with massive infections which may render one or both lungs almost inoperable (Fig. 7.1b). Hydatid disease in small populations of endangered brush-tailed rock wallabies has been identified as an important threatening process (Barnes et al. 2008). Even if infected animals do not die, their respiration is seriously impaired, rendering infected animals more susceptible to predation than uninfected animals. Hydatid cysts appear to grow much faster in macropods than in sheep. In an experimental infection of tammar wallabies and sheep (Barnes et al. 2007, 2011) cysts became fertile in the wallabies within 9 months compared to the sheep where their cysts had not grown sufficiently to become fertile during the life of the study (Barnes et al. 2011). Hydatid cysts in sheep may not become fertile for about 2 years and up to 6 in some cases (Slais 1980; Gemmell et al. 1986).

7.2.6 Wombats

Wombats (*Vombatus ursinus*) have rarely been reported infected with hydatid cysts. The only report to date has been that of Grainger and Jenkins (1996) who found two animals infected from 10 examined. These animals had been collected near Lake Eildon in eastern Victoria. All cysts in the wombats, as with macropodid marsupials, were located in the lungs (Fig. 7.1) and contained many microlitres of protoscoleces (total volume not recorded). Wild dogs predate on wombats but since there are so little data on the prevalence of infection in different wombat populations their importance as a wildlife intermediate host remains undetermined.

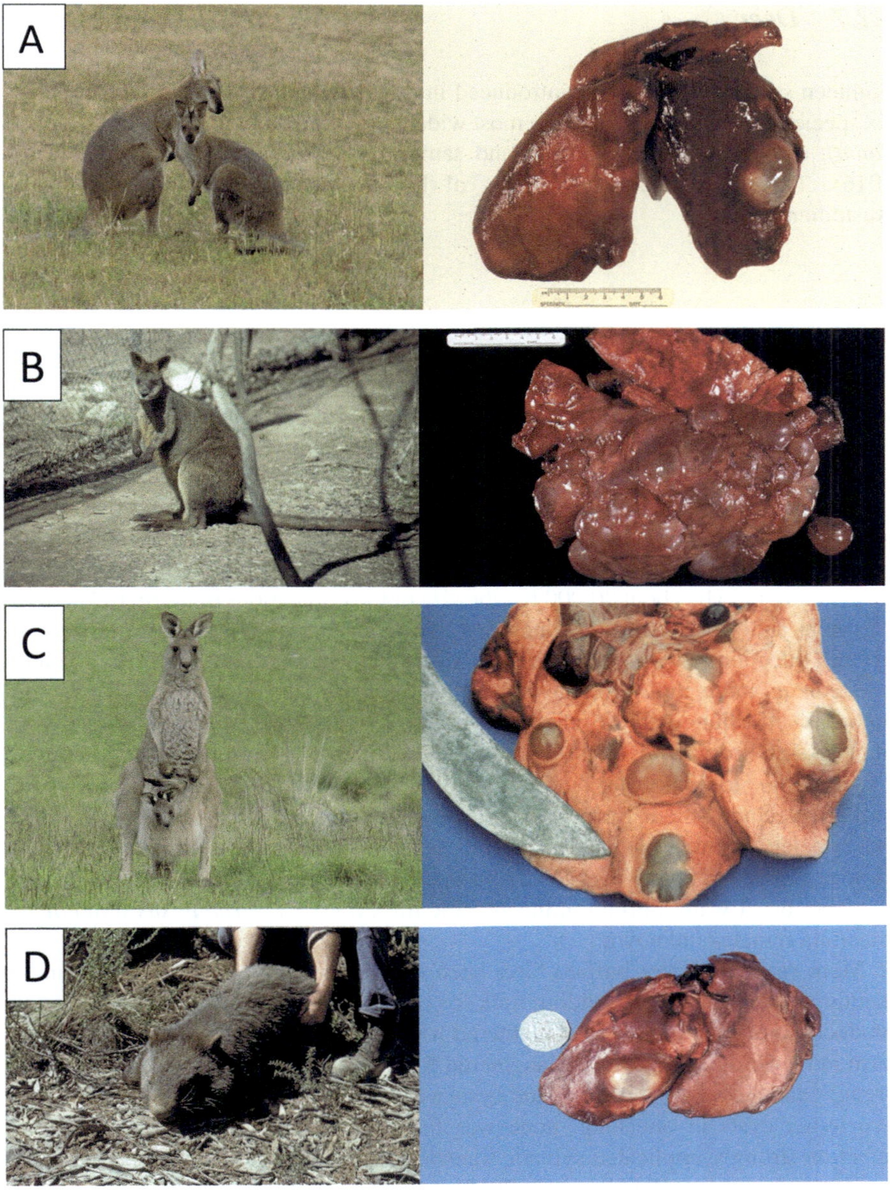

Fig. 7.1 Common marsupial wildlife intermediate hosts of *Echinococcus granulosus* in south-eastern Australia. (**a**) Red-necked wallaby (*Macropus rufogriseus*); (**b**) swamp wallaby (*Wallabia bicolour*); (**c**) eastern grey kangaroo (*Macropus giganteus*) and (**d**) common wombat (*Vombatus ursinus*)

7.2.7 Deer

Eighteen species of deer were introduced into Australia during the 1800s, but only six species became established, the most widely distributed being fallow deer (*Dama dama*), red deer (*Cervus elephus*) and samba deer (*Rusa unicolor*) (Davis et al. 2016). There have been no reports of hydatid infection in farmed or wild Australian deer.

7.2.8 Water Buffalo

Water buffalo (*Bubalus bubalus*) occur widely across northern Australia. Water buffalo were introduced into Australia from Indonesia in 1825 and 1843 to provide a source of meat, hides and milk for new remote northern Australia settlements. The settlements were unsuccessful and the buffalo released (Jenkins et al. 2019). Currently, feral water buffalo occur widely across northern Western Australia, the Northern Territory and northern Queensland. The population in the Northern Territory is thought to be about 70,000 but the size of the populations of northern Western Australia and Queensland are unknown. To date, there have been no reports of hydatid infection in wild populations of Australian water buffalo. However, hydatid infection in cattle that have never left the Northern Territory has been recently reported, infection in these animals was thought to have occurred via sylvatic transmission from wild dogs (Wilson et al. 2020b). Therefore, transmission of *E. granulosus* to feral water buffalo may also be occurring. The large size of water buffalo precludes them from wild dog predation, even by a pack of wild dogs. Nevertheless, animals eventually die and could be scavenged by wild dogs. It is not known if hydatid cysts ever reach patency in Australian water buffalo so the contribution of water buffalo in the sylvatic life cycle of *E. granulosus* in northern Australia remains unknown.

More recently, water buffalo have been re-domesticated and are being farmed, mainly as dairy animals providing milk for mozzarella cheese and gelato ice cream production, but some are also slaughtered for meat. A 3-year-old animal on a dairy farm in New South Wales was slaughtered for human consumption and the internal organs examined. Four small hydatid cysts were found on the surface of the liver and two larger cysts in one lung, none was fertile. Investigation into the source of infection strongly implicated sylvatic transmission from wild dogs that periodically traverse the property (Jenkins et al. 2019).

7.2.9 Goats

In many countries around the world goats are commonly infected with *E. granulosus* (Romig et al. 2017). Goats were introduced into Australia during settlement (Rolls 1969), over time some escaped and were able to survive, large populations of feral goats are now found in many parts of Australia. Despite there being large populations of wild rangeland and domesticated goats living in a variety of climate types in Australia, none has been reported infection with hydatid disease (Thompson and Jenkins 2014). Data from a recent study on 374,580 slaughtered and inspected wild rangeland goats sourced from Victoria, New South Wales and Western Australia have recently become available (Jenkins et al. 2018). The study focused on metacestodes of taeniid cestodes. Metacestodes of *Taenia hydatigena* and *T. ovis* were found in 8923 (2.4%) and 857 (0.24%) of carcasses, respectively, and one hydatid cyst was reported in one animal. This single report was suspected to be a false positive. Therefore, from an *E. granulosus* transmission perspective, goats in Australia appear to be unimportant.

7.2.10 Wild Camels, Donkeys and Horses

Camels, donkeys and horses were introduced into Australia as beasts of burden. Over time some escaped or were released, currently large populations of feral camels, donkeys and horses occur in various parts of Australia (Anon 2011, 2017; Brough 2006; Rolls 1969). Despite camels, donkeys and horses acting as intermediate hosts for *E. granulosus* elsewhere (Romig et al. 2017) naturally acquired hydatid disease in camels, donkeys and horses have never been reported in Australia. A contributing factor for this, particularly in respect of camels and donkeys, is that they tend to inhabit the hotter, more arid areas where eggs of *E. granulosus* are more likely to become desiccated. However, several thousand horses live in and around the Kosciuszko National Park, an area of high transmission for *E. granulosus* between wild dogs and macropodid marsupials (Jenkins and Morris 2003). Until studies have been undertaken the status of wild horses as hosts for *E. granuulosus* in Australia is unknown. Hydatidosis has been identified in domesticated horses at autopsy in Australia but these were animals imported into Australia from overseas, mainly the United Kingdom (Kumaratalake and Thompson 1982). It is generally accepted that camels, donkeys and horses have no role in the transmission *of E. granulosus* in Australia.

7.2.11 Feral Pigs

Occurrence of hydatid disease in feral pigs in Australia/or the lungs. Infection in feral pigs has been reviewed in Jenkins and Macpherson (2003) and Thompson and Jenkins (2014) and nothing new has been published. Feral pigs infected with hydatid disease have cysts both in the liver and lungs have been reported from the eastern states of Australia (Jenkins and Morris 2003; Lidetu and Hutchinson 2007) and Western Australia (Thompson et al. 1988). Cyst fertility is variable appearing to be higher in Queensland compared to New South Wales. However, this observation was only based on two available studies (Jenkins and Morris 2003; Lidetu and Hutchinson 2007). The role of feral pigs in the transmission of hydatid disease is probably limited since wild dogs predate mainly on piglets that are least likely to be infected with fertile cysts. Older pigs are less likely to be predated, especially by a single wild dog, because of their large size. However, wild dogs and foxes may scavenge carcasses of feral pigs that have died of natural causes or been shot and butchered by hunters.

7.2.12 Lagomorphs

Rabbits (*Oryctolagus cuniculus*) were introduced into Australia during settlement for recreational hunting (Rolls 1969). Rabbits now occur commonly in most parts of Australia and are a serious agricultural and environmental pest and a major food item for wild dogs and foxes. Apart from two historic reports of hydatid infection in Australian rabbits (Johnson 1909; Sweet 1909), natural infection with *E. granulosus* in Australian rabbits has never been reported. It is thought the two historic reports were misidentification of *Taenia serialis* infection that occurs commonly in rabbits in many areas (Ross 1926; Gemmell 1959b; Kumaratalake and Thompson 1982). Australian wild rabbits are susceptible to infection with *E. granulosus* as shown through experimental infection (Jenkins and Thomson 1995). However, these animals were infected with a large dose of eggs on one occasion, something highly unlikely to happen in the wild. Hares (*Lepus europeus*) also occur widely in Australia but, as with rabbits, have never been found naturally infected with hydatid cysts. However, hares of the same species have been reported as intermediate hosts for *E. granulosus* in Argentina (Schantz et al. 1976) with South American red foxes (*Dusicyon culpaeus*) acting as the definitive host. Nevertheless, rabbits and hares in Australia are not contributing to the sylvatic transmission of *E. granulosus*.

7.3 The Role of Domestic Animals in the Transmission of *E. granulosus* in Australia

7.3.1 Domestic Dogs

There have been only two recent studies reporting on the occurrence of *E. granulosus* in Australian rural dogs (Jenkins et al. 2006, 2014). In both these studies determining infection relied on the detection or *E. granulosus* copro antigens in faeces, using an enzyme-linked immunosorbent assay (ELISA). In the study by Jenkins et al. (2014) copro-PCR was also included to provide added diagnostic support for ELISA-positive faeces. However, in both studies neither adult *E. granulosus* nor released terminal segments were recovered from any of the ELISA positive or ELISA/copro-PCR positive faecal samples and there was no opportunity to purge any of the dogs with arecoline hydrobromide to irrefutably confirm infection. Jenkins et al. (2006) detected *E. granulosus* copro-antigens in 24% (137/561) rural dogs from New South Wales and Victoria. The study of Jenkins et al. (2014) detected far fewer coproantigen-positive dogs, 21 of 1119 (2.0%) of mainland dogs and 45 of 306 (7.8%) dogs from Tasmania. These findings were unexpected particularly in view of the highly successful hydatid control program previously undertaken in Tasmania and need further investigation.

The prevalence of taeniid cestodes in Australian rural dogs has been steadily falling in recent years (Jenkins personal observations). This fall in the prevalence is thought to be because the patent on praziquantel has lapsed and the drug is now included in many cheap generic “all wormers” and nutritious, palatable dry dog food, that is cheap and convenient to use, has been developed.

7.3.2 Cattle

Early reports of hydatidosis in cattle on the Australian mainland and Tasmania have been reviewed in Schantz et al. (1995). Following the concerted hydatid control campaign in Tasmania between 1962 and 1996 and the announcement of provisional freedom from *E. granulosus* in 1996 (Beard et al. 2001), hydatidosis in cattle has almost disappeared. Nevertheless, a few cases are reported annually in cattle previously imported for the mainland, mainly eastern Victoria. Curiously, each year there are also a few cases reported in young cattle that have remained on Tasmania their whole life. The pathway for the infection of these cattle is unclear (Jenkins et al. 2014).

Hydatid infection in cattle remains common on the Australian mainland, mainly through transmission from the sylvatic transmission cycle of *E. granulosus* because cattle are commonly grazed in areas of Australia where populations of *E. granulosus*-infected wild dogs also reside (Banks et al. 2006b). The most recent study of bovine hydatidosis on the Australian mainland has been undertaken over the last 4 years (2016–2020). The study took place in an abattoir in northern New South Wales that processes over 300,000 animals per year, sourced from all states and territories of Australia. During this study infection data from 1,178,329 cattle, slaughtered between 2010 and 2018, were analysed and hydatid disease was present in 33% (true prevalence, adjusted for imperfect detection at meat inspection) (Wilson et al. 2019a). Although many of the infected animals originated in eastern Australia, particularly in areas associated with the Great Dividing Range (GDR) and the coastal areas east of the GDR, infected cattle occurred widely throughout much of eastern Australia (Wilson et al. 2019a). One of the reasons why large numbers of cattle infected with hydatid disease occur along the GDR is that these areas are also inhabited by large populations of *E. granulosus*-infected wild dogs.

To obtain values of true prevalence, the sensitivity and specificity of the standard abattoir meat inspection process were investigated (Wilson et al. 2019b). The study was based on 636 pre-inspected bovine livers, either identified as infected with *E. granulosus* or not infected by meat inspectors. When these livers were sliced into approximately 5-mm slices the prevalence jumped from an apparent prevalence of 8.8% (meat inspection) to a true prevalence of 33%. Many of the animals where infection had been missed were in the younger age groups with cysts less than 10 mm in diameter, located deep in the liver. These cysts are completely impossible to detect by current standard inspection methods because of their small size and location (Wilson et al. 2019b). Cyst burden with respect to sex, age and other risk factors were also investigated in the same study (Wilson et al. 2019c). Animals that were male and grass-fed in all age groups were more likely to be infected. The economic impact of infection from the abattoir where the study was undertaken was also analysed (Wilson et al. 2020a). The annual financial loss for the abattoir was approximately $(Au)94,000 per year, equating to $(Au)6.70 per infected animal slaughtered. Also during this study Wilson et al. (2020b) showed that hydatid disease is being naturally transmitted to cattle in the Northern Territory. In a previous abattoir study in the Northern Territory by Small and Pinch (2003) they found that all hydatid-infected cattle were animals imported from Queensland. They concluded there was no natural transmission of *E. granulosus* to cattle in the Northern Territory. From the study of Wilson et al. (2020b), this appears not to be the case. It is most likely infection seen in the cattle described by Wilson et al. (2020b) originated via sylvatic transmission from dingoes.

7.3.3 Sheep

There have been no recent reports on the prevalence of hydatid disease in Australian sheep. Currently, the status of sheep in the domestic transmission of *E. granulosus* is unknown. Nevertheless, there has been a marked decline in hydatid-infected sheep passing through abattoirs during the last 30 years (Jenkins, personal observations). Infection is still seen periodically, most commonly in older sheep, particularly those coming from properties that abut national and state parks, state forests and vacant

Crown Land inhabited by wild dogs. The wildlife reservoir for *E. granulosus* composed of wild dogs and macropodid marsupials is large and extends right up the eastern side of Australia (Thompson and Jenkins 2014), providing a constant, potential source of infection for sheep (Grainger and Jenkins 1996).

7.3.4 Alpacas and Llamas

Alpacas and llamas have been introduced into Australia from South America to provide an alternative fibre to wool. They are also used on sheep, goat and free-range poultry farms as guards to provide protection, particularly from foxes but are also used by some farmers to protect sheep from wild dogs (Jenkins 2003). South American camelids have been reported with fertile cysts of *E. granulosus ss*. To date, hydatid disease has not been reported in either alpacas or llamas in Australia.

7.4 Human Infection

Human hydatidosis is no longer a nationally notifiable disease in Australia but remains notifiable in some states and territories (Northern Territory, South Australia and Tasmania). It is not known how many cases are diagnosed annually as no recent studies have been undertaken and human hydatid disease remains a notoriously under-reported zoonosis (Jenkins and Power 1996; O'Hern and Coolie 2013). In the 5 years to 2000, an annual average of 40 cases per year were diagnosed in Australia (Annon 2000). Since that time official data on human hydatid disease in Australia has become increasingly difficult to find (Harriott et al. 2019). Harriott et al. (2019) attempted to find data on human infection through a Brisbane pathology laboratory and found they tested 8–26 cases per year which, in their opinion, were positive cases, some of which were Aboriginal people. An added complication in determining the prevalence of human hydatid disease in Australia is that a proportion of the diagnosed cases occur in people not born in Australia who became infected in their country of origin.

Following the completion of the Tasmanian hydatid control campaign (Beard et al. 2001) human hydatidosis in Tasmania seems to now be a public health issue of the past. Human hydatidosis in Tasmania has been comprehensively reviewed by O'hern and Coolie (2013) who reported there had been no new incidents of transmission of *E. granulosis* to humans since the declaration of "provisional" eradication in 1996, up to the time they undertook their study in 2012. Each year a few new cases continue to be reported but these are in older people who contracted infection prior to 1996. Hydatid disease remains a notifiable disease in Tasmania.

7.5 Hydatid Control in Domestic and Wild Canids in Australia

Currently, there are no programmes to control hydatid disease underway anywhere in Australia. The most well-documented programme that focused on domestic transmission between sheep and domestic dogs was in Tasmania between 1962 and 1969 (Beard et al. 2001). On mainland Australia, there have been no state or territory-wide hydatid control programmes focusing on domestic or sylvatic transmission. There have been nine small hydatid control programmes since 1969 that have run for a few years with modest funding, usually from state departments of health and/or agriculture and then ceased. These programmes have focused on a limited geographical area and none has published an official record of their activities. It has been shown that control of *E. multilocularis* is possible in rural wild fox populations (Schelling et al. 1997) and also in urban fox populations (Hegglin et al. 2003). Although there have never been any attempts to use baits containing praziquantel to control *E. granulosus* in Australian fox or wild dog populations, this strategy has been suggested by Jenkins (2005, 2006) as a potential strategy in certain circumstances.

7.6 Conclusion

During the time of settlement, and for several decades thereafter, transmission of *E. granulosus* was mainly in a domestic transmission pattern (domestic dogs/sheep), and human hydatidosis was a major public health problem. Progressively, since that time, the parasite has taken advantage of the presence of the suite of susceptible wildlife hosts present in Australia and firmly established itself in a wildlife transmission cycle in much of the country. This wildlife transmission cycle poses a constant risk of transmission to livestock, domestic dogs and humans. However, with an increased understanding of the epidemiology of the parasite, the development of a highly effective cestocidal drug (praziquantel) to treat infected dogs, development of sophisticated surgical techniques for cyst removal in humans and good hygiene, hydatid disease is no longer the public health problem it used to be. Nevertheless, Australian authorities should not become complacent. Wild dogs and foxes infected with *E. granulosus* are encroaching into urban environments, there is a massive biomass of *E. granulosus* in wildlife in many rural areas popular with the general public, such as recreation sites in national parks. There is an apparent lack of interest in monitoring infection in domestic livestock, dogs and humans. The use of praziquantel-medicated baits should also be explored in urban environments and around popular camping grounds in national parks where populations of *E. granulosus*-infected wild dog and fox populations live. Also, the understanding of the general public regarding the importance of hydatid disease control is rapidly being lost. The apparent lack of interest in this important parasite by the public and authorities is of concern and needs to be addressed.

References

Allen BL, Goullet M, Allen LR et al (2013) Dingoes at the doorstep: preliminary data on the ecology of dingoes in urban areas. Landsc Urban Plan 119:131–135

Annon (2000) Communicable diseases intelligence, vol 15. Australian Government Department of Health

Anon (2011) Feral horse (*Equus caballus*) and feral donkey (*Equus asinus*). Australian Government, Department of Sustainability, Environment, Water, Population and communities. https://www.environment.gov.au/system/files/resources/b32a088c-cd31-4b24-8a7c-70e1880508b5/files/feral-horse.pdf

Anon (2017) Camels in Western Australia. Parks and Wildlife Service, Department of Biosecurity, Conservation and Attractions, Government of Western Australia. https://www.dpaw.wa.gov.au/management/pests-diseases/202-camels-in-western-australia?showall=1

Banks DJD, Copeman DB, Skerratt LF (2006a) *Echinococcus granulosus* in northern Queensland. 2. Ecological determinants of infection in beef cattle. Aust Vet J 84:308–311

Banks DJD, Copeman DB, Skerratt LF et al (2006b) *Echinococcus granulosus* in northern Queensland. 1. Prevalence in cattle. Aust Vet J 84:303–307

Barnes TS, Hinds LA, Jenkins DJ et al (2007) Precocious development of hydatid cysts in a macropod host. Int J Parasitol 37:1379–1389

Barnes TS, Goldizen AW, Morton JM et al (2008) Cystic echinococcosis in a wild population of the brush-tailed rock-wallaby (*Petrogale penicillata*), a threatened macropodid. Parasitology 135:715–723

Barnes TS, Hinds LA, Jenkins DJ et al (2011) Comparative pathology of pulmonary hydatid cysts in macropods and sheep. J Comp Pathol 144:113–122

Beard TC, Bramble AJ, Middleton MJ (2001) Eradication in our lifetime: a log book of the Tasmanian hydatid control campaign, 1962–1969. Department of Primary Industry, Water and Environment, Hobart

Breckwoldt R (1988) A very elegant animal, the dingo. Angus and Robertson, North Ryde

Brough J (2006) From value to vermin: a history of the donkey in Australia. Aust Zool 33:388–397

Claridge AW, Mills DJ, Hunt R et al (2009) Satellite tracking of wild dogs in south-eastern Australian forests: implications for management of a problematic top-order carnivore. Forest Ecol Manage 258:814–822

Claridge AW, Spencer RJ, Wilton AN et al (2014) When is a dingo not a dingo? Hybridisation with domestic dogs. In: Glen AS, Dickman CR (eds) Carnivores of Australia: past, present and future. CSIRO, Collingwood

Coman BJ (1973) The diet of red foxes *Vulpes vulpes* L., in Victoria. Aust J Zool 21:391–401

Corbett L (1995) The dingo in Australia and Asia. University of New South Wales Press, Sydney

Cursino MS, Harriott L, Allen BL, Gentle M, Leung LKP (2017) Do female dingo–dog hybrids breed like dingoes or dogs? Aust J Zool 65:112–119

Davis NE, Bennett A, Forsyth DM et al (2016) A systematic review of the impacts and management of introduced deer (family Cervidae) in Australia. Wildl Res 43:515–532. https://doi org/10.1071/WR16148

Durie PH, Riek RF (1952) The role of the dingo and wallaby in the infestation of cattle with hydatids (*Echinococcus granulosus* (Batsch 1786) Rudolphi, 1805) in Queensland. Aust Vet J 28:249–254

Gemmell MA (1959a) Hydatid disease in Australia VI. Observations on the carnivore of New South Wales as definitive hosts of *Echinococcus granulosus* (Batsch 1786), (Rudolphi, 1801), and their role in the spread of hydatidosis in domestic animals. Aust Vet J 35:450–455

Gemmell MA (1959b) Hydatid disease in Australia VII. An appraisal of the present position and some problems of control. Aust Vet J 35:505–514

Gemmell MA (1990) Australasian contributions to an understanding of the epidemiology and control of hydatid disease caused by *Echinococcus granulosus* – past, present and future. Int J Parasitol 20:431–456

Gemmell MA, Lawson JR (1986) Epidemiology and control of hydatid disease. In: Thompson RCA (ed) The biology of *Echinococcu*s and hydatid disease. George Allen and Unwin, London, pp 189–216

Gemmell MA, Lawson JR, Roberts (1986) Population dynamics in echinococcosis and cysticercosis: biological parameters of *Echinococcus granulosus* in dogs and sheep. Parasitology 92:599–620

Grainger HJ, Jenkins DJ (1996) Transmission of hydatid disease to sheep from wild dogs in Victoria, Australia. Int J Parasitol 26:1263–1270

Harriott L, Gentle M, Traub R, Cobbald R, Magalhães RS et al (2019) Geographical distribution and risk factors for *Echinococcus granulosus* infection in peri-urban wild dog populations. IJP: PAW 10:112–119

Hegglin D, Ward PI, Deplazes P (2003) Anthelmintic baiting of foxes against urban contamination with Echinococcus multilocularis. Emerg Inf Dis 9:1266–1272

Hornsby PE (1982) Predation of the euro (*Macropus robustus*) (Marsupialia: Macropodidae) by the European fox (*Vulpes vulpes*) (Placentalia: canidae). Aust Mammal 5:225–227

Howkins AB (1966) The role of macropididae in Tasmania as intermediate hosts in hydatid disease. Aust Vet J 42:240–241

Jenkins D (2003) Guard animals for livestock protection: existing and potential use in Australia. In: Noad B (ed) NSW Agriculture, Orange

Jenkins DJ (2005) Hydatid control in Australia: where it began, what has been achieved and where to from here. IJP 35:733–740

Jenkins DJ (2006) *Echinococcus granulosus* in Australia, widespread and doing well! Parasitol Int 55:S203–S206

Jenkins DJ, Craig NA (1992) The role of foxes *Vulpes vulpes* in the epidemiology of *Echinococcus granulosus* in urban environments. Med J Aust 157:754–756

Jenkins DJ, Macpherson CNL (2003) Transmission ecology of *Echinococcus* in wildlife in Australia and Africa. Parasitology 127:S63–S72

Jenkins DJ, Morris B (1991) Unusually heavy infections of *Echinococcus granulosus* in wild dogs in south-east Australia. Aust Vet J 68:36–37

Jenkins DJ, Morris B (2003) *Echinococcus granulosus* in wildlife in and around the Kosciuszko National Park, South-Eastern Australia. Aust Vet J 81:81–85

Jenkins DJ, Power K (1996) Human hydatidosis in New South Wales and the Australian Capital Territory. Med J Aust 164:18–21

Jenkins DJ, Thomson RCA (1995) Hydatid cysts in an experimentally infected rabbit. Vet Record 137(6):148–149

Jenkins DJ, Murray AJ, Claridge AW et al (2005a) The contribution of spotted-tailed quolls in the transmission of *Echinococcus granulosus* in the Byadbo wilderness area of the Kosciuszko National Park, Australia. Wildlf Res 32:37–41

Jenkins DJ, Romig T, Thompson RCA (2005b) Emergence/re-emergence of *Echinococcus* spp – a global update. Int J Parasitol 35:1205–1219

Jenkins DJ, McKinlay A, He D (2006) Detection of *Echinococcus granulosus* coproantigens in faeces from naturally infected rural domestic dogs in south eastern Australia. Aust Vet J 84:12–16

Jenkins DJ, Allen L, Goullet M (2008) Encroachment of *Echinococcus granulosus* into urban areas in eastern Queensland, Australia. Aust Vet J 86:294–300

Jenkins DJ, Lievaart JJ, Boufana B et al (2014) *Echinococcus granulosus* and other intestinal helminths in rural dogs from eastern Australia, the current situation. Aust Vet J 92:292–298

Jenkins DJ, Cowled B, Brooks V (2018) The role of feral goats in the transmission of taeniid cestodes in Australia. Vet Parasitol 255:1–9

Jenkins DJ, Williams T, Raidal S et al (2019) The first report of hydatid disease (*Echinococcus granulosus*) in an Australian water buffalo (*Bubalus bubalis*). IJP:PAW 8:256–259

Johnson TH (1909) Notes on some Australian parasites. Agric Gazette NSW 20:581–584

Johnson PM, Spear R, Beveridge I (1998) Mortality in captive and wild rock-wallabies and nailtailed wallabies due to hydatid disease by *Echinococcus granulosus*. Aust Mammal 20:419–423

King HWH (1959) Sheep movements to snow leases and permissive occupancies, 1945-1955. The AustGeog 7:129–140

Kumaratalake LM, Thompson RCA (1982) Hydatidosis/Echinosis in Australia. Helminthol Abstr – Ser A 51:233–252

Lidetu D, Hutchinson GW (2007) The prevalence, organ distribution and fertility of cystic echinococcosis in feral pigs in tropical Queensland, Australia. Onderst J Vet Res 74:73–79

Mackenstedt U, Jenkins D, Romig T (2015) The role of wildlife in the transmission of parasitic zoonoses in peri-urban and urban areas. IJP:PAW 4:71–79

Meek PD, Wishart J (2017) Camera trap evidence of redfox (*Vulpesvulpes*) predation attempts on adult macropods. Pac Conserv Biol 23:302–305

O'Hern JA, Coolie L (2013) A description of human hydatid disease in Tasmania in the post eradication era. Med J Aust 199:117–120

Obendorf DL, Matheson MJ, Thompson RCA (1989) *Echinococcus granulosus* infection in foxes in South-Eastern New South Wales. Aust Vet J 66:123–124

Reichel MP, Lyford RA, Gasser RB (1994) Hyperendemic focus of echinococcosis in north-eastern Victoria. Med J Aust 160:499–501

Roberts MW, Dexter N, Meek PD et al (2006) Does baiting influence the relative composition of the diet of foxes? Wildl Res 33:481–488

Rolls EC (1969) They all ran wild. The animals and plants that plague Australia. Angus and Robertson, Sydney

Romig T, Deplazes P, Jenkins D et al (2017) Ecology and life cycle patterns of *Echinococcus* species. In: Thompson RCA, Deplazes P, Lymbery AJ (eds) Advances in parasitology *Echinococcus* and *Echinococcosis*, Part A Pub. Academic, New York, pp 213–314

Ross IC (1926) A survey of the incidence of *Echinococcus granulosus* or hydatid disease in New South Wales. J Aust Vet Assoc 2:56–37

Schantz PM, Colli C, Cruz-Reyes A (1976) Sylvatic echinococcosis in Argentina. II. Susceptibility of wild carnivores to Echinococcus granulosus (Batsch 1786) and host-induced morphological variation. Tropmed Parasitol 27:70–78

Schantz PM, Chai J, Craig PS et al (1995) Epidemiology and control of hydatid disease. In: Thompson RCA, Lymbery AJ (eds) *Echinococcus* and hydatid disease. CAB International, pp 233–331

Schelling U, Frank W, Will R et al (1997) Chemotherapy with praziquantel has the potential to reduce the prevalence of *Echinococcus multilocularis* in wild foxes (*Vulpes vulpes*). Ann Trop Med Parasitol 91:179–186

Shima AL, Constantinoiu CC, Johnson L et al (2018) *Echinococcus granulosus* infection in two free-ranging Lumholtz's tree kangaroo (*Denrolagus lumholtzi*) from the Atherton Tablelands, Queensland. Trop Med Infect Dis 3:47. https://doi.org/10.3390/tropicalmed3020047

Slais J (1980) Experimental infection of sheep and pigs with *Echinococcus granulosus* (Batsch 1786) and the origin of pouching in hydatid cysts. Acta Vet Acad Sci Hung 28:375–387

Small LM, Pinch DS (2003) Survey for hydatidosis in cattle bred in the northern region of the Northern Territory of Australia. Aust Vet J 81:355–358. https://doi.org/10.1111/j.1751-0813.2003.tb11513.x

Sweet G (1909) The endo parasites of Australian stock and native fauna. Part II, New and unrecorded species. Proc R Soc Vic 12:503–527

Thevenet PS, Jensen O, Drut R, Cerrone G et al (2005) Viability and infectioness of eggs of *Echinococcus granulosus* aged under natural conditions of inferior arid climate. Vet Parasitol 133:71–77

Thompson RCA (1983) The susceptibility of the red fox to infection with *Echinococcus granulosus* of Australian sheep origin. Ann Trop Med Parasitol 77:75–82

Thompson RCA, Jenkins DJ (2014) *Echinococcus* as a model system: biology and epidemiology. Int J Parasitol 44:865–877

Thompson RCA, Nicholas WL, Howell MJ et al (1985) *Echinococcus granulosus* in a fox. Aust Vet J 62:200–201

Thompson RCA, Lymbery AJ, Hobbs RP et al (1988) Hydatid disease in urban areas of Western Australia: and unusual cycle involving western grey kangaroos (*Macropus fuliginosus*), feral pigs and domestic dogs. Aust Vet J 65:188–190

Thomson PC (1992) The behavioural ecology if dingoes in North-Western Australia, IV. Social and spatial organisation, and movements. Wildl Res 19:543–563

Turni C, Smales LR (2001) Parasites of the bridled nailtail wallaby (*Onychogalea fraenata*) (Marsupialia: Macropodidae). Wildl Res 28:403–411

von Siebold CTE (1853) Ueber die Verwandlung der *Echinococcus* Brut in Taenien. Z Wissenschaft Zool 5:409–424

Wilson C, Jenkins D, Brookes V, Barnes T (2019a) An eight-year retrospective study of hydatid disease (*Echinococcus granulosus* sensustricto) in beef cattle slaughtered at an Australian abattoir. Pre Vet Med:173, 104806. https://doi.org/10.1016/j.prevetmed.2019.104806

Wilson C, Brookes V, Jenkins D, Barnes T (2019b) Evaluation of the diagnostic sensitivity of meat inspection for hepatic hydatid disease in beef cattle in an Australian abattoir. Prev Vet Med 167:9–15. https://doi.org/10.1016/j.prevetmed.2019.03.014

Wilson CS, Brookes VJ, Barnes TS et al (2019c) Revisiting cyst burden and risk factors for hepatic hydatid disease (*Echinococcus granulosus* sensustricto) in Australian beef cattle. Prev Vet Med 172:104791. https://doi.org/10.1016/j.prevetmed.2019.104791

Wilson CS, Jenkins DJ, Brookes VJ, Barnes TS, Budke CM (2020a) Assessment of the direct economic losses associated with hydatid disease (*Echinococcus granulosus* sensustricto) in beef cattle slaughtered at an Australian abattoir. Prev Vet Med 176:104900. https://doi.org/10.1016/j.prevetmed

Wilson CS, Brookes V, Jenkins DJ (2020b) Transmission of *Echinococcus granulosus* in the Northern Territory, Australia: hydatid disease reported in beef cattle from the region. Aust Vet J 98:100. https://doi.org/10.1111/avj.12916

Chapter 8
Dipylidium caninum

Heinz Mehlhorn

Abstract Adult tapeworms of this species parasitize mainly in the intestine of dogs and cats and are only rarely found in the intestine in humans. The reason is, that dogs ingest much more common potentially infected adult fleas or portions of them than humans. Thus, mainly children become infected due to hand-mouth contacts after touching/caressing the dog's fur.

Keywords Dog tapeworm · Cysticercoids

1. *Name*: Greek: *di* = two, double; *pyle* = opening. Latin: *canis* = dog. English: Cucumber seed tapeworm; rice seed tapeworm. The Latin name of the worm was given describing the fact that both lateral sides of the proglottids have a genital opening.
2. *Geographic distribution/epidemiology*: Worldwide, common among dogs and cats—therefore, increased transmission risks for holders of dogs and cats exist.
3. *Biology, morphology*: *D. caninum* parasitizes as adult worm mainly in dogs and cats and only rather seldom in humans (Figs. 8.1 and 8.2a–c). It reaches a length of 20–80 cm and a width of 3–4 mm. The scolex is 0.5-mm wide, has a rostellum with 3–5 rows of small (7 μm) and large (13 μm) hooks and is provided with four suckers. Each proglottid includes two sets of male and female sexual organs, which are situated opposite to each other at the lateral sites possessing separate excretion channels. The terminal proglottids (Fig. 8.2b, c) reach a length of 8–20 mm and are closely filled with 120 × 200 μm sized egg packages containing 8–18 eggs with diameters of about 30–50 μm (Fig. 8.3). These thin-walled eggs contain the infectious so-called 6-hooklarva (oncosphaera) (Figs. 8.3 and 8.4). The shape of the whitish terminal proglottids is similar to

H. Mehlhorn (✉)
Department of Parasitology, Heinrich Heine University, Düsseldorf, Nordrhein-Westfalen, Germany
e-mail: mehlhorn@uni-duesseldorf.de

C. Strube, H. Mehlhorn (eds.), *Dog Parasites Endangering Human Health*, Parasitology Research Monographs 13,
https://doi.org/10.1007/978-3-030-53230-7_8

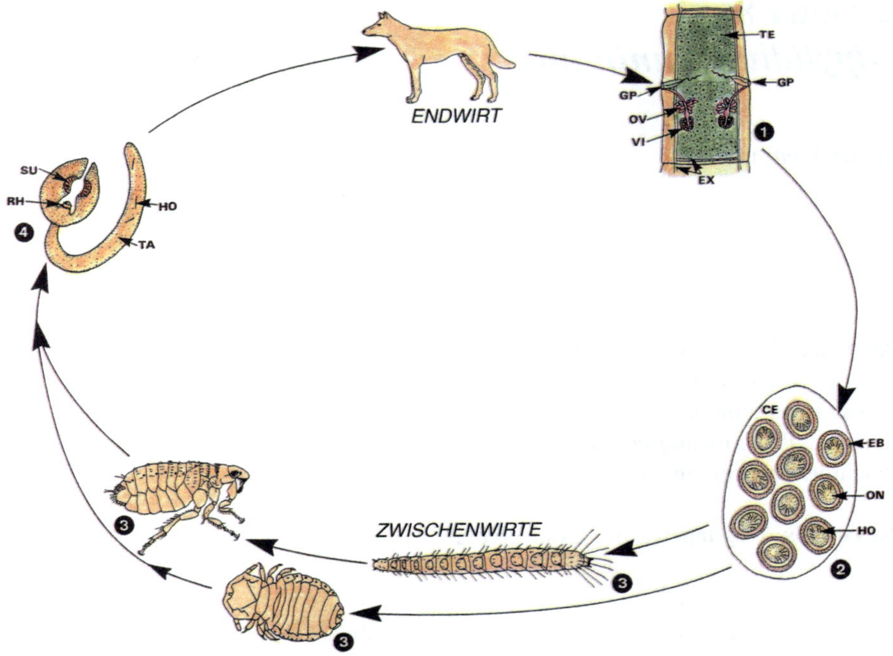

Fig. 8.1 Diagrammatic representation of the developmental cycle of one of the tapeworms of dogs (*Dipylidium caninum*). (**1–3**) The dog excretes proglottids (**1**), which contain egg packages (**2**). If larvae of mallophages or fleas (**3**) ingest such egg packages, the cysticercoid larvae finally occur also inside adult fleas and mallophages. If these are ingested by the same dog a reinfection occurs, while infections of other dogs occur as soon as the fleas or mallophages have entered another host. CE = egg package; EB = embryophore; ES = egg shell; EX = excretion vessel; GP = genital pore; HO = hook; ON = oncosphere; OV = ovary; RH = hook at the rostellum; SU = sucker; TA = tail of cysticercoid larva; RE = testes; VI = vitellarium; Endwirt = final host; Zwischenwirt = engl. intermediate host. Reprinted by permission from Springer Nature: Springer International Publishing "Animal Parasites: Diagnosis, Treatment, Prevention" by Heinz Mehlhorn © 2016

that of cucumber seeds. Thus, this tapeworm got its trivial name "cucumber tape worm". Due to heavy contractions, these proglottids are able to leave the intestine and start creeping around the anus and finally are also found inside the fur of dogs and cats. During these movements, the typical egg packages were pressed out from the proglottids and thus they may become ingested by intermediate hosts such as flea larvae or mallophages. Inside the body cavity of these insects, the so-called *oncosphaera* larva is transformed into the infectious *cysticercoid* larva, which grows up into the adult worm, as soon as the intermediate host has been ingested by a final host. This obligatory need to ingest flea larvae or at least portions of them in order to become infected shows that with respect to humans mainly children might become infected when playing with dogs and cats and touching their fur. After ingestion of such cysticercoid larvae, it takes about 3 weeks until these stages are grown up into fertile adult

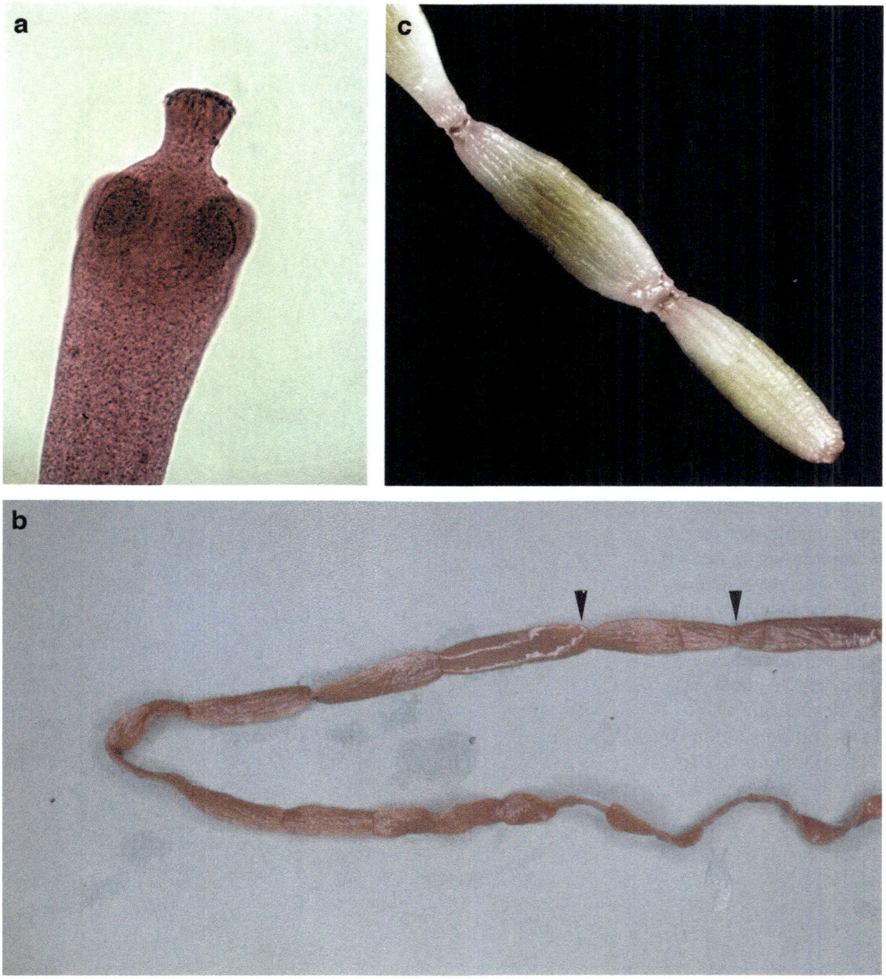

Fig. 8.2 a-c Light micrographs of *Dipylidium caninum*. (**a**) Anterior end with protruded rostrum. (**b**) Rice grain-like terminal proglottids obtained from dog feces. (**c**) Portions of the strobila being excreted after treatment with praziquantel. Reprinted by permission from Springer Nature: Springer International Publishing "Animal Parasites: Diagnosis, Treatment, Prevention" by Heinz Mehlhorn © 2016

tapeworms inside their final hosts. Besides *D. caninum*, other species have the same type of life cycle: *D. sexcoronatum*, *Diplophylidium acanthotetra,* and *D. noelleri* as well as several *Joyeuxiella* species.

4. *Symptoms of the disease (Dipylidiasis)*: In the case of low-grade infections, the symptoms of disease are unspecific. However, as soon as 100–200 adult worms are present inside a host a typical bloody-slimy diarrhea may occur accompanied

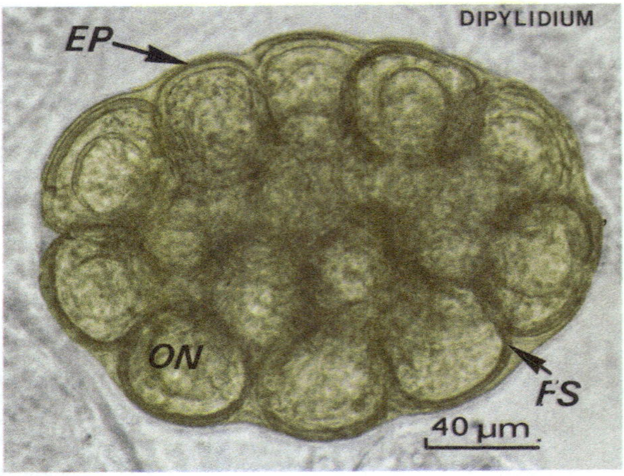

Fig. 8.3 Light micrograph of an egg package (EP) of *Dipylidium caninum*. FS = outer layer of each single egg, which surrounds the enclosed oncosphaera larva. Adapted by permission from Springer Nature: Springer International Publishing "Animal Parasites: Diagnosis, Treatment, Prevention" by Heinz Mehlhorn © 2016

Fig. 8.4 Diagrammatic representation of a dog sliding its anus on the soil in order to decrease itching. Reprinted by permission from Springer Nature: Springer International Publishing "Animal Parasites: Diagnosis, Treatment, Prevention" by Heinz Mehlhorn © 2016

by cramps. In addition, signs of urticaria, loss of body weight, and blood eosinophilia have been reported.

5. *Diagnosis*: Feces of infected persons contain whitish-reddish appearing proglottids, which may reach in fresh feces a length of up to 1.5 cm (Fig. 8.2b, c), but which appear after drying rice-grain-like and thus can be easily observed by the naked eye as whitish dots, e.g., in underwear. In addition, infected persons may show symptoms of a slight eosinophilia. Heavily infected children may be hit by restlessness due to itching along the anus region initiated by creeping proglottids. Infected dogs in households show sledging movements on the floor due to anal itching. Microscopical investigation of the feces reveals the typical egg packages (Fig. 8.3).
6. *Pathway of infection*: Oral uptake of portions of fleas, flea larvae, or mallophages containing the infectious cysticercoid larvae while stroking or playing with infected dogs or cats.

7. *Prophylaxis*: In the case of cats or dogs in households, these animals must regularly become dewormed at fixed intervals and kept free from fleas by application repellents or insecticides onto the body, on the floor and onto the hair of animals. In cases dogs and cats are infected they suffer from anus itching and thus they slide their anus along the floor (Fig. 8.4).
8. *Incubation period*: 10–25 days.
9. *Prepatent period*: 18–25 days.
10. *Patency*: Up to 1 year (however, for dogs and cats remains the danger of repeated self-infections by uptake of infected insects.
11. *Therapy*: For humans: see *Taenia solium*; in the case of dogs and cats: use of praziquantel and treatment with insecticides, in addition, to use of repellents.

Parasite details have been revised and are fully discussed in Mehlhorn (2016b).

Further Reading

Beugnet F et al (2013) Preventive efficacy of Frontline® Combo and Certified® against *Dipylidium caninum* infestation of cats and dogs using a natural flea infestation model. Parasite 20:7. https://doi.org/10.1051/parasite/2013006

Beugnet F et al (2014) Occurrence of *Dipylidium caninum* in fleas from client owned cats and dogs. Vet Parasitol 205:300–306

Beugnet F, Labuschagne M, Vos C, Crafford D, Fourie J (2018) Analysis of *Dipylidium caninum* tapeworms from dogs and cats, or their respective fleas – Part 2. Distinct canine and feline host association with two different *Dipylidium caninum* genotypes. Parasite 25:31

Beugnet F, Meyer L, Fourie J, Larsen D (2017) Preventive efficacy of NexGard spectra® against *Dipylidium caninum* infection in dogs using a natural flea (*Ctenocephalides felis*) infestation model. Parasite 24:16

Gopinath D, Meyer L, Smith J, Armstrong R (2018) Topical or oral fluralaner efficacy against flea (*Ctenocephalides felis*) transmission of *Dipylidium caninum* infection to dogs. Parasit Vectors 11:557

JesudossChelladurai J, Kifleyohannes T, Scott J, Brewer MT (2018) Praziquantel resistance in the zoonotic cestode *Dipylidium caninum*. Am J Trop Med Hyg 99(5):1201–1205

Labuschagne M, Beugnet F, Rehbein S, Guillot J, Fourie J, Crafford D (2018) Analysis of *Dipylidium caninum* tapeworms from dogs and cats, or their respective fleas – Part 1. Molecular characterization of *Dipylidium caninum*: genetic analysis supporting two distinct species adapted to dogs and cats. Parasite 25:30

Mehlhorn H (2016a) Animal parasites. Diagnosis, treatment, prevention, 1st edn. Springer International, Cham

Mehlhorn H (2016b) Human parasites. Diagnosis, treatment, prevention, 1st edn. Springer International, Cham

Portokalidou S, Gkentzi D, Stamouli V, Varvarigou A, Marangos M, Spiliopoulou I, Dimitriou G (2019) *Dipylidium caninum* infection in children: clinical presentation and therapeutic challenges. Pediatr Infect Dis J38:e157–e159

Raza A, Rand J, Qamar AG, Jabbar A, Kopp S (2018) Gastrointestinal parasites in shelter dogs: occurrence, pathology, treatment and risk to shelter workers. Animals (Basel) 8(7):E108

Saini VK, Gupta S, Kasondra A, Rakesh RL, Latchumikanthan A (2016) Diagnosis and therapeutic management of *Dipylidium caninum* in dogs: a case report. J Parasit Dis 40:1426–1428
Wani ZA, Allaie IM, Shah BM, Raies A, Athar H, Junaid S (2015) *Dipylidium caninum* infection in dogs infested with fleas. J Parasit Dis 39:73–75
Zhu GQ, Li L, Ohiolei JA, Wu YT, Li WH, Zhang NZ, Fu BQ, Yan HB, Jia WZ (2019) A multiplex PCR assay for the simultaneous detection of *Taenia hydatigena*, *T. multiceps*, *T. pisiformis*, and *Dipylidium caninum* infections. BMC Infect Dis 19:854

Chapter 9
Ancylostoma caninum and Other Canine Hookworms

John M. Hawdon and Kira A. Wise

Abstract The hookworm *Ancylostoma caninum* is the most common nematode parasite of dogs. Cosmopolitan in distribution, *A. caninum* is a serious canine pathogen, causing blood loss, anemia, and sometimes death, especially in puppies. In addition, *A. caninum* and related hookworms are zoonoses, and are capable of causing disease in humans. This chapter will first review the life cycle of *A. caninum*, its pathology and treatment in dogs. Next, the diseases caused by *A. caninum* and closely related canine hookworms in humans will be discussed, including cutaneous larva migrans, eosinophilic enteritis, and diffuse unilateral subacute neuroretinitis. Finally, evidence of the emergence of multidrug-resistant *A. caninum* and its possible impact on diseases in both dogs and humans will be explored.

Keywords Hookworm · *Ancylostoma caninum* · Cutaneous larva migrans · Diffuse unilateral subacute neuroretinitis · DUSN · Eosinophilic enteritis · Anthelmintic resistance

9.1 Introduction

Domestic dogs are among the most common companion animal, with an estimated 700 million worldwide (Hughes and Macdonald 2013). In the United States, 38.4% of households own a dog, totaling approximately 76.8 million (American Veterinary Medical Association 2018). In poor rural areas, community dogs are ubiquitous and function as de facto pets to much of the population. This popularity comes with the potential for exposure of people to canine parasites and risk of disease from these parasites. Hookworms are one of the greatest concerns, and present significant risk to human health (Jiraanankul et al. 2011; Otranto et al. 2017; Traub et al. 2005). There

J. M. Hawdon (✉) · K. A. Wise
George Washington University, Washington, DC, USA
e-mail: jhawdon@gwu.edu

C. Strube, H. Mehlhorn (eds.), *Dog Parasites Endangering Human Health*, Parasitology Research Monographs 13,
https://doi.org/10.1007/978-3-030-53230-7_9

are four canine hookworms with zoonotic potential. The most common species, *Ancylostoma caninum,* is found in both wild and domestic canids and felids around the world. It is the most prevalent and significant intestinal nematode of dogs in the United States (Little et al. 2009), and its prevalence in the United States has increased 47% from 2015 to 2018 (Drake and Carey 2019). *Ancylostoma braziliense* and *A. ceylanicum* parasitize dogs and cats primarily in the tropics and subtropics, whereas *Uncinaria stenocephala* infects dogs primarily in the northern latitudes of North America and Europe (Bowman et al. 2010). While all of these species are zoonotic, this chapter will concentrate on the human diseases caused by *A. caninum* because of its ubiquity, with the other species addressed when relevant.

9.2 Life Cycle

Ancylostoma caninum is a zoonotic hookworm whose definitive hosts are canids and occasionally felids (Liu et al. 2013). Typically, the *A. caninum* life cycle begins with the adult male and female hookworms in the small intestine of their host (Fig. 9.1). Following mating, females release eggs that are shed in the host's feces. Female

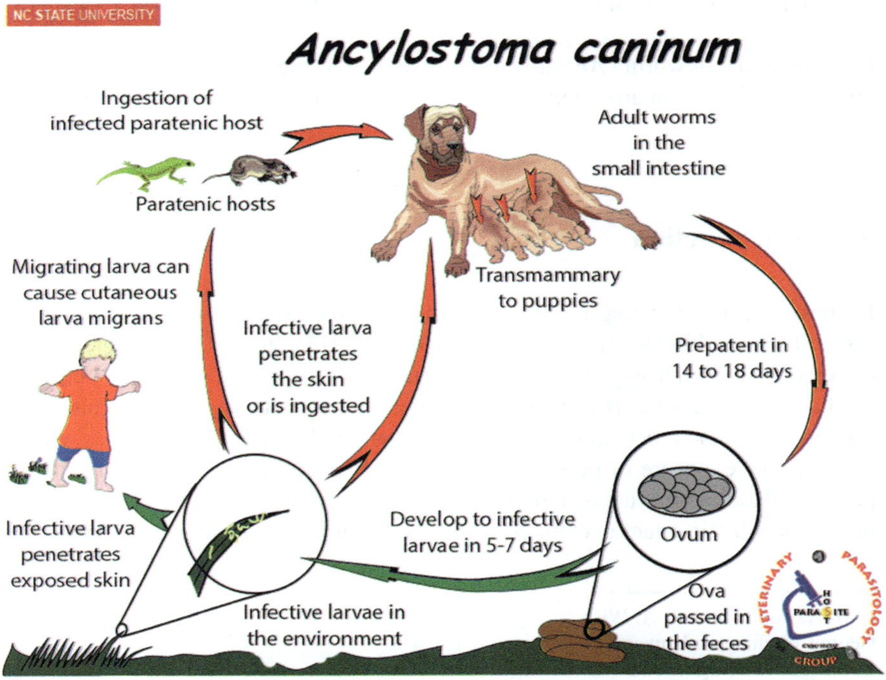

Fig. 9.1 Life cycle of the canine hookworm *Ancylostoma caninum*. Image courtesy of Flowers et al. (2015)

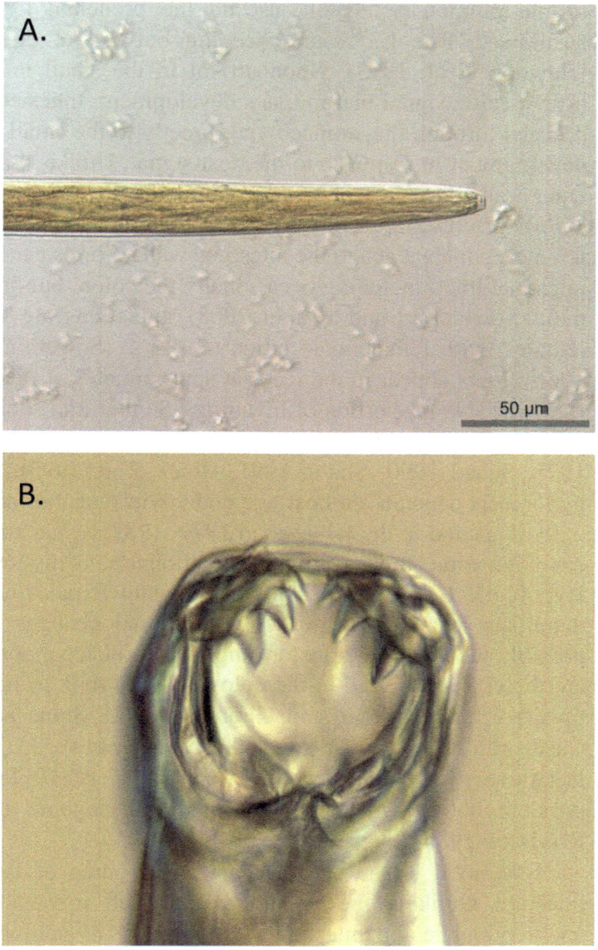

Fig. 9.2 Life cycle stages of *Ancylostoma caninum*. (**a**) Anterior of infective third-stage larva showing filariform esophagus. Magnification 400×. (**b**) Adult showing three pairs of ventral teeth in the buccal capsule, a diagnostic character of the species

A. caninum egg production has been estimated to range from 2000 to 17,000 eggs per female per day (McCoy 1931; Sowemimo and Asaolu 2008; Herrick 1928). In suitable conditions—that is, a warm, moist, and shaded environment—the larvae hatch within 1–2 days. The rhabditiform first-stage larvae (L1) feed on bacteria in the stool and soil, undergoing two molts to become the developmentally arrested, infective filariform third-stage larvae (iL3) after 5–10 days (Fig. 9.2a). Under favorable environmental conditions, these infective larvae can survive up to 3 or 4 weeks. Dogs most often contract *A. caninum* orally while eating contaminated soil, feces, or grass, while grooming, or when drinking contaminated water. Less frequently, L3 can infect their host by penetrating the skin, usually through the footpad or abdomen. Larvae that infect via the skin enter the bloodstream and are carried through the blood vessels to the heart and then to the lungs. Here the larvae break out of the pulmonary alveoli, ascend the bronchial tree to the pharynx, and are swallowed. The L3 resume feeding but fail to develop during this migration (Hawdon et al. 1993). Upon arrival in the small intestine, the L3 encounters a host-specific signal that initiates development. Ingested larvae forgo migration and proceed through the stomach and directly to the small intestine where they resume development in response to the host signal. Unlike skin penetrating larvae, L3 that enter orally do not resume feeding until after they molt to the L4 stage (Hawdon et al. 1993). In either case, the L3 undergoes two molts to reach the adult males and females, which attach to the intestinal wall. Worms can be distributed throughout the intestinal tract, including occasionally the colon, but prefer the jejunum and duodenum (Sowemimo and Asaolu 2008). Infections are slightly biased toward more females, with female:male ratios of 1 to 2 (Sowemimo and Asaolu 2008; Roche 1966). Eggs appear in the feces at approximately 14 days post-infection.

A variable proportion of invading L3 enter into skeletal or intestinal muscle and prolong developmental arrest in the host, a condition known as hypobiosis (Lee et al. 1975; Schad 1990; Shoop 1991; Hotez et al. 1993). The fraction that undergoes hypobiosis depends on host age and environmental factors, with more arresting in older dogs and in the fall (Schad 1979, 1982; Schad and Page 1982). This alternate developmental strategy provides additional transmission options for the hookworm. Hypobiotic L3 in the muscles of pregnant females reactivate and migrate to the mammary gland in response to hormonal changes associated with parturition, possibly mediated by transforming growth factor-β upregulation (Arasu 2001), and are thereby transferred in the colostrum and milk to newborn puppies for approximately 18 days (Stone and Girardeau 1960; Stone and Peckham 1970; Jimenez Castro et al. 2019). Once infected, a female can shed L3 in the milk for at least three lactations (Stoye and Krause 1976; Stoye 1973). Prenatal transmission is also possible, but much less frequent than the lactogenic route (Stoye 1973; Burke and Roberson 1985).

Hypobiotic larvae can also act as a source of larvae to repopulate the host intestine. Unlike some nematodes in which arrested larvae resume development and repopulate the intestine after removal of adults (Gibbs 1986), *A. caninum* hypobiotic larvae reactivate idiosyncratically and spontaneously (Schad and Page 1982), a phenomenon sometimes referred to as "larval leak" (Jimenez Castro et al. 2019; Bowman 2014). In this phenomenon, hypobiotic larvae in the intestine and skeletal muscle reactivate and migrate to the small intestine via the lungs or directly from the intestinal muscle, and mature in the intestine. These hypobiotic larvae function as a reservoir for continuous reinfection, manifested by chronic low levels of egg shedding in older dogs. The mechanism of larval reactivation is unknown.

Hypobiosis also functions to protect infective larvae from harsh environmental conditions. Infective larvae of numerous nematode species arrest development prior to seasonal periods of suboptimal conditions for the development of their offspring, such as cold winters in the north and dry summers in the south (Gibbs 1986). Larvae of the closely related human hookworm *A. duodenale* enter into hypobiosis prior to the dry inter-monsoon season in West Bengal, India, when soil conditions are too dry for development of infective L3 (Schad et al. 1973). The larvae resume development and mature prior to the monsoon season, ensuring that their eggs are released into an environment more conducive to development. This is reflected in seasonal reductions in egg output associated with the dry season. The period of arrest is genetically programmed, as larvae brought to the United States arrested for a similar length of time as those in India (Nawalinski and Schad 1974). While seasonal variation in egg output has not been conclusively demonstrated, *A. caninum* infections in northern Australia appear to show seasonality (Prociv et al. 1994; Croese 1995), and larvae can be induced to undergo hypobiosis by sudden chilling, as would be expected in late fall in the northern hemisphere (Schad 1982).

Finally, hypobiosis increases transmission by enabling paratenesis. Hookworm L3 remains in developmental arrest in the muscles of nonpermissive hosts, including rodents and insects, which can function as paratenic or transport hosts (Matsusaki 1951; Little 1961; Miller 1970). Canids ingesting such a host will develop patent infections due to reactivation of hypobiotic larvae in the prey.

9.3 Pathology in Dogs

Like most hookworms, *A. caninum* is a bloodsucking parasite that uses a cutting apparatus in its buccal cavity to affix to the intestinal mucosa and submucosa. The buccal capsule of *A. caninum* has three prominent pairs of ventrally located teeth that can be used to distinguish this species from other hookworms (Fig. 9.2b). The contraction of their muscular esophagus causes negative pressure that sucks tissue into their buccal capsules, which is hydrolyzed with secreted proteases, causing capillary lysis and bleeding. Adult worms are voracious bloodsuckers that have esophageal pump rates of 120–250 per minute (Wells 1931; Roche et al. 1962). Hookworms change attachment sites frequently, as often as every 4–6 h, thereby leaving multiple lesions that ooze blood for some period (Walker et al. 1995). This is due to anticoagulants released during feeding. *Ancylostoma caninum* secretes several anticoagulants (Eiff 1966) targeting multiple steps in the coagulation pathway to keep blood flowing, including factor Xa, factor VIIa/tissue factor (Cappello et al. 1995; Stassens et al. 1996; Mieszczanek et al. 2004; Li et al. 2010; Jiang et al. 2011), as well as a platelet inhibitor (Del Valle et al. 2003). While the worm consumes much of the released blood, a significant portion leaks out surrounding the bite wound. This blood loss and resultant iron deficiency anemia is the major clinical manifestation of hookworm disease. The disease varies by severity, ranging from asymptomatic infections to fatal exsanguination, depending on host age, resistance, and worm burden (Bowman 2014).

Hookworm disease can present as four different forms: peracute, acute, chronic (compensated), and secondary (decompensated) hookworm disease (Bowman 2014). Peracute hookworm disease typically occurs in neonatal puppies infected via transmammary (lactogenic) transmission. Depending on the number of larvae transmitted, rapid anemia can develop in the second week of life, prior to the appearance of eggs in the feces at day 14 post infection. This is because the immature fourth stage larva (L4) feeds on blood, but is not yet reproducing and releasing eggs. As few as 50–100 worms can be fatal. Therefore, diagnosis of peracute hookworm disease must be made based on clinical signs of anemia (pale mucous membranes) and dark, semifluid feces (the color is caused by partial digestion of the intestinal lining). Prognosis is poor in these cases. Affected pups usually require a blood transfusion, which may keep them alive long enough for anthelmintics to remove the adult worms. Because of the widespread prevalence of *A. caninum* and the possibility of lactogenic transfer of hypobiotic larvae, veterinarians should assume that every nursing puppy is in danger of hookworm anemia, and should be cognizant of the signs of the disease.

Acute hookworm disease results from sudden exposure of older puppies to a large number of infective larvae. Anemia may develop, but is less sudden and dramatic than in peracute disease. Normally, eggs are present in the feces, however, in heavy infections clinical signs of anemia may manifest prior to the emergence of the eggs by about 4 days. Chronic (compensated) hookworm infection occurs in older dogs, and is typically asymptomatic due to age-associated resistance to infection. Diagnosis is contingent on the presence of eggs in the feces and decreases in erythrocyte count, blood hemoglobin levels, or packed cell volume. Occasionally the infection can lead to chronic ill health if left untreated and/or the host fails to compensate sufficiently. Compensated hookworm disease can progress to secondary (decompensated) hookworm disease, usually occurring in older dogs that are ailing with other illnesses. The primary sign is severe anemia, commonly in a malnourished animal. Because of the frailty of the animal, the disease may lead to death, but it is important to note that hookworms play a secondary role in these cases (Bowman 2014).

9.4 Treatment for *Ancylostoma caninum* Infection

There are multiple efficacious and safe drugs, known as anthelmintics, for deworming dogs infected with hookworms. Drugs belonging to four major classes, as well as formulations containing drug combinations, are used depending on the specific situation and whether coinfections are present. The benzimidazoles, including fenbendazole and febantel, kill nematodes by binding to tubulin and preventing polymerization of microtubules (Lacey 1988). They may be given alone or in combination with other drugs. The tetrahydropyrimidines (oxantel, pyrantel, morantel) are nicotinic acetylcholine receptor (nAChR) agonists. Formulations containing febantel, pyrantel, and praziquantel (a cestodicide) were approximately 99.9% effective against *A. caninum* (Hellmann et al. 2003; Miro et al. 2007), and a chewable formulation containing pyrantel, oxantel, and praziquantel was similarly effective (Schmid et al. 2010).

The cyclooctadepsipeptide emodepside induces a flaccid paralysis of nematode muscles in the pharynx and body wall (Harder et al. 2005). Combinations with praziquantel are greater than 98% effective against larval and adult *A. caninum*

(Schimmel et al. 2009). Emodepside is available in Europe, but is not approved for use in dogs in the United States.

The macrocyclic lactone class includes ivermectin, selamectin, moxidectin, and milbemycin oxime. They bind to glutamate-gated chloride ion channels in muscle cells, increasing cell membrane permeability and hyperpolarization that causes paralysis and death (Laing et al. 2017). Each of the drugs, alone or in combination with praziquantel, are highly efficacious against *A. caninum*, with a cure rate greater than 90% (Traversa 2012).

The US Companion Animal Parasite Council (CAPC) and the European Scientific Counsel Companion Animal Parasites (ESCCAP) both recommend aggressive deworming of puppies, with treatment every 2 weeks for the first 8 weeks, followed by monthly preventative (Traversa 2012). This schedule is targeted at controlling perinatal transmission of *A. caninum* and the canine roundworm *Toxocara canis* because of their zoonotic potential. Treatment intervals of 4–6 weeks can prevent most patent infections, but frequencies less 3–4 times per year do not decrease hookworm prevalence (Traversa 2012; Epe 2009). Monthly preventatives typically contain a benzimidazole and a macrocyclic lactone that targets circulatory system parasites (i.e., *Dirofilaria immitis*). A review of treatment guidelines is available (Traversa 2012).

9.5 Human Diseases Caused by Dog Hookworms

9.5.1 Cutaneous Larva Migrans

Hookworm-related cutaneous larva margin (CLM or HrCLM), also known as "creeping eruption," is a self-limited, parasitic skin disease characterized by erythematous and highly pruritic serpiginous tracks (Bowman et al. 2010; Hochedez and Caumes 2007) (Fig. 9.3). The slightly elevated, tortuous linear tracks are caused by the penetration and migration of zoonotic hookworm infective larvae, primarily *Ancylostoma braziliense* but also *A. caninum* and *Uncinaria stenocephala*, all of which infect dogs but generally do not reach maturity in humans (Bowman et al. 2010). Another zoonotic species, *A. ceylanicum*, readily infects humans in addition to dogs and cats, causing patent infections usually without CLM (Bowman et al. 2010; Traub 2013).

Lee first described "creeping eruption" in 1874, and the first organism incriminated was the botfly *Gasterophilus*, whose larvae migrate through the skin causing myiasis (Bowman et al. 2010). In 1926, CLM was attributed to unknown nematode larvae, which was given the placeholder name *Agamonematodum migrans* until the adult stage of the nematode could be determined (Kirby-Smith et al. 1926). Subsequently, epidemiological and volunteer infections demonstrated that hookworms, specifically *A. braziliense*, caused most cases of creeping eruption in the southern United States (White and Dove 1929; Dove 1932). The terms "creeping eruption" and "cutaneous larva migrans" have been used interchangeably since to describe the

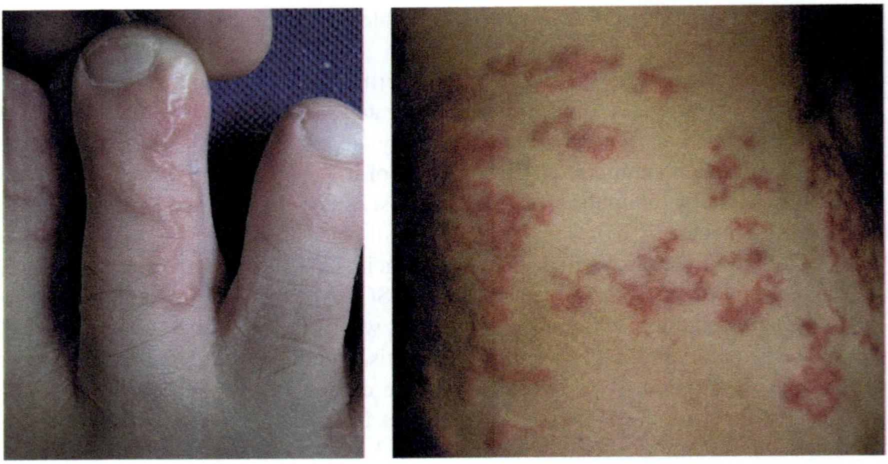

Fig. 9.3 Examples of hookworm-related cutaneous larva migrans (HrCLM). Left panel, creeping eruption on the toes. Used with permission (Rosh and McStay 2012). Right panel, a severe case of CLM on the hip (from the public domain)

dermatosis associated with zoonotic hookworm infection. Pointing out that creeping eruption is a clinical sign, and that cutaneous larva migrans is a syndrome, Caumes and Danis proposed that the disease should be renamed HrCLM (Caumes and Danis 2004; Caumes 2006). This name is generally now accepted for the disease.

HrCLM is among the most frequently seen skin disease in travelers returning from tropical and subtropical climates, where they are exposed to the iL3 during sunbathing or walking barefoot on the beach (Rosh and McStay 2012; Sow et al. 2017). In the United States, the distribution of HrCLM in coastal Atlantic and Gulf states overlaps that of *A. braziliense* (Bowman et al. 2010; Traversa 2012). CLM is also common in endemic countries, where the epidemiology and clinical presentation can differ (Heukelbach et al. 2004). Infection is usually self-limiting, although it causes significant discomfort for the patient, with pain and intense itching for up to several months that can be sufficiently severe to disrupt sleep (Heukelbach et al. 2004; Fuller 1966; Shelmire 1928; Davies et al. 1993). Cases lasting 14–18 months have been reported (Richey et al. 1996). Lesions from *A. braziliense* and *U. stenocephala* may persist for many weeks to a year before the larvae die, while lesions caused by *A. caninum* are small and transient (Mackenstedt et al. 2015).

CLM is transmitted through direct contact with contaminated soil and/or sand. Because *A. braziliense* larvae lack the collagenase enzymes required to penetrate the basal membrane of the human skin (Sandground 1939; Vetter and Leegwater-v.d. Linden 1977), it is typically confined to the upper epidermis/dermis layer of the skin, where it primarily migrates (Rosh and McStay 2012; Jones 1993). Larval migration can begin within minutes or months after initial skin penetration, and results in an allergic reaction that manifests variably. Lesions may include papules, nonspecific dermatitis, vesicles, or narrow (2–4 mm) elevated serpiginous (snakelike) or linear inflamed tracks with intense pruritus (Murphy and Spickler 2013; Veraldi et al. 2013a). There is usually no eosinophilia in HrCLM, so lab tests are not helpful for diagnosing HrCLM but can exclude other tropical infections in travelers (Blackwell and Vega-Lopez 2001). While the disease can affect any area exposed to the soil, the most common site for the lesions is the extremities (especially the dorsal side of the feet and toes), followed by the hands, arms, buttocks, and genitalia (Murphy and Spickler 2013; Belizario et al. 2016). Also, colloquially referred to as "sandworms," "sandworm eruption," or (erroneously) ground itch, any activity that involves skin contact with contaminated soil, including gardening, beach activities, and picnicking can result in infection. A common presentation, especially in the coastal US south, is "plumber's itch" acquired while working on exposed plumbing in crawl spaces underneath buildings (Bowman et al. 2010; Kirby-Smith et al. 1926; Fuller 1966). Secondary bacterial infection arising from scratching lesions is a frequent complication. Without treatment, larvae can migrate 1–5 cm per day before dying, typically within 2–8 weeks, although occasionally infections may persist for a year (Veraldi et al. 2013a; Chaudhry and Longworth 1989). Because humans are accidental, nonpermissive hosts, the larvae are unable to complete their life cycle and eventually die (Bowman et al. 2010; Sow et al. 2017). However, larvae of *A. caninum* may enter and persist in deeper tissue, or enter the human intestine and develop further (Bowman et al. 2010; Wang et al. 2017; Little et al. 1983) (see below).

Both *A. caninum* and *A. ceylanicum* can cause CLM, but it is usually a macropapular dermatitis instead of the classic "creeping eruption" caused by *A. braziliense* (Kirby-Smith et al. 1926; White and Dove 1929; Dove 1932; Shelmire 1928). Infection of human volunteers with *A. caninum* caused a papular dermatitis that lacked the classic serpiginous tracks seen with *A. braziliense* induced CLM (White and Dove 1929). They used *A. caninum* iL3 from dogs and cats for the experiments. However, their infections with *A. caninum* isolated from cats are suspect, as the parasite might have actually been *A. tubaeforme*, which had yet to be described as a separate species (Burrows 1962). Nonetheless, additional reports suggest that macropapular dermatosis is most likely caused by *A. caninum*, whereas the classic serpiginous tracks are the result of *A. braziliense* infection (Dove 1932; Shelmire 1928; Caumes 2000; Haydon and Bearup 1963; Wijers and Smit 1966).

Other hookworms produce skin lesions that differ significantly from the classic creeping eruption produced by *A. braziliense*. Experimental infections with the northern canine hookworm *Uncinaria stenocephala* produced very limited creeping lesions that disappeared spontaneously after several days (Shelmire 1928). There are no confirmed cases of CLM caused by *U. stenocephala* in the recent literature, although reports of multiple autochthonous CLM cases in the United Kingdom, Germany, and northern France, where *A. caninum* and *A. braziliense* are absent, implicate *U. stenocephala* (Baple and Clayton 2015; Diba et al. 2004; Roest and Ratnavel 2001; Beattie and Fleming 2002; Patterson and Kersey 2003; Tamminga et al. 2009; Müller-Stöver et al. 2010; Kienast et al. 2007; Herrmann et al. 2004; Klose et al. 1996).

Ground itch is the term for the papular and vesicular lesions caused by penetration of the skin by the anthroponotic hookworms *A. duodenale* and *Necator americanus*.

Typically there are no tracks or lines associated with skin penetration by these species, and the lesions resolve within a few weeks (Shelmire 1928; Diemert et al. 2018).

Atypical Presentations of HrCLM

Veraldi and Carrera (1999) and Veraldi et al. (2013b) described atypical presentations of HrCLM and noted an increase in their frequency. The reason for the increase is unknown and could be caused by increased travel to zoonotic hookworm endemic areas, increased awareness by physicians, or a true increase caused by abiotic or biotic changes (Veraldi et al. 2013b). Eczematous CLM appears as inflamed, vesicular plaques, with very short tracks when they are present. Tinea pedis-like CLM manifests as a highly pruritic vesicular rash without tracks, whereas bullous CLM presents as blisters containing clear fluid along the tracks (Veraldi et al. 2017). A previously uncommon form that is increasing in frequency is follicular HrCLM (Caumes and Danis 2004; Veraldi et al. 2013b). Originally described in 1978, the causative organism was identified as *Pelodera strongyloides*, a saprophytic nematode (Pasyk 1978). The first case of hookworm folliculitis was reported in 1991 (Miller et al. 1991). Subsequently, several cases have been linked to hookworms, usually based on symptoms resolving in response to anthelmintic treatment (Richey et al. 1996; Veraldi et al. 2013a, b; Caumes et al. 1995; Rivera-Roig et al. 2008; Lockmann et al. 2018).

Follicular HrCLM is characterized by small, inflamed follicular papules, often topped with pustules or vesicles (Veraldi et al. 2005, 2013b). Occurring primarily in adults, the lesions typically occur on the back, trunk, abdomen, inguinal area, and buttocks, and are usually intensely pruritic. Tracks are rare, and when present are usually short and fragmented (Veraldi et al. 2013b; Caumes et al. 2002; Ezzedine and Pistone 2013). Histology typically shows a mild edema in the dermis and cellular infiltration surrounding the follicles, which on occasion contain hookworm larvae (Richey et al. 1996; Miller et al. 1991; Rivera-Roig et al. 2008; Vanhaecke et al. 2014a).

Chronic or persistent HrCLM is a folliculitis characterized by an absence of pruritus and lesions that persist for months or years (Veraldi et al. 2013a). Richey described a case of HrCLM of 22 months duration (Richey et al. 1996). A larva thought to be an *Ancylostoma* species was found deep in a hair follicle. Since then, more than 20 cases of chronic HrCLM have been described, including cases of 5, 6, 8, 9, 10, 12, and 14 months (Veraldi et al. 2013a; Esser et al. 1999; Jelinek et al. 1994), as well as one of 5 years duration (Cayce et al. 2007). Veraldi reported 13 persistent cases ranging from 5 to 14 months duration, all of which eventually resolved with albendazole treatment (Veraldi et al. 2013a).

Follicular HrCLM can often be difficult to diagnose. In one case, pruritic folliculitis early in the disease course was misdiagnosed as herpes zoster infection, and treated with acyclovir and topical steroids (Malvy et al. 2006). The lesion continued to expand, causing extreme pain and itching that disrupted the patient's sleep. After 3 weeks, the patient was examined at a travel clinic, by which time multiple serpiginous and linear tracks had developed, along with secondary bacterial infection. A single 12-mg dose of ivermectin resolved the symptoms within 10 days. The uncommon presentation in which classic creeping eruption lesions were absent early probably led to the misdiagnosis of this case (Malvy et al. 2006).

Identification of the hookworm species causing HrCLM is often difficult, and in many cases depends on the specific presentation. Larvae are rarely recovered from biopsies of the tracks, as the larva is thought to be well ahead of the inflamed track. This is supported by successful treatment of CLM by aiming a carbon dioxide laser ahead of an advancing track (see below). Aiming at the leading edge failed to stop the larva (Soriano and Piansay-Soriano 2017). Reflectance confocal microscopy has been used to visualize a larva in situ (Gao and Liu 2019).

Larvae have been recovered from at least eight cases of HrCLM by skin scraping or cutaneous biopsy. In three cases, the larvae were identified as uncharacterized *Ancylostoma* spp. (Kim et al. 2006; Purdy et al. 2011; Manning et al. 2006). In another case, a larva found by cutaneous biopsy of a follicular lesion was identified as *A. caninum* (Miller et al. 1991) by morphological characteristics (Nichols 1956). In three cases, they were not identified to species (Vanhaecke et al. 2014a). More recently, three larvae were recovered from skin scrapings of follicular lesions that co-presented with serpiginous tracks (Le Joncour et al. 2012). DNA was isolated from two of the larvae and the ITS2 region amplified by polymerase chain reaction (PCR). Sequencing revealed a 100% identity to *A. braziliense* sequences present in the GenBank DNA sequence database. Although the larvae were recovered from follicular lesions, there were extensive serpiginous tracts typically seen with HrCLM caused by *A. braziliense*. This represents the first molecular confirmation of a hookworm-induced case of CLM.

A case of HrCLM presenting as a creeping eruption was diagnosed as *A. caninum* infection by ELISA (Kwon et al. 2003). The patient acquired the infection in Cambodia, but denied participating in behavior that would have resulted in skin contact with soil. Dot ELISA against a panel of helminth antigens (although without hookworm) was negative, but a microplate ELISA using *A. caninum* antigen was positive. However, the lack of *A. braziliense* antigen as a control calls the diagnosis into question, especially since creeping eruption is more frequently associated with infection with *A. braziliense* than with *A. caninum* (Bowman et al. 2010), and there could be antibody cross-reaction between the closely related hookworms. While there are no reports of *A. braziliense* in Cambodia, it is endemic in neighboring countries of Thailand (Areekul and Tipayamontri 1974) and Vietnam (Ng-Nguyen et al. 2015), and therefore likely present in Cambodia as well.

A recent case of HrCLM reported from a child in Ecuador was attributed to *A. caninum* (Coello et al. 2019). Filariform larvae were identified in soil samples from the environment and hookworm eggs were present in local dogs. However, no molecular testing or morphological differentiation was performed on the larvae or ova, so *A. braziliense* cannot be ruled out as the causative species, especially since the lesion was serpiginous.

Complications of CLM

Secondary bacterial infections of skin lesions due to scratching are relatively common complications, with rates up to 8% of cases reported (Davies et al. 1993; Bouchaud et al. 2000). Systemic complications are rare. When they occur, they are usually pulmonary (Hochedez and Caumes 2007). One such uncommon complication is eosinophilic pneumonia or Löffler's syndrome (LS) secondary to CLM. Löffler's syndrome, also known as pulmonary infiltration with eosinophilia syndrome (PIE), is a transient, migratory eosinophilic infiltration of the lungs with an associated blood eosinophilia (Akuthota and Weller 2012). Usually associated with pulmonary migration of *Ascaris* larvae, LS can be caused by both infectious and noninfectious agents (Akuthota and Weller 2012). Löffler's syndrome is an uncommon complication of CLM, with 44 cases reported in the literature (Table 9.1). The first report of LS in association with CLM was published in 1945, describing nine cases (Wright and Gold 1945). A second paper from the same authors describing 17 additional cases of LS followed in 1946 (Wright and Gold 1946). All of the patients had classic serpiginous lesions associated with CLM caused by *A. brazilienze* infection. Aside from a mild cough and the skin lesions, none of the patients reported feeling ill. All had some level of blood eosinophilia, as well as a high number of eosinophils in their sputum. All were afebrile. No ova were ever seen in patient feces, indicating that the larva could not establish in the intestine despite likely undergoing a lung migration following skin penetration. Radiographs revealed an often migratory and transient patchy pulmonary infiltration appearing after 7 days of skin symptoms. If cases where the skin lesions were allowed to progress untreated, the lung infiltration continued for several weeks. There was a nearly complete absence of signs of systemic disease in all the patients (Wright and Gold 1946).

The etiology of LS in unclear. In *Ascaris*, LS is caused by the migration of larvae through the lungs on the way to the intestine. However, larvae have rarely been recovered from sputum (Muhleisen 1953) in HrCLM cases, leading some authors to propose that hookworm larvae in the skin cause a generalized sensitization and reaction to soluble antigens in the lung that results in eosinophilic infiltration (Tan and Liu 2010; Butland and Coulson 1985), or a combination of larval migration and allergic reaction causes LS (Gao and Liu 2019; Wong-Waldamez and Silva-Lizama 1995). Support for the idea that the eosinophilic infiltration of LS is associated with larval pulmonary migration comes from a case in which *Ancylostoma* larvae were found in the sputum of an HrCLM patient (Muhleisen 1953). The patient acquired the infection while working under his house effecting plumbing repairs. Two days later an intensely pruritic rash developed, and a week later he developed dyspnea, wheezing, and a productive cough. There was no pulmonary infiltrate as is seen in LS. Larvae were found in daily sputum samples collected over 24 days. Soil samples taken from the area under his house revealed hookworm infective larvae. None of the larvae were identified beyond genus level, but the authors suspected *A. braziliense* because of the serpiginous appearance of the rash. Ova were never detected in the patient's feces.

Table 9.1 Cases of Löffler's syndrome associated with hookworm-related cutaneous larva migrans

Year	Location	Cases	Symptoms	Treatment	References
1943–1944	Camp Blanding, Florida	9	First report of CLM with LS; eosinophilia, cough	Not reported	Wright and Gold (1945)
1943–1944	Florida	17	Classic CLM with linear tracks; mild respiratory symptoms; eosinophilia	Ethyl chloride freezing,	Wright and Gold (1946)
1948	Jacksonville, Florida	1	CLM, cough, eosinophilia	Home remedies; Ethyl chloride freezing, fuadin	Horton (1949)
1951	Oklahoma	1	CLM; cough; eosinophilia; egg positive stool	Ethyl chloride freezing, tetrachloroethylene	Kalmon (1954)
1959	Texas	1	CLM, fever, respiratory symptoms, eosinophilia	O_2 therapy, cryotherapy, DEC, antibiotics	Falconer and Lea (1961)
1978	Yucatan, Mexico	2	CLM, cough, elevated IgE	Oral thiabendazole, 0.1% triamcinolone acetonide cream	Guill and Odom (1978)
1963	Barbados	1	CLM, cough, eosinophilia	Topical thiabendazole	Butland and Coulson (1985)
1988	West Indies, Venezuela, or Guianas	1	CLM, eosinophilia, pulmonary infiltrates	Cigarette burns; topical thiabendazole	Ambrus and Klein (1988)
1995	Guatemala	1	Bullous CLM; eosinophilia, respiratory symptoms	400 mg oral albendazole	Wong-Waldamez and Silva-Lizama (1995)
2001	Thailand	1	CLM, cough, eosinophilic alveolitis	Oral thiabendazole, corticosteroids	Del Giudice et al. (2002)
2009	Thailand	1	CLM, pulmonary infiltrates, eosinophilia, aggravated asthma	18 mg ivermectin, clobetasol-17 propionate, Symbicort	Te Booij et al. (2010)
2010	Bali	1	CLM, respiratory symptoms, pulmonary infiltrates	Mebendazole, albendazole (x5), IV hydrocortisone, O_2 supplementation	Tan and Liu (2010)
2011	Africa or Southeast Asia	1	Classic and follicular CLM, cough, wheeze, eosinophilia	Ivermectin (x2)	Pasricha (2011)
2016	India	1	CLM, intense cough, eosinophilia, pulmonary infiltrates, first case of CLM and LS in India	Albendazole (x5)	Podder et al. (2016)
2019	Malaysia	1	CLM, eosinophilia, pulmonary infiltrates	Albendazole (x7)	Gao and Liu (2019)

Another case often used to support pulmonary migration by larvae in LS was the finding of hookworm life cycle stages in localized hemorrhagic lesions of the submucosa of the small intestine of five people in Java in 1937 (Bonne 1937). In two cases immature and mature adults were found, and in two other cases an adult worm was found with eggs and larvae nearby. In the final case, eggs and larvae were present. In all cases, the worms were identified as *A. braziliense.* If true, this would indicate that *A. braziliense* larvae undergo pulmonary migration in humans, thereby implicating the migrating larva as the cause of LS. However, *A. braziliense* has never since been found to develop in humans. Furthermore, at the time, *A. braziliense* and *A. ceylanicum* were thought to be the same species, and the names used synonymously, including by the author. Not until 1951 were the two species recognized as separate, 14 years after this description (Biocca 1951). *Ancylostoma ceylanicum* is known to infect humans, and is more abundant than *A. duodenale* in many areas of Southeast Asia (Traub 2013). More likely, the five cases reported by Bonne represent aberrant infection with *A. ceylanicum* rather than *A. braziliense* (Bonne 1937).

An enigmatic case of HrCLM complicated by LS was reported in 1954 (Kalmon 1954). A 23-year-old male in Oklahoma contracted CLM while working under a house where the family dog frequented. Creeping eruption lesions developed and worsened over 10 days and were accompanied by fits of annoying, nonproductive cough and mild dyspnea. The patient was admitted to the hospital and had no signs of illness other than the skin lesions, wheezing, and eosinophilia. The skin lesions were frozen by ethyl chloride spray under general anesthesia. Chest radiographs indicated scattered patches of pneumonia. After 18 days, the rash had improved enough to permit discharge, but a nagging cough continued for nearly 9 months. Interestingly, numerous hookworm-type ova were found in his feces, indicating a patent hookworm infection. The authors took this as evidence that larvae had traversed the lungs, but failed to realize the larger implications, namely that this would indicate that a zoonotic *Ancylostoma* hookworm could establish and reproduce in a human host, which had never been reported to occur. More likely, the patient had a patent infection with the anthroponotic hookworm *N. americanus* coincident with the HrCLM caused by *A. braziliense*. Human hookworm was endemic in parts of the southern US until the 1980s (Martin 1972; Gloor et al. 1970; Starr and Montgomery 2011), and has been found as recently as 2017 in Alabama (McKenna et al. 2017). Hookworm eggs are practically indistinguishable morphologically. Therefore, coincident *A. braziliense* and *N. americanus* coinfection might explain the particularly persistent nature of this case of LS (Kalmon 1954).

A case of optic disc edema associated with CLM was reported in a UK patient who had traveled to Kenya (Dhir et al. 2010). In addition to skin lesions, the patient complained of blurred vision and floaters in his left eye, pain with eye movement, and left side headache. Examination revealed a swollen left optic disc. No other abnormalities were noted, and the condition resolved following ivermectin treatment. The authors postulate that an *A. caninum* larva migrated into or close to the optic nerve, causing neuritis. No larva was recovered, and the skin lesions were described as typical, suggesting the causative hookworm might have been

A. braziliense instead of *A. caninum*. This appears to be the first case of peri-optic neuritis associated with HrCLM and differs from the more common diffuse unilateral subacute neuroretinitis condition, another ocular disease related to zoonotic nematode infection discussed next.

Diffuse Unilateral Subacute Neuroretinitis

Diffuse unilateral subacute neuroretinitis (DUSN) is a serious infectious eye disease that can cause visual impairment and blindness (Mazzeo et al. 2019). The disease is characterized by vision loss in one eye, vitreous inflammation, and focal pigment epithelial loss in the early stages, progressing to retinal vessel stenosis, optic atrophy, and permanent vision loss. First described in 1978 (Gass and Scelfo 1978), a single nematode wandering in the subretinal space of the eye is thought to cause inflammation and degeneration of the retina and retinal pigment epithelium. The worm is found in fewer than half of the cases (Mazzeo et al. 2019). If treated early with oral anthelmintics or laser mediated destruction of the larva (photocoagulation), vision usually improves and the prognosis is favorable. However, many cases present in more advanced stages of the disease, when prognosis is poor even with treatment.

When the nematode is found, it seems to be one of two different sizes. In the southeastern US, the Caribbean, and South America, the worms vary from 400 to 600 microns in length. This size larva is thought to be a hookworm. In other cases, the nematode is longer, from 1500 to 2000 microns, and believed to be an ascarid larva, most likely the raccoon roundworm *Baylisascaris procyonis* or perhaps the dog roundworm *Toxocara canis*. The larger worm is most frequently found in the northern Midwestern states, although a case has been reported from South America (Moraes et al. 2002; de Souza et al. 1992; Cortez et al. 2005). The disease is rare in Asia, with only three cases reported: two involving the larger size worm in India and China (Venkatesh et al. 2005; Cai et al. 2000), and one due to the smaller worm seen in Korea (Kang and Lee 2015).

The nematode is typically found in the subretinal space (Kang and Lee 2015). As the larva moves, distinctive whitish inflammatory patches develop in its wake, probably in reaction to secretions or excretions from the worm (Mazzeo et al. 2019). A characteristic of DUSN is sequential appearance of these patches as the worm advances. The lesions are not seen when the larva is inactive. Early in the disease, there is active inflammation of the optic nerve, vitreous, retina and choroid, and optic disc edema may accompany the lesions (Kang and Lee 2015). As the disease progresses, further damage occurs, including vessel narrowing and pigment epithelial atrophy over large parts of the fundus. Vision loss may be due to damage to retinal ganglion cells or optic nerve disorders, and may appear suddenly early in the disease, or more slowly when the optic vessel narrowing occurs later in the disease course (Mazzeo et al. 2019; de Souza et al. 1992).

Finding the worm in suspected cases of DUSN is often difficult, and the worm is actually seen in fewer than half of the cases. Recently, a scanning laser ophthalmoscope (SLO) was used to visualize motile nematodes in three DUSN cases (Moraes et al. 2002). SLO allows better visualization of the retina in real time, facilitating detection of a migrating larva. Because the results are instantaneous, laser photocoagulation of the worm can be performed immediately, thereby halting further damage and ensuring that the worm is not missed during a subsequent visit. In addition, SLO imaging after laser treatment provides confirmation that the larva has been destroyed (Moraes et al. 2002).

Because the causative worm is seen infrequently, identification of the culprit species had not been accomplished until recently. In this case, a patient presenting with acute vision loss was found to have retinal lesions and a detached retina (Poppert et al. 2017). During surgery to repair the detached retina, a nematode purported to be approximately 10 mm long was seen and removed. The worm was destroyed during removal, but DNA was isolated from solutions used to rinse the eye during surgery. The nematode ITS 2 and 3 regions were amplified and sequenced, and found to be 99% and 87% identical to *A. ceylanicum* and *A. caninum*, respectively (Poppert et al. 2017). This case is perplexing for several reasons. First, the origin of the infection is unclear. The patient was adopted at the age of four from Colombia and moved to Germany, where he lived for 6 years prior to vision loss. During that time, he never traveled outside Germany other than on family vacations to Spain. *Ancylostoma ceylanicum* has not been reported from any of these countries, but has been reported from South America (Traub 2013; Rep and Heinemann 1976). If the infection was acquired in Colombia, the nematode remained in the patient for at least 6 years. Furthermore, the reported size of nearly 10 mm is the size of an adult hookworm, and well outside the size ranges typically reported for DUSN of 400–600 or 1500–2000 micrometers (Mazzeo et al. 2019). Adult hookworms are not known to leave the small intestine, suggesting that if the reported size is correct, the worm developed to the adult stage elsewhere in the body, perhaps even the eye. In any case, this represents the first molecular identification of the causal nematode in a DUSN case, and confirms the role of zoonotic hookworms in at least some cases (Poppert et al. 2017).

When the worm is found, laser photocoagulation is the preferred treatment, as it does not exacerbate inflammation and quickly inactivates the DUSN, preventing further damage (de Souza et al. 1992; Kang and Lee 2015; Garcia et al. 2005). Extended course (30 days) of high dose albendazole (400 mg) is efficacious and safe, and should be used when the worm cannot be located and destroyed (Mazzeo et al. 2019; Souza et al. 2005; Gass et al. 1992). However, anthelmintics including thiabendazole and ivermectin have also been ineffective in some cases (Souza et al. 2005; Casella et al. 1998). In general, the response to anthelmintic treatment is variable, and the optimal dose and schedule has yet to be established. However, some physicians recommend anthelmintic treatment even after successful laser photocoagulation, especially if inflammation continues, which suggests another worm is present (Mazzeo et al. 2019; Cortez et al. 2005). If DUSN is diagnosed and treated in the early stage, vision may improve (Garcia et al. 2005; Garcia et al. 2004).

DUSN caused by a hookworm likely begins with a case of CLM, although a prior history of skin lesions is rarely reported (Moraes et al. 2002). The more subtle manifestations of HrCLM caused by *A. caninum* and *A. ceylanicum* could be missed easily, whereas *A. braziliense* CLM is difficult to overlook or ignore. The only case with a definitive worm identification was caused by *A. ceylanicum* (Poppert et al. 2017), but aspects of this case are questionable. Until more worms are identified to species, the origin of this infection will remain enigmatic.

Miscellaneous CLM

A purported *Ancylostoma* larva was identified in a section of a sebaceous gland on the scalp of a young girl in Brazil (Guimarães et al. 1999), however, the identification is suspect. CLM in the oral mucosal has been reported on several occasions (Capuano et al. 2006; Damante et al. 2011; Andre et al. 1988; Lopes et al. 1994), and there is a report in which CLM triggered a case of erythema multiforme, an acute hypersensitivity skin reaction (Vaughan and English 1998). One case of visceral larva migrans was attributed to *A. caninum* (Gandullia et al. 1981). In another case, an *Ancylostoma* larva was found in a skeletal muscle biopsy following a severe case of CLM (Little et al. 1983). The patient also had pulmonary involvement. The authors postulate that the offending nematode was *A. caninum* because of its ability to undergo hypobiosis and remain viable in muscle of paratenic hosts.

Finally, an outbreak of HrCLM in Naples Italy was linked to dried floral arrangements (Galanti et al. 2002), and another in a group of 140 vacationers in Barbados (Tremblay et al. 2000).

Treatment of CLM

Prior to the advent of effective anthelmintics, physical treatments such as cryotherapy or surgical excision were standard (Blackwell and Vega-Lopez 2001). However, these treatments are usually ineffective because the larva is rarely on the leading edge of the track, and therefore difficult to locate. For example, in one study cryotherapy with repeated use of liquid nitrogen was unsuccessful in all six patients that were treated, and resulted in severe blistering or ulceration in some cases (Jelinek et al. 1994). Thus, the pain and ineffectiveness associated with liquid nitrogen cryotherapy and surgery recommend against them, as other treatments with fewer side effects and higher efficacy are readily available (Davies et al. 1993; Jelinek et al. 1994). Local remedies can be even worse, including attempting to kill the larvae by burning the lesions with a cigarette, which in addition to being painful and causing scarring, was ineffective (Ambrus and Klein 1988).

Anthelmintics are now regarded as first choice for the treatment of CLM, and several are highly efficacious. Thiabendazole, albendazole, flubendazole, and ivermection given orally are all excellent options. Thiabendazole, which has been used the longest, is most effective when given as multiple doses. In one study, cure rates ranged from 68% with a single 50 mg/kg dose to 89% after four weekly doses (Katz et al. 1965). Thiabendazole is also associated with more side effects (Caumes 2000). Albendazole, a newer member of the benzimidazole class of anthelmintics, is currently the drug of choice for treatment of hookworm and other nematode intestinal infections. As with that indication, cure rates against HrCLM can be variable. Albendazole is often given as a single 400-mg dose, but efficacy increases if it is administered for 3–5 days or combined with ivermectin (Prickett and Ferringer 2015; Veraldi et al. 2012). Albendazole is generally well tolerated. The macrocyclic lactone ivermectin is also well tolerated and is highly effective at a single dose of

12 mg, with cures of 81–100% (Caumes 2000; Prickett and Ferringer 2015), although in some cases multiple treatments are required (Bouchaud et al. 2000). At this dose, pruritus completely resolves, and progress of the tracks is halted within 48 h (Dourmishev et al. 2005; Caumes et al. 1992). One study found a single dose of ivermectin to be more effective than a single 400-mg dose of albendazole (Caumes et al. 1993). When oral ivermectin or albendazole are contraindicated, topical thiabendazole, albendazole, or ivermectin ointments used daily for 10–15 days are effective (Hochedez and Caumes 2007; Davies et al. 1993; Blackwell and Vega-Lopez 2001; Patterson and Kersey 2003; Robert et al. 2019; Magri et al. 2019). Pruritus often responds to oral antihistamines and topical corticosteroids, and antibiotics are recommended for secondary bacterial infections (Te Booij et al. 2010; Ghosh and Bandyopadhyay 2009).

Complicated HrCLM cases, including those with Löffler's syndrome, often require combined or longer treatment periods (Tan and Liu 2010; Prickett and Ferringer 2015). Follicular cases of CLM appear more resistant to treatment than classic CLM cases (Davies et al. 1993; Veraldi et al. 2005; Vanhaecke et al. 2014a, b; Le Joncour et al. 2012; Heukelbach and Feldmeier 2008). In one case, a single dose of ivermectin achieved a cure rate of only 41%, compared with 81–100% in typical CLM cases (Caumes et al. 2002). Another study found that single-dose ivermectin had a cure rate of 98% in creeping eruption cases, but 66% in follicular cases (Vanhaecke et al. 2014b). These cases may require repeated courses of anthelmintics.

Recently, a carbon dioxide laser was successfully used to treat cases of CLM in the Philippines due to the unavailability of currently approved treatments in national pharmacies (Soriano and Piansay-Soriano 2017). Eight patients previously diagnosed with CLM were treated for their lesions. The laser was used to trace the serpiginous tracts first in continuous mode for two passes. Next, the laser was shifted to fractional mode and several passes were made over a 1–2 cm perimeter of the normal skin surrounding the lesions because the larvae are typically found ahead of the erythematous part of the tract (Jelinek et al. 1994). Initially, each track received a single session of carbon dioxide laser treatment. However, when the number of passes of fractional CO_2 laser was increased to three to four, larval migration ceased. During the 4-week follow-up period, patients were given antihistamines on a need basis. All the patients reported relief from the pruritus immediately after treatment, and no recurrences were seen during follow-up. While costly, this study suggests carbon dioxide laser treatment as an efficacious alternative treatment for CLM.

Prevention of CLM

Prevention of CLM requires avoiding contact with contaminated soil (Hochedez and Caumes 2007). In tropical countries, dogs and cats frequent beaches, and high prevalence of hookworm infections ensures deposition of large numbers of ova there. Wearing footwear while walking on tropical beaches, as well as lying on a mat rather than a towel when sunbathing, are effective. Avoiding dry areas of sand above the high tide line will also prevent exposure, as wave washed areas will have fewer larvae.

9.5.2 Eosinophilic Enteritis

Eosinophilic gastrointestinal disorders (EGID) are a group of rare, poorly understood conditions characterized by eosinophilic infiltration in the gastrointestinal tract (Munjal et al. 2017). The entire gastrointestinal tract from the esophagus to the rectum can be affected, but the stomach and the duodenum are most commonly involved (Triantafillidis et al. 2002). The disorders result in various gastrointestinal manifestations depending on the location, depth, and severity of eosinophilic invasion (Triantafillidis et al. 2002). Eosinophilic enteritis (EE) is characterized by abdominal pain, peripheral blood eosinophilia (PE) in 75% of patients, and a severe, segmental eosinophilic infiltration of the muscular layer of the small intestinal mucosa, leading to thickening of the intestinal wall and occasional obstruction of the intestinal lumen (Uenishi et al. 2003; Talley et al. 1990; Yun et al. 2007). Eosinophilic gastroenteritis (EG), first describer in 1937, involves invasion of eosinophils in one or more layers of the intestinal wall (Talley et al. 1990). Since then, approximately 300 cases have been reported in literature. The cause of EG is still unknown, although several epidemiological and clinical features suggest an allergic component, possibly a food allergy (Talley et al. 1990). On the other hand, EE occurs as abdominal pain associated with PE, but without involvement of the stomach (Walker et al. 1995).

The rarity of EE cases (approximately 100 prior to 1979) made an outbreak in Townsville, a city in northern Queensland, Australia, in the mid-1980s stand out (Cello 1979; Croese 1988). Between 1983 and 1987, 33 cases were reported, with an additional 60 by 1990 (Prociv and Croese 1990). In the original group, patients presented with gastrointestinal symptoms, including severe, transient abdominal pain resembling obstruction, weight loss, and sometimes diarrhea, melena, eosinophilia, and elevated IgE (Croese 1988). Laparotomy on nine patients revealed prominent segmented inflammation and thickening of the jejunum or ileum. Hookworm was implicated when a single worm was recovered from one patient. Positive identification of the worm could not be obtained; however, modern sanitation in Queensland had eliminated the transmission of anthroponotic hookworm infection, suggesting a zoonotic species as the causal agent. The answer came in 1990, when a single intact immature worm was recovered during colonoscopy of a patient presenting with abdominal pain, eosinophilia, and high IgE levels (Prociv and Croese 1990). The worm was identified as *A. caninum* by the distinctive three pairs of buccal teeth and other morphological characteristics.

Unsurprisingly, the association of *A. caninum* infection with EE generated considerable skepticism. To further investigate this phenomenon, Walker et al. reported the pathological findings of 79 different patients with biopsy-proven EE from northeastern Australia, seventy of whom were seen since 1987 (Walker et al. 1995). Prior to this, the most biopsy-proven cases reported were a series of 38 identified in records of the Mayo Clinic between 1950 and 1987 (Talley et al. 1990). The Australian cases occurred between 1971 and 1993, and involved patients primarily from coastal Queensland. Nearly all of the cases (77) affected the ileum or colon, unlike typical cases involving the stomach and proximal small intestine seen previously (Talley et al. 1990). Of these cases, 10 were associated with the presence of a single, immature adult *A. caninum*. Over the remaining 69 cases, 65% had positive serology for hookworm, and some responded to anthelmintics, thereby implicating *A. caninum* in these cases also (Walker et al. 1995). An enzyme-linked immunosorbent assay (ELISA) and Western blot measuring IgG and IgE antibodies against Ac68, an excretory–secretory (ES) antigen from adult *A. caninum*, was used for serological testing (Loukas et al. 1992, 1994, 1996). All of the sera from confirmed hookworm-induced EE were positive using these tests, as were 92% of EE patients (Loukas et al. 1994).

From the mid-1980s until present, 93 cases of eosinophilic gastroenteritis were linked to *A. caninum* infection in northern Australia (Loukas et al. 2016). The infections had unique, interesting characteristics. Clinically, patients presented with symptoms mimicking appendicitis or a small intestinal obstruction, frequently with moderate to severe pain. Blood eosinophilia was unreliable to differentiate between those conditions and hookworm, as it was not always present. In chronic cases, diarrhea and occasionally weight loss were common. When laparotomies were performed, segmental mid or distal intestinal inflammation was generally seen, sometimes with ascites (Croese et al. 1994a).

Only single, immature worms were ever recovered, so they did not reproduce, and consequently no ova were ever seen in patients' feces. Most of the EE symptoms resolved within 24–48 h of anthelmintic administration, further supporting the role of hookworm as the causal agent. No skin symptoms or CLM from penetrating larvae were reported in these cases. Worms were rarely recovered or visualized from prospective patients, as standard endoscopy and colonoscopy could not reach the mid to distal small intestine where the worms typically reside. These procedures could only detect worms attached in the terminal ileum or colon. Even during laparotomies, worms were usually not recovered. In most cases, diagnosis of hookworm induced EE relied on positive serology and resolution of symptoms following anthelmintic treatment.

The difficulty diagnosing *A. caninum* induced EE is underscored by two early cases reported in 1995 (Walker et al. 1995). The first case occurred in 1976, and involved an 18-year-old male from outside Brisbane. He sought treatment for abdominal pain and vomiting after consuming shellfish. Laparotomy revealed an inflamed, constricted ileum, dilation of the intestine proximal to the constriction, mesenteric lymphadenopathy, and ascites. Samples were taken from the resected intestine for histology, on which the diagnosis of EE was made. The remaining portion of the intestine was mounted and used for teaching in the Department of Pathology at the University of Queensland. Over 3400 medical students and 30 surgical pathology instructors studied this specimen over the ensuing 17 years, until a medical student noticed a worm within the constricted area in 1993. After demounting and clearing, the worm was identified as an immature female *A. caninum* based on its buccal dentition.

In 1982, a 12-year-old male student from the same area presented with symptoms of small intestine obstruction. Laparotomy revealed a 2.5-cm inflamed and constricted distal ileum, with thickening of the intestinal wall. Histology showed eosinophilic infiltration, indicative of EE. Again, part of the specimen was mounted for teaching. After a hookworm was discovered in the first case in 1993, this specimen, together with another in the teaching collection, was carefully examined. A worm was found in the constricted section of the specimen. The worm was not removed from the specimen, but given its size, appearance, and the fact that the patient had not left Queensland, it was assumed to be *A. caninum*. These cases highlight the difficulty in diagnosing EE even when the affected tissue is available, and stress the need for careful examination to find the worm.

Cultural, environmental, and epidemiological conditions in Queensland were ideal for exposure and transmission of *A. caninum* to humans. Dog ownership was high, and the prevalence of canine hookworm infection ranged from 22 to 50% (Croese et al. 1994b). Most yards were not fenced in, thereby allowing both roaming and pet dogs free access. Rather than removing the feces, homeowners frequently hosed them into the lawn. The warm, tropical climate allowed people to walk barefoot on the lawn during gardening and other activities, thereby exposing them to *A. caninum* iL3. These factors all contributed to the high rate of zoonotic hookworm infection in this population.

While the assumption that the warm climate and the "tropical lifestyle" led to exposure and infection through the skin, results of a series of experiments in which human volunteers were infected with *A. caninum* challenge this dogma. A volunteer infected orally with 100 iL3 showed a peak in eosinophilia associated with mild abdominal pain at approximately 4 weeks post-infection (Landmann and Prociv 2003). After the peak, the eosinophilia slowly resolved over the next 5 weeks, reaching a plateau at a level slightly higher than normal by 8–9 weeks post-infection. It remained slightly elevated until approximately 14 weeks, when eosinophil levels rose again, peaking at 20 weeks, after which it slowly returned to the earlier levels. The second peak was similar to the first, and was associated with intermittent mild abdominal pain as well. After treatment with anthelmintic, the volunteer was exposed to 200 iL3 on the skin. This elicited a painful, pruritic skin reaction at the infection site, but only a slight rise in eosinophils. Following a third, oral dose of 20 iL3, eosinophil levels again rose sharply, with similar kinetics as the first oral infection (Landmann and Prociv 2003). Similar results were seen in additional volunteer studies, and skin infections with large numbers of larvae had to be repeated numerous times to induce a serum antibody response and eosinophilia (Landmann and Prociv 2003; Prociv and Croese 1996).

In contrast, oral infection with small doses of iL3 provoked marked eosinophilia in the volunteers. This response was thought to be a response to adult worms in the intestine (Landmann and Prociv 2003). Oral infection of dogs with *A. caninum* is probably the most common route of entry, and leads to robust infections without migration outside of the intestinal tract. The volunteer studies suggest that oral infection of humans with *A. caninum* is more pathogenic than percutaneous infection, and may be responsible for the more clinically serious cases of EE reported in Queensland (Landmann and Prociv 2003).

Further evidence supports a role for oral infection of humans with *A. caninum*. Neither CLM nor "ground itch," cutaneous manifestations commonly seen at the site of hookworm larval penetration, were ever reported in any of the Australian EE cases (Prociv and Croese 1996). Ground itch associated with hookworm infection is thought to only develop on repeat exposure, suggesting host hypersensitivity to hookworm molecules encountered during infection (Prociv and Croese 1996). In the volunteer studies, the skin rash appeared following percutaneous infection, suggesting the necessity of previous exposure also (Landmann and Prociv 2003), although this is not the case with anthroponotic hookworms, where first exposure leads to pruritus and a papulovesicular rash at the infection site (Diemert et al. 2018). On the other hand, severe EE is thought to be a Type 1 hypersensitivity response provoked by re-exposure to hookworms in the intestine (Walker et al. 1995; Loukas et al. 1992). The lack of skin pathology in any EE cases, despite the obvious re-exposure in severe intestinal cases, suggests that ingestion of *A. caninum* L3 might be the predominant route of exposure in the Australian EE cases. Contaminated water or food represents the most likely vehicle for oral entry of iL3, although consumption of hypobiotic larvae in undercooked meat cannot be ruled out (Schad et al. 1984).

An interesting phenomenon associated with EE infections in Australia was an apparent seasonality of cases, suggesting the potential for *A. caninum* to undergo hypobiosis in humans. Two lines of evidence support this. First, a retrospective study of 156 patients revealed that EE cases in Queensland remained similar throughout much of the year, but decreased in June and remained low during the winter months of June, July, and August (Croese 1995). Cases increased in the spring, prior to the monsoon season from December through March. This paralleled egg output in infected dogs in the region, which also decreases in winter and increases in spring (Prociv et al. 1994). This seasonal variation ensures that the eggs will land in soil that is moist and conducive to development, and occurs in *A. caninum* infections elsewhere, as well as in humans infected with the congeneric *A. duodenale* in India (Schad 1982; Schad and Page 1982; Schad et al. 1973). Second, multiple cases of recurring EE following anthelmintic treatment were reported. These were judged not to be from re-exposure, as these patients typically altered their behavior to avoid contact with larvae after treatment. Some patients had multiple recurrences, each of which would resolve after treatment (Croese et al. 1994a). These individual bouts are likely caused by the emergence and development of iL3 from a tissue reservoir of hypobiotic larvae, analogous to the "larval leak phenomenon described in *A. caninum* infected dogs (Jimenez Castro et al. 2019; Bowman 2014). Further support comes from volunteer infections with *A. caninum* (discussed previously), in which two peaks of eosinophilia occurred 17 weeks apart in the absence of reinfection (Landmann and Prociv 2003). In the Townsville/Brisbane area, *A. caninum* infections undergo hypobiosis and seasonal variation in egg output (Prociv et al. 1994), and they appear to maintain this behavior when they infect aberrant hosts such as humans.

EE Outside Australia

There have been two reports of EE likely caused by *A. caninum* in the United States, both from Louisiana. In the early 1990s, a 7-year-old child presented with bloody stool, elevated IgE, but a normal eosinophil count (Khoshoo et al. 1994). Colonoscopy revealed multiple small nodules in the terminal ileum and mild erythema of the colonic mucosa. Histology showed tissue eosinophilia in the large intestine and terminal ileum. Serology was positive for *Toxocara* (dog roundworm) and *A. caninum* Ac68 by both ELISA and Western blot, and the family dog was infected with *A. caninum*. Symptoms resolved within 72 h of anthelmintic treatment, but returned approximately 6 weeks later. A second treatment cured the infection permanently. The case was somewhat atypical, as there was no associated abdominal pain and the infection occurred in a child. The Australian cases were all reported from adults.

The second case was an 11-year-old child who presented with a 3-month history of abdominal pain, poor appetite with weight loss, intermittent vomiting, and frank blood in her stool (Khoshoo et al. 1995). Initial blood work was normal except for eosinophilia. Colonoscopy revealed mild inflammation and moderate to severe eosinophilia in the terminal ileum. Serology was positive for *A. caninum* by ELISA and Western blot. The symptoms resolved after two treatments with anthelmintic. These two cases indicated that *A. caninum* infection should be considered in cases of EE or abdominal pain, and suggest that *A. caninum* induced EE is not unique to Australia. However, cultural differences, as well as possible biological differences in virulence may account for the higher incidence in Australia.

Ancylostoma caninum was implicated in 11 cases of unexplained abdominal pain in patients in Egypt (Bahgat et al. 1999). All of the patients were positive for Ac68 antigen by ELISA and IgG/IgG4 Western blot. Three patients were initially diagnosed with appendicitis, and the removed appendices showed eosinophilic infiltration, but no worms were recovered. Symptoms resolved with mebendazole treatment.

Despite widespread infection in dogs and high levels of likely exposure, no *A. caninum* induced EE has been reported in India (Traub et al. 2005). Large populations of stray and semi-domesticated infected dogs living in close association with people, especially in rural areas, where EE may be common, under-reported, and therefore under-investigated.

Canine hookworms have been recovered sporadically from humans in the past, often without symptoms of EE. Prior to the Queensland outbreak, *A. caninum* have been recovered from humans following deworming or at autopsy in the Philippines, Indonesia, South America, the United States, and Israel (Nichols 1956; Muhleisen 1953; Croese et al. 1994b; Carrias and de Vargas 1985). Recently, a single female *A. caninum* was recovered from the descending colon of a patient in South Korea during colonoscopy (Jung et al. 2020). The patient had moderate eosinophilia but no signs associated with EE. The worm was confirmed to be *A. caninum* by morphology and sequencing of the internal transcribed spacer 1 (ITS-1) and 5.8S rDNA region, which had a 100% identity with an *A. caninum* sequence in GenBank. These reports suggest that *A. caninum* infections in humans may be more widespread than commonly believed. It is unclear why these infections are not associated with EE. Strain differences in virulence in various populations may account for cryptic infections in some areas. Another possibility is that in endemic areas, the immune response to hookworm is generally dampened from repeated exposure to anthroponotic species beginning at an early age, and *A. caninum* may be seen by the immune system as "just another hookworm."

Interestingly, molecular sampling of hookworm eggs collected from humans in Tamil Nadu, India, found *A. caninum* DNA in at least one patient in seven of nine sampled villages, with an overall prevalence of 16.8% (George et al. 2016). This exceeded that of the human hookworm *A. duodenale*, which was only found in people from four of the nine villages, with an overall prevalence of 8.4%. They also found that DNA from animal hookworms including *A. caninum* were more prevalent than human species in soil samples. Detecting *A. caninum* DNA in human stools suggests patent hookworm infections, as it is usually assumed that fecal DNA is due to eggs. However, the authors could not conclusively demonstrate *A. caninum* eggs from any humans, as all hookworm-infected people also had *N. americanus* eggs in their feces, nor could they rule out contamination during collection despite instructions to patients to avoid soil contamination. They considered pass through of ingested hookworm eggs or DNA unlikely (George et al. 2016).

Finally, a survey of hookworms from stray dogs, selected wildlife, and humans was recently conducted in South Africa (Ngcamphalala et al. 2019). Fifty students in two primary schools were tested, and three were found to have hookworm eggs, as identified by microscopy. Molecular analysis and sequencing indicated that all three patients were infected with *A. caninum*. No human hookworm species were found. These indicate the first confirmed patent cases of *A. caninum* in humans reported to date. More recently, individual *A. caninum* eggs were identified from a person in Brazil by molecular methods (Furtado et al. 2020). Together with the results from India (George et al. 2016), these findings suggest that *A. caninum* can infect and reach sexual maturity and patency in humans, at least under certain conditions or in certain locales. It also suggests that more surveys using molecular diagnostics are likely to reveal a wider incidence of human infections with *A. caninum* and other zoonotic hookworms.

Aberrant *A. caninum* Infections

Ancylostoma caninum and other zoonotic hookworm larvae have been identified in tissues outside of the skin and intestine, including muscle, lungs, and eye (Little et al. 1983; Muhleisen 1953; de Souza et al. 1992; Nadbath and Lawlor 1965; Kwiatkowska-Kawecka 1973). Intestinal, dermal, and extradermal cases of *A. caninum* in humans may be under- or mis-diagnosed. High levels of dog ownership and close association of dogs with humans worldwide likely leads to frequent exposure. Perhaps the most bizarre case of zoonotic hookworm in humans was the recovery of a dead adult *A. tubaeforme* female from the eye of a human in Sri Lanka (Dissanaike et al. 2000). *Ancylostoma tubaeforme* is a cat hookworm closely related to *A. caninum*.

9.6 Other Canine Hookworms Affecting Humans

Ancylostoma ceylanicum is a hookworm of dogs, cats, and humans found primarily in southeast Asia, China, Australia, India, Tanzania, and South Africa (Traub 2013; Haydon and Bearup 1963; Ngcamphalala et al. 2019; Wang et al. 2019; Liu et al. 2015; Smout et al. 2013, 2017; Ngui et al. 2012; Sato et al. 2010; Bradbury et al. 2017; Conlan et al. 2011; Setasuban et al. 1976; Chowdhury and Schad 1972; Margono et al. 1979; Choo et al. 2000; Merino-Tejedor et al. 2019; Pa Pa Aung et al. 2017). It has also been reported from South America (Rep and Heinemann 1976). Originally described by Loos in 1911 (Looss 1911) from a civet cat in Ceylon (now Sri Lanka), it was considered synonymous with *A. braziliense*, described a year earlier. By mid-century, however, the species were confirmed to be separate (Biocca 1951). The life cycle is similar to *A. caninum*, except that paratenic hosts and hypobiotic larvae are believed not to occur (Ray et al. 1972; Yoshida et al. 1974). While it feeds on blood, it is less virulent than *A. caninum*, but more so than *A. braziliense*, which is not a vigorous blood feeder (Miller 1966, 1968; Rep 1980). However, heavy infections will result in significant anemia and gastrointestinal symptoms in dogs (Traub 2013).

Among the zoonotic hookworms, *A. ceylanicum* is the only one capable of maturing and reproducing in humans. First documented from prisoners in Calcutta in 1913 (Lane 1913), *A. ceylanicum* infections in humans have subsequently been reported throughout Southeast Asia and parts of Africa and South America (Traub 2013; Rep and Heinemann 1976; Chowdhury and Schad 1972; Inpankaew et al. 2014; Areekul et al. 1970; Yoshida et al. 1968; Velasquez and Cabrera 1968). More recently, molecular epidemiological surveys have found that the most common hookworm in humans after *N. americanus* in Asia is *A. ceylanicum*, with an estimated 19–73 million cases (Traub 2013). Most natural *A. ceylanicum* infections in humans are light, although heavy infections and anemia can occur (Anten and Zuidema 1964; Chung et al. 2012).

Molecular diagnostics have highlighted the previously unknown zoonotic risk of *A. ceylanicum*. Several studies have found monospecific and mixed infections of *A. ceylanicum* in humans, especially in areas with high prevalence of the hookworm in dogs, including Laos (Sato et al. 2010; Conlan et al. 2012), Thailand (Jiraanankul et al. 2011), Cambodia (Inpankaew et al. 2014), and Malaysia (Ngui et al. 2012). Relationships with companion animals in rural and poor areas are often close but semi-domestic, with community dogs and cats far outnumbering owned pets. This makes controlling parasites like hookworm difficult in these areas, and consequently increases human exposure to infective stages (Otranto et al. 2017). Therefore, greater vigilance is required in *A. ceylanicum* endemic areas, especially in areas where the prevalence of anthroponotic hookworms is low or control programs are not undertaken. The emergence of *A. ceylanicum* as an important parasite of humans has been reviewed elsewhere (Traub 2013).

The ability of *A. ceylanicum* to reproduce in hamsters has made it an important model for hookworm infection and disease (Ray et al. 1972, 1975; Carroll and Grove

1985, 1986; Bungiro et al. 2001; Alkazmi et al. 2008; Behnke et al. 1986, 1997; Ray and Bhopale 1972).

Ancylostoma braziliense was described in 1910 by Gomez da Faria (Faria 1910), one year before Loos described *A. ceylanicum* (Looss 1911), and were shortly thereafter a subject of contention. For the first half of the twentieth century, the two were thought to be synonymous, often referred to as *A. braziliense* var. *ceylanicum* (Wijers and Smit 1966). In 1951, Biocca (1951) finally separated them into two species, but this was not universally accepted (Rep 1963) until attempts to crossbreed the species failed to produce offspring (Rep et al. 1968). Additional morphological evidence cemented their status as two species (Yoshida 1971).

A parasite of dogs and cats, *A. braziliense* is most common in tropical and subtropical regions. In the United States, it is most common in the Atlantic and Gulf coastal states, especially Florida, but it has been seen as far north as New Jersey, and from bobcats (*Felis rufus*) in West Virginia (Bowman et al. 2010; Dove 1932; Watson et al. 1981; Anderson et al. 2003). Elsewhere in the Western hemisphere, *A. braziliense* has been found in the Caribbean, Surinam, Brazil, Uruguay, and Cuba (Davies et al. 1993; Rep and Heinemann 1976; Anderson et al. 2003; Malgor et al. 1996; Liotta et al. 2012; Labarthe et al. 2004; Coelho et al. 2011; Oliveira-Arbex et al. 2017; Prieto Fernández et al. 1978; Rep 1975). In Asia, *A. braziliense* is restricted to Malaysia, Borneo, Indonesia, Vietnam, and Australia (Haydon and Bearup 1963; Ng-Nguyen et al. 2015; Margono et al. 1979; Choo et al. 2000; Yoshida et al. 1973). It has also been reported from South Africa, Kenya, Tanzania, and India (Ngcamphalala et al. 2019; Merino-Tejedor et al. 2019; Traub et al. 2004; Mulinge et al. 2020; Minnaar and Krecek 2001). While it apparently prefers more tropical climates, increased application of molecular diagnostics during surveys is likely to expand its range. Over its distribution, it is usually sympatric with the other hookworms of canids and felids, *A. caninum*, *A. tubaeforme*, and *A. ceylanicum*. Of these, *A. braziliense* consumes the least blood, and is considered the least pathogenic to its host (Miller 1966; Rep 1980).

Ancylostoma braziliense is the primary etiological agent of CLM, causing the classic "creeping eruption" characterized by long, serpiginous tracks that can persist for up to 100 days (Bowman et al. 2010; Dove 1932; Chaudhry and Longworth 1989; Haydon and Bearup 1963; Malgor et al. 1996; Traub et al. 2008). Unlike *A. caninum*, which causes a less severe macropapular or follicular dermatitis (White and Dove 1929; Haydon and Bearup 1963), *A. braziliense* is unable to penetrate through the deeper layers of human skin, although it rapidly penetrates canine skin (Sandground 1939; Vetter and Leegwater-v.d. Linden 1977). Consequently, no confirmed cases of EE or adults have been reported. A single case of an adult *A. braziliense* in a person was reported in 1928 (Dove 1928), but this was subsequently contested (Beaver 1956). On one occasion, a larva of *A. braziliense* was recovered from the sputum of a man with CLM (Muhleisen 1953), indicating that larvae can leave the skin and undergo somatic migration but are unable to reach or develop in the intestine.

Warming associated with anthropogenic climate change has extended the range of *A. braziliense* to regions where it was formerly absent, as evidenced by recent cases in India (Kaur et al. 2015; Siddalingappa et al. 2015) and multiple recent autochthonous CLM cases in Europe (Baple and Clayton 2015; Tamminga et al. 2009; Robert et al. 2019; Okulewicz 2017; Del Giudice et al. 2019; Blaizot et al. 2017; Panés-Rodríguez et al. 2016; Gutiérrez García-Rodrigo et al. 2017). Furthermore, the high prevalence of *A. braziliense* in low-income settings increases the risk of CLM (Malgor et al. 1996; Oliveira-Arbex et al. 2017; Minnaar and Krecek 2001; Minnaar et al. 2002; Reichert et al. 2018). Transmission increases in the rainy season (Reichert et al. 2018; Heukelbach et al. 2008), and increased precipitation is predicted in the future due to climate change (Weaver et al. 2010). Together, HrCLM caused by *A. braziliense* is likely to increase on a warmer planet of the future.

9.7 Anthelmintic Resistance in *Ancylostoma caninum*

Several anthelmintics are quite efficacious against canine hookworms including *A. caninum*, and are widely used for controlling infections in puppies and older dogs, as well as for zoonotic infections in humans. However, as with other drugs targeting pathogenic organisms, the pathogen is not a static player, and is capable of evolving genetic resistance to the drug, especially if used incorrectly. This is evident by the emergence of anthelmintic resistance in many of the parasitic nematodes that infect livestock and horses globally. Anthelmintic resistance is the heritable ability of a subpopulation of worms to survive a dose of anthelmintic that they were unable to survive previously (Wolstenholme et al. 2004). Widespread resistance to multiple drugs in multiple nematode species that infect livestock have been reported worldwide. Mutations that confer resistance to specific anthelmintics reside in natural worm populations, typically at low frequencies. Heavy use of anthelmintics selects for these resistant worms, which survive treatment and reproduce (Prichard 1990). Over time, as resistant individuals predominate and the frequency of their resistance alleles increases due to natural selection, phenotypic resistance to the drug will begin to emerge (Kaplan 2004). Strongylid nematodes infecting livestock have genetic features that favor the development of anthelmintic resistance, such as rapid rates of nucleotide sequence evolution and extremely large effective population sizes (Jimenez Castro et al. 2019; Blouin et al. 1995, 1999). In some areas, formerly effective drugs like ivermectin, pyrantel, and especially the benzimidazole anthelmintics, are no longer able to eliminate the infections they once did (Wolstenholme et al. 2004; Kaplan 2004). For example, resistance of the sheep parasite *Haemonchus contortus* to benzimidazoles is 100% among 34 farms in the southern US, with 30% of the farms having total failure against all anthelmintics (Kaplan 2020).

Parasitic nematodes of companion animals, such as *A. caninum*, differ from livestock strongylids in several important ways that should limit development of anthelmintic resistance. Pets are kept and treated for parasites individually or in small numbers, allowing a large proportion of the hookworm metapopulation to remain untreated (Traversa 2012; Irwin 2002). The untreated worms act as *refugia* of susceptible alleles, which are available to dilute the resistance alleles (Hodgkinson et al. 2019). This prevents the frequency of resistance alleles from increasing to the level required for phenotypic resistance to appear. However, several authors have pointed out that improper or over use of anthelmintics in concentrated groups of animals, such as kennels, breeding facilities, or shelters, could lead to resistance in species like *A. caninum* (Traversa 2012; Irwin 2002). Indeed, this has now occurred, as outlined below.

The first report of anthelmintic resistance in *A. caninum* occurred in 1987 to the drug pyrantel (Jackson et al. 1987). A greyhound in poor condition was treated with a combination of pyrantel and oxantel prior to importation from Australia into New Zealand, as is standard practice. After a second treatment, the animal was tested and found to have an extremely high fecal hookworm egg count. Larvae raised from the eggs were used to infect two puppies, which were subsequently treated with up to five time the recommended dose of pyrantel or pyrantel/oxantel. There was no decrease in fecal egg count during the treatment period. Examination of adult worms from a third infection confirmed that they were *A. caninum* (Jackson et al. 1987).

Several anthelmintic trials conducted in Brisbane, Australia further suggested that pyrantel resistance was becoming established in the *A. caninum* population of the area. Three formulations containing pyrantel, oxantel, and praziquantel were 25–40% less effective than a pyrantel/fenbendazole product (Hopkins et al. 1998; Kopp et al. 2008a; Thompson and Roberts 2001). A subsequent trial indicated that pyrantel alone reduced hookworm worm burdens by only 75% (Kopp et al. 2008a; Hopkins and Gyr 1991). Using the local Brisbane strain of *A. caninum*, a third trial found a high level of resistance to pyrantel, with only 25.1% efficacy (Kopp et al. 2007). Furthermore, egg counts rose in the treated animals more than in controls, despite fewer adult worms, suggesting a compensatory mechanism to increase egg production in response to a reduction in adult worms (Kopp et al. 2007). The same group subsequently isolated a strain with low resistance to pyrantel, in which efficacy was 71% (Kopp et al. 2008b, c). These trials indicate that pyrantel resistant *A. caninum* is circulating in Brisbane area pet dogs.

The mechanism of pyrantel resistance is unknown. Pyrantel is an agonist of the nicotinic AChR channel (Robertson et al. 1994). Binding to the receptor opens the channel, causing cell depolarization and Ca^{++} entry, resulting in spastic muscle contraction and the inability of the worm to maintain its position in the intestine (Kopp et al. 2008a; Martin et al. 2012). The AChR is composed of α- and β-subunits that form a ring of five subunits with a central pore (Martin et al. 2012; Williamson et al. 2007; Neveu et al. 2010). The cholinergic anthelmintics, which include pyrantel, oxantel, morantel, and levamisole, target the nematode AChR channel, which is composed of five individual subunits. There are three types of AChR that differ in their sensitivity to specific drugs. N-type channels are most sensitive to nicotine and oxantel, L-type are most sensitive to pyrantel and levamisole, and B-subtype are most sensitive to bephenium (Kopp et al. 2007; Martin et al. 2004). There are at least two subtypes of L-AChR, composed of various subunit combinations, that differ in their relative sensitivities to pyrantel and levamisole (Boulin et al. 2011). L-type AChR channels from resistant nematodes are less abundant on muscle cells and open less frequently (Robertson et al. 1999, 2000). Efforts to determine the resistance mechanism in *A. caninum* have been directed toward polymorphisms and expression differences in channel subunits. Expression of three subunits was reduced significantly in a highly pyrantel-resistant strain of *A. caninum*, but no polymorphisms were detected (Kopp et al. 2009). Therefore, the genetic mechanism of resistance to pyrantel remains unclear in *A. caninum* and other parasitic nematodes.

Until recently, reports of pyrantel resistance from Australia were the only confirmed cases of anthelmintic resistance in hookworms. In 2019, an isolate of *A. caninum* resistant to benzimidazoles and ivermectin was identified (Kitchen et al. 2019). The isolate was recovered from a retired racing greyhound raised in Kansas, and was refractory to several rounds of anthelmintic treatment with fenbendazole. Shortly thereafter, three additional resistant isolates were reported, each of which was resistant to pyrantel in addition to benzimidazoles and ivermectin (Jimenez Castro et al. 2019). One of these isolates originated in greyhounds, while the other two were from different breeds.

The mechanism of benzimidazole resistance has been studied in livestock parasitic nematodes, and is generally accepted to be caused by single nucleotide polymorphisms (SNPs) in the β-tubulin isotype 1 gene (Schwenkenbecher et al. 2007). SNPs in codons 200 and 167 change phenylalanine to tyrosine (TTC→TAC), while an SNP in codon 198 changes glutamate to alanine (GAG→GCG). All of these mutations can confer resistance independently, but the 200 and 167 SNPs are more common (Silvestre and Cabaret 2002; Kwa et al. 1994; Ghisi et al. 2007; Vercruysse et al. 2011). All of the recently identified multidrug-resistant *A. caninum* isolates had the codon 167 SNP, which is at least several cases was fixed at 100% frequency in the population (Jimenez Castro et al. 2019; Kitchen et al. 2019). Furthermore, introducing the codon 167 SNP into the homologous β-tubulin gene of the free-living nematode *Caenorhabditis elegans* using CRISPR/Cas9 conferred thiabendazole resistance, but not ivermectin or pyrantel resistance, on the edited worms (Kitchen et al. 2019). This is the first direct proof that any of the tubulin SNPs are responsible for benzimidazole resistance. The mechanisms of resistance to ivermectin and pyrantel are currently unknown. Changes in expression levels of nAChR subunits have been associated with pyrantel resistance in *A. caninum* (Kopp et al. 2009), but the underlying genetic mutation is not known.

Interestingly, two of the four resistant *A. caninum* isolates were isolated from greyhounds. While not formally reported previously, it is nonetheless unsurprising that anthelmintic resistance arose in greyhounds. Greyhound breeding farms have long been suspected of harboring anthelmintic resistant hookworms (Ridley et al. 1994). Anecdotal reports from greyhound adoption groups on the Internet bemoan the inability to deworm adopted dogs using standard protocols. Long standing, chronic infections that are not resolved with anthelmintics are frequently attributed to the "larval leak" phenomenon, in which dormant, developmentally arrested L3 in tissues sporadically reactivate to repopulate the intestine following removal of resident adult hookworms (Schad and Page 1982; Bowman 2014; Epe 2009). These infections are characterized by chronic shedding of low numbers of hookworm eggs, with treatment only providing a short egg-free period until the next batch of dormant larvae are mobilized (Jimenez Castro et al. 2019). More likely, however, greyhound husbandry and management practices have led to the development of anthelmintic resistance. Greyhound breeding farms are similar to livestock farms in practice, with a relatively large number of animals concentrated in a small area, and therefore are more likely to generate the conditions necessary for high transmission levels and development of resistance than in a typical companion animal setting. Following weaning, a litter of puppies is allowed unrestricted access to an exercise run of grass, dirt, or sand. As they get older, they are moved to sequentially larger exercise areas. Feces are not routinely removed from these runs, presenting ideal conditions for parasite transmission.

Little information about parasitism in greyhound farms is available, with only a single published study examining the epidemiology of parasites at greyhound breeding farms in the United States found (Ridley et al. 1994). The studies, performed by investigators at Kansas State University, found a high prevalence of parasites in greyhounds from three breeding farms in Kansas, with the most common species being *A. caninum*, at prevalence of 16.1% despite an intense deworming schedule. Ova (*Toxocara* and *Trichuris*) and hookworm larvae recovered from soil samples at several of the breeding farms indicated that environmental contamination was a major problem, leading to nearly continuous exposure of the animals to parasites. Because of this high level of exposure, anthelmintics are administered frequently in an effort to control hookworm and other nematode infections. At the farms in the study, deworming typically started at 2–3 weeks of age. This is followed by treatment every 1–3 weeks until age one, then every 3–4 weeks for their entire racing career. Treatments are typically rotated among several different anthelmintics under the assumption that any worms not killed by a treatment will be killed by a subsequent treatment (Ridley et al. 1994). The combination of high prevalence, heavy environmental contamination with infective stages, and excessive and improper use of anthelmintics has provided an ideal environment for the emergence of multidrug-resistant strains of *A. caninum*.

As of this writing, there is little evidence that resistant *A. caninum* has become widespread outside greyhound breeding facilities. However, identification of two resistant isolates in other breeds suggests that it has begun to spread to the pet population (Jimenez Castro et al. 2019). Anthelmintic resistant *A. caninum* would present serious problems for management of hookworm disease in dogs, as well as for treatment of HrCLM and other zoonotic disease resulting from *A. caninum* infection. Options are limited for treatment of multidrug-resistant strains. The combination of topical moxidectin (Advantage Multi) and an oral formulation of pyrantel, praziquantel, and febantel (Drontal-Plus) was used to eliminate infection in eight greyhounds with persistent hookworm infections that were refractory to standard treatments (Hess et al. 2019), although it is not known if the worms were genetically resistant. Moxidectin is generally effective against ivermectin-resistant nematodes (Craig et al. 1992; Oosthuizen and Erasmus 1993; Pankavich et al. 1992), however, regular use in these populations can lead to high levels of resistance, as was seen with *H. contortus* (Kaplan et al. 2007) and one of the multidrug-resistant *A. caninum* isolates (Jimenez Castro et al. 2019). The relatively new octadepsipeptide drug emodepside is effective against canine hookworms (Schimmel et al. 2009) and is likely still effective against multidrug-resistant isolates. However, it is not approved for use in dogs in the United States, and therefore must be used off label or experimentally.

9.8 Conclusion

The canine hookworm *A. caninum* is a longtime companion of our favorite companion animal, and represents a potential threat to humans. Diseases ranging from CLM to DUSN to eosinophilic enteritis have been reported in humans when infected with *A. caninum*. However, given the cosmopolitan distribution of both dogs and their hookworm *A. caninum*, it is somewhat surprising that the incidence of these diseases is so low. Humans are likely exposed to infective *A. caninum* larvae multiple times throughout the course of their lives, yet relatively few serious cases of disease resulting from such exposure are seen. The already low risk of contracting zoonotic hookworm can be lowered further by simple practices like disposing of canine feces, proper deworming, and exercising care when contacting sand or soil while vacationing in the tropics or in areas where dogs roam freely. The benefits of dog ownership should not be forsaken on the off chance of acquiring a zoonotic hookworm infection.

References

Akuthota P, Weller PF (2012) Eosinophilic pneumonias. Clin Microbiol Rev 25(4):649–660. https://doi.org/10.1128/CMR.00025-12

Alkazmi LM, Dehlawi MS, Behnke JM (2008) The mucosal cellular response to infection with *Ancylostoma ceylanicum*. J Helminthol 82(1):33–44

Ambrus JL, Klein E (1988) Löffler syndrome and ancylostomiasis brasiliensis. NY State J Med 88 (9):498–499

American Veterinary Medical Association (2018) U.S. pet ownership & demographics sourcebook. Veterinary Economics Division, American Veterinary Medical Association, Schaumburg, IL

Anderson TC, Foster GW, Forrester DJ (2003) Hookworms of feral cats in Florida. Vet Parasitol 115(1):19–24. https://doi.org/10.1016/s0304-4017(03)00162-6

Andre J, Bernard M, Ledoux M, Achten G (1988) Larva migrans of the oral mucosa. Dermatologica 176(6):296–298. https://doi.org/10.1159/000248739

Anten JF, Zuidema PJ (1964) Hookworm infection in Dutch servicemen returning from West New Guinea. Trop Geogr Med 16:216–224

Arasu P (2001) In vitro reactivation of *Ancylostoma caninum* tissue-arrested third-stage larvae by transforming growth factor-beta. J Parasitol 87(4):733–738

Areekul S, Tipayamontri U (1974) Experimental infection of *Ancylostoma braziliense* in dogs and cats in Thailand. I. Morphology, route of infection, worm burden and egg output. Southeast Asian J Trop Med Public Health 5(1):31–38

Areekul S, Radomyos P, Viravan C (1970) Preliminary report of *Ancylostoma ceylanicum* infection in Thai people. J Med Assoc Thai 53(5):315–321

Bahgat MA, El Gindy AE, Mahmoud LA, Hegab MH, Shahin AM (1999) Evaluation of the role of *Ancylostoma caninum* in humans as a cause of acute and recurrent abdominal pain. J Egypt Soc Parasitol 29(3):873–882

Baple K, Clayton J (2015) Hookworm-related cutaneous larva migrans acquired in the UK. BMJ Case Rep 2015. https://doi.org/10.1136/bcr-2015-210165

Beattie PE, Fleming CJ (2002) Cutaneous larva migrans in the west coast of Scotland. Clin Exp Dermatol 27(3):248–249. https://doi.org/10.1046/j.1365-2230.2002.09852.x

Beaver PC (1956) The record of *Ancylostoma braziliense* as an intestinal parasite of man in North America. Am J Trop Med Hyg 5(4):737–738. https://doi.org/10.4269/ajtmh.1956.5.737

Behnke JM, Paul V, Rajasekariah GR (1986) The growth and migration of *Necator americanus* following infection of neonatal hamsters. Trans R Soc Trop Med Hyg 80:146–149

Behnke JM, Guest J, Rose R (1997) Expression of acquired immunity to the hookworm *Ancylostoma ceylanicum* in hamsters. Parasite Immunol 19(7):309–318. https://doi.org/10.1046/j.1365-3024.1997.d01-213.x

Belizario V Jr, Delos Trinos JP, Garcia NB, Reyes M (2016) Cutaneous manifestations of selected parasitic infections in Western Pacific and Southeast Asian regions. Curr Infect Dis Rep 18 (9):30. https://doi.org/10.1007/s11908-016-0533-x

Biocca E (1951) On *Ancylostoma braziliense* (de Faria, 1910) and its morphological differentiation from *A. ceylanicum* (Looss, 1911). J Helminthol 25(1–2):1–10. https://doi.org/10.1017/S0022149X00018903

Blackwell V, Vega-Lopez F (2001) Cutaneous larva migrans: clinical features and management of 44 cases presenting in the returning traveller. Br J Dermatol 145(3):434–437. https://doi.org/10.1046/j.1365-2133.2001.04406.x

Blaizot R, Goiset A, Caumes E, Gabriel F, Milpied B (2017) Cutaneous larva migrans: a case in Bordeaux, France and a systematic review of locally acquired cases in Europe. Eur J Dermatol 27(4):426–429. https://doi.org/10.1684/ejd.2017.3043

Blouin MS, Yowell CA, Courtney CH, Dame JB (1995) Host movement and the genetic structure of populations of parasitic nematodes. Genetics 141(3):1007–1014

Blouin MS, Liu J, Berry RE (1999) Life cycle variation and the genetic structure of nematode populations. Heredity 83(Pt 3):253–259

Bonne C (1937) Invasion of the submucosa of the human small intestine by *Ancylostoma braziliense*. Am J Trop Med Hyg s1–17(4):587–594. https://doi.org/10.4269/ajtmh.1937.s1-17.587

Bouchaud O, Houze S, Schiemann R, Durand R, Ralaimazava P, Ruggeri C, Coulaud JP (2000) Cutaneous larva migrans in travelers: a prospective study, with assessment of therapy with ivermectin. Clin Infect Dis 31(2):493–498. https://doi.org/10.1086/313942

Boulin T, Fauvin A, Charvet CL, Cortet J, Cabaret J, Bessereau JL, Neveu C (2011) Functional reconstitution of *Haemonchus contortus* acetylcholine receptors in *Xenopus* oocytes provides mechanistic insights into levamisole resistance. Br J Pharmacol 164(5):1421–1432. https://doi.org/10.1111/j.1476-5381.2011.01420.x

Bowman DD (2014) Georgis' parasitology for veterinarians, 10th edn. Saunders, St. Louis

Bowman DD, Montgomery SP, Zajac AM, Eberhard ML, Kazacos KR (2010) Hookworms of dogs and cats as agents of cutaneous larva migrans. Trends Parasitol 26(4):162–167

Bradbury RS, Hii SF, Harrington H, Speare R, Traub R (2017) *Ancylostoma ceylanicum* Hookworm in the Solomon Islands. Emerg Infect Dis 23(2):252–257. https://doi.org/10.3201/eid2302.160822

Bungiro RD Jr, Greene J, Kruglov E, Cappello M (2001) Mitigation of hookworm disease by immunization with soluble extracts of *Ancylostoma ceylanicum*. J Infect Dis 183(9):1380–1387

Burke TM, Roberson EL (1985) Prenatal and lactational transmission of *Toxocara canis* and *Ancylostoma caninum*: experimental infection of the bitch before pregnancy. Int J Parasitol 15:71–75

Burrows RB (1962) Comparative morphology of *Ancylostoma tubaeforme* (Zeder, 1800) and *Ancylostoma caninum* (Ercolani, 1859). J Parasitol 48(5):715–718. https://doi.org/10.2307/3275261

Butland RJ, Coulson IH (1985) Pulmonary eosinophilia associated with cutaneous larva migrans. Thorax 40(1):76–77. https://doi.org/10.1136/thx.40.1.76
Cai J, Wei R, Zhu L, Cao M, Yu S (2000) Diffuse unilateral subacute neuroretinitis in China. Arch Ophthalmol 118(5):721–722
Cappello M, Vlasuk GP, Bergum PW, Huang S, Hotez PJ (1995) *Ancylostoma caninum* anticoagulant peptide: a hookworm-derived inhibitor of human coagulation factor Xa. Proc Natl Acad Sci USA 92(13):6152–6156
Capuano ACT, de Sousa S, de Carvalhosa AA, dos Santos Pinto Junior D (2006) Larva migrans in the oral mucosa: report of a case. Quintessence Int 37(9):721–723
Carrias VR, de Vargas PR (1985) Infection of the human intestine by *Ancylostoma caninum* Ercolani 1859. Rev Soc Brasil Med Trop 18:57
Carroll SM, Grove DI (1985) *Ancylostoma ceylanicum*: immunization with soluble worm extract and responses to challenge infection of dogs. Exper Parasitol 60(3):263–269
Carroll SM, Grove DI (1986) Experimental infection of humans with *Ancylostoma ceylanicum*: clinical, parasitological, haematological and immunological findings. Trop Geogr Med 38 (1):38–45
Casella AM, Farah ME, Belfort R Jr (1998) Antihelminthic drugs in diffuse unilateral subacute neuroretinitis. Am J Ophthalmol 125(1):109–111. https://doi.org/10.1016/s0002-9394(99)80248-7
Caumes E (2000) Treatment of cutaneous larva migrans. Clin Infect Dis 30(5):811–814. https://doi.org/10.1086/313787
Caumes E (2006) It's time to distinguish the sign 'creeping eruption' from the syndrome 'cutaneous larva migrans'. Dermatology 213(3):179–181. https://doi.org/10.1159/000095032
Caumes E, Danis M (2004) From creeping eruption to hookworm-related cutaneous larva migrans. Lancet Infect Dis 4(11):659–660. https://doi.org/10.1016/S1473-3099(04)01178-8
Caumes E, Datry A, Paris L, Danis M, Gentilini M, Gaxotte P (1992) Efficacy of ivermectin in the therapy of cutaneous larva migrans. Arch Dermatol 128(7):994–995
Caumes E, Carriere J, Datry A, Gaxotte P, Danis M, Gentilini M (1993) A randomized trial of ivermectin versus albendazole for the treatment of cutaneous larva migrans. Am J Trop Med Hyg 49(5):641–644. https://doi.org/10.4269/ajtmh.1993.49.641
Caumes E, Carrière J, Guermonprez G, Bricaire F, Danis M, Gentilini M (1995) Dermatoses associated with travel to tropical countries: a prospective study of the diagnosis and management of 269 patients presenting to a tropical disease unit. Clin Infect Dis 20(3):542–548. https://doi.org/10.1093/clinids/20.3.542
Caumes E, Ly F, Bricaire F (2002) Cutaneous larva migrans with folliculitis: report of seven cases and review of the literature. Br J Dermatol 146(2):314–316. https://doi.org/10.1046/j.0007-0963.2001.04531.x
Cayce KA, Scott CM, Phillips CM, Frederick C, Park HK (2007) What is your diagnosis? Cutaneous larva migrans. Cutis 79(6):429, 435–426
Cello JP (1979) Eosinophilic gastroenteritis – a complex disease entity. Am J Med 67 (6):1097–1104. https://doi.org/10.1016/0002-9343(79)90652-1
Chaudhry AZ, Longworth DL (1989) Cutaneous manifestations of intestinal helminthic infections. Dermatol Clin 7(2):275–290
Choo J, Pang E, Prociv P (2000) Hookworms in dogs of Kuching, Sarawak (north Borneo). Trans R Soc Trop Med Hyg 94(1):21–22. https://doi.org/10.1016/s0035-9203(00)90425-5
Chowdhury AB, Schad GA (1972) *Ancylostoma ceylanicum*: a parasite of man in Calcutta and environs. Am J Trop Med Hyg 21(3):300–301
Chung C-S, Lin C-K, Su K-E, Liu C-Y, Lin C-C, Liang C-C, Lee T-H (2012) Diagnosis of *Ancylostoma ceylanicum* infestation by single-balloon enteroscopy (with video). Gastrointest Endosc 76(3):671–672. https://doi.org/10.1016/j.gie.2012.05.010
Coelho WMD, Amarante AFT, Apolinário JC, Coelho NMD, Bresciani KDS (2011) Occurrence of *Ancylostoma* in dogs, cats and public places from Andradina city, São Paulo state, Brazil. Rev Inst Med Trop Sao Paulo 53:181–184

Coello RD, Pazmiño BJ, Reyes EO, Rodríguez EX, Rodas EI, Rodas KA, Dávila AX, Rodas JP, Cedeño PP (2019) A case of cutaneous larva migrans in a child from Vinces, Ecuador. Am J Case Rep 20:1402–1406. https://doi.org/10.12659/AJCR.915154

Conlan JV, Sripa B, Attwood S, Newton PN (2011) A review of parasitic zoonoses in a changing Southeast Asia. Vet Parasitol 182(1):22–40

Conlan JV, Khamlome B, Vongxay K, Elliot A, Pallant L, Sripa B, Blacksell SD, Fenwick S, Thompson RC (2012) Soil-transmitted helminthiasis in Laos: a community-wide cross-sectional study of humans and dogs in a mass drug administration environment. Am J Trop Med Hyg 86 (4):624–634. https://doi.org/10.4269/ajtmh.2012.11-0413

Cortez R, Denny JP, Muci-Mendoza R, Ramirez G, Fuenmayor D, Jaffe GJ (2005) Diffuse unilateral subacute neuroretinitis in Venezuela. Ophthalmology 112(12):2110–2114. https://doi.org/10.1016/j.ophtha.2005.05.029

Craig TM, Hatfield TA, Pankavich JA, Wang GT (1992) Efficacy of moxidectin against an ivermectin-resistant strain of *Haemonchus contortus* in sheep. Vet Parasitol 41(3–4):329–333. https://doi.org/10.1016/0304-4017(92)90090-v

Croese TJ (1988) Eosinophilic enteritis – a recent north Queensland experience. Aust N Z J Med 18 (7):848–853. https://doi.org/10.1111/j.1445-5994.1988.tb01643.x

Croese J (1995) Seasonal influence on human enteric infection by *Ancylostoma caninum*. Am J Trop Med Hyg 53(2):158–161. https://doi.org/10.4269/ajtmh.1995.53.158

Croese J, Loukas A, Opdebeeck J, Prociv P (1994a) Occult enteric infection by *Ancylostoma caninum*: a previously unrecognized zoonosis. Gastroenterology 106(1):3–12. https://doi.org/10.1016/s0016-5085(94)93907-1

Croese J, Loukas A, Opdebeeck J, Fairley S, Prociv P (1994b) Human enteric infection with canine hookworms. Ann Intern Med 120(5):369–374. https://doi.org/10.7326/0003-4819-120-5-199403010-00003

Damante JH, Chinellato LE, Oliveira FT, Soares CT, Fleury RN (2011) Larva migrans in the oral mucosa: report of two cases. Braz Dent J 22(2):166–170. https://doi.org/10.1590/s0103-64402011000200014

Davies HD, Sakuls P, Keystone JS (1993) Creeping eruption. A review of clinical presentation and management of 60 cases presenting to a tropical disease unit. Arch Dermatol 129(5):588–591. https://doi.org/10.1001/archderm.129.5.588

de Souza EC, da Cunha SL, Gass JD (1992) Diffuse unilateral subacute neuroretinitis in South America. Arch Ophthalmol 110(9):1261–1263. https://doi.org/10.1001/archopht.1992.01080210079029

Del Giudice P, Desalvador F, Bernard E, Caumes E, Vandenbos F, Marty P, Le Fichoux Y, Dellamonica P (2002) Loeffler's syndrome and cutaneous larva migrans: a rare association. Br J Dermatol 147(2):386–388. https://doi.org/10.1046/j.1365-2133.2002.48102.x

Del Giudice P, Hakimi S, Vandenbos F, Magana C, Hubiche T (2019) Autochthonous cutaneous larva migrans in France and Europe. Acta Derm Venereol 99(9):805–808. https://doi.org/10.2340/00015555-3217

Del Valle A, Jones BF, Harrison LM, Chadderdon RC, Cappello M (2003) Isolation and molecular cloning of a secreted hookworm platelet inhibitor from adult *Ancylostoma caninum*. Mol Biochem Parasitol 129(2):167–177

Dhir L, O'Dempsey T, Watts MT (2010) Cutaneous larva migrans with optic disc edema: a case report. J Med Case Rep 4(1):209. https://doi.org/10.1186/1752-1947-4-209

Diba VC, Whitty CJM, Green T (2004) Cutaneous larva migrans acquired in Britain. Clin Exp Dermatol 29(5):555–556. https://doi.org/10.1111/j.1365-2230.2004.01592.x

Diemert D, Campbell D, Brelsford J, Leasure C, Li G, Peng J, Zumer M, Younes N, Bottazzi ME, Mejia R, Pritchard DI, Hawdon JM, Bethony JM (2018) Controlled human hookworm infection: accelerating human hookworm vaccine development. Open Forum Infect Dis 5(5):ofy083–ofy083. https://doi.org/10.1093/ofid/ofy083

Dissanaike AS, Ihalamulla RL, de Silva D, Pathirana S, Weerakoon U (2000) On a dead female hookworm, probably *Ancylostoma tubaeforme*, from the vitreous of a patient in Sri Lanka. Ceylon J Med Sci 43:25–30

Dourmishev AL, Dourmishev LA, Schwartz RA (2005) Ivermectin: pharmacology and application in dermatology. Int J Dermatol 44(12):981–988. https://doi.org/10.1111/j.1365-4632.2004.02253.x

Dove WE (1928) An intestinal infection of *Ancylostoma braziliense* in a boy and skin lesions produced with larvae from this strain. J Parasitol 15:136–137

Dove WE (1932) Further studies on *Ancylostoma braziliense* and the etiology of creeping eruption. Am J Epidemiol 15(3):664–711. https://doi.org/10.1093/oxfordjournals.aje.a117837

Drake J, Carey T (2019) Seasonality and changing prevalence of common canine gastrointestinal nematodes in the USA. Parasit Vectors 12(1):430–430. https://doi.org/10.1186/s13071-019-3701-7

Eiff JA (1966) Nature of an anticoagulant from the cephalic glands of *Ancylostoma caninum*. J Parasitol 52(5):833–843. https://doi.org/10.2307/3276520

Epe C (2009) Intestinal nematodes: biology and control. Vet Clin North Am Small Anim Pract 39 (6):1091–1107. https://doi.org/10.1016/j.cvsm.2009.07.002

Esser AC, Kantor I, Sapadin AN (1999) Souvenir from the Hamptons – a case of cutaneous larva migrans of six months' duration. Mt Sinai J Med 66(5–6):334–335

Ezzedine K, Pistone T (2013) Hookworm folliculitis. CMAJ 185(4):E213–E213. https://doi.org/10.1503/cmaj.120266

Falconer HS, Lea WA Jr (1961) Cutaneous larva migrans associated with Loeffler's syndrome, report of a case. Tex State J Med 57:373–376

Faria JG (1910) Contribution towards the classification of Brazilian Entozoa. III. *Ancylostoma braziliense* n. sp. Parasite of cats and dogs. Mem Inst Oswaldo Cruz 2(2):286–293

Flowers JR, Hammerberg B, Qurollo B (2015) Life cycle: *Ancylostoma caninum*. https://parasitology.cvm.ncsu.edu/life_cycles/nematodes/ancylcan.html

Fuller CE (1966) A common source outbreak of cutaneous larva migrans. Public Health Rep 81 (2):186–190

Furtado LFV, Dias LTO, Rodrigues TO, Silva VJD, Oliveira V, Rabelo EML (2020) Egg genotyping reveals the possibility of patent *Ancylostoma caninum* infection in human intestine. Sci Rep 10(1):3006. https://doi.org/10.1038/s41598-020-59874-8

Galanti B, Fusco FM, Nardiello S (2002) Outbreak of cutaneous larva migrans in Naples, southern Italy. Trans R Soc Trop Med Hyg 96(5):491–492. https://doi.org/10.1016/s0035-9203(02)90415-3

Gandullia E, Lignana E, Rabagliati AM, Penna R (1981) Visceral larva migrans caused by *Ancylostoma caninum* ("Larva migrans viscerale" determinata da *Ancylostoma caninum*). Minerva Pediatr 33(18):917–923

Gao YL, Liu ZH (2019) Cutaneous larva migrans with Loeffler's syndrome. Am J Trop Med Hyg 100(3):487–488. https://doi.org/10.4269/ajtmh.18-0406

Garcia CAA, Gomes AHB, Garcia Filho CAA, Vianna RNG (2004) Early-stage diffuse unilateral subacute neuroretinitis: improvement of vision after photocoagulation of the worm. Eye 18 (6):624–627. https://doi.org/10.1038/sj.eye.6700742

Garcia CAA, Gomes AHB, Vianna RNG, Souza Filho JP, Garcia Filho CAA, Oréfice F (2005) Late-stage diffuse unilateral subacute neuroretinitis: photocoagulation of the worm does not improve the visual acuity of affected patients. Int Ophthalmol 26(1–2):39–42. https://doi.org/10.1007/s10792-005-0078-8

Gass JD, Scelfo R (1978) Diffuse unilateral subacute neuroretinitis. J R Soc Med 71(2):95–111

Gass JD, Callanan DG, Bowman CB (1992) Oral therapy in diffuse unilateral subacute neuroretinitis. Arch Ophthalmol 110(5):675–680. https://doi.org/10.1001/archopht.1992.01080170097031

George S, Levecke B, Kattula D, Velusamy V, Roy S, Geldhof P, Sarkar R, Kang G (2016) Molecular identification of hookworm isolates in humans, dogs and soil in a tribal area in Tamil Nadu, India. PLoS Negl Trop Dis 10(8):e0004891. https://doi.org/10.1371/journal.pntd.0004891

Ghisi M, Kaminsky R, Maser P (2007) Phenotyping and genotyping of *Haemonchus contortus* isolates reveals a new putative candidate mutation for benzimidazole resistance in nematodes. Vet Parasitol 144(3–4):313–320. https://doi.org/10.1016/j.vetpar.2006.10.003
Ghosh SK, Bandyopadhyay D (2009) Dermacase: can you identify this condition? Cutaneous larva migrans. Can Fam Physician 55(5):489, 491
Gibbs HC (1986) Hypobiosis in parasitic nematodes-an update. Adv Parasit 25:129–174
Gloor RF, Breyley ER, Martinez IG (1970) Hookworm infection in a rural Kentucky county. Am J Trop Med Hyg 19(6):1007–1009
Guill MA, Odom RB (1978) Larva migrans complicated by Loeffler's syndrome. Arch Dermatol 114(10):1525–1526
Guimarães LC, Silva JH, Saad K, Lopes ER, Meneses AC (1999) Larva migrans within scalp sebaceous gland. Rev Soc Bras Med Trop 32(2):187–189. https://doi.org/10.1590/s0037-86821999000200011
Gutiérrez García-Rodrigo C, Tous Romero F, Zarco Olivo C (2017) Cutaneous larva migrans, welcome to a warmer Europe. J Eur Acad Dermatol Venereol 31(1):e33–e35. https://doi.org/10.1111/jdv.13621
Harder A, Holden-Dye L, Walker R, Wunderlich F (2005) Mechanisms of action of emodepside. Parasitol Res 97(Suppl 1):S1–S10. https://doi.org/10.1007/s00436-005-1438-z
Hawdon JM, Volk SW, Rose R, Pritchard DI, Behnke JM, Schad GA (1993) Observations on the feeding behaviour of parasitic third-stage hookworm larvae. Parasitology 106(Pt 2):163–169
Haydon GAM, Bearup AJ (1963) *Ancylostoma braziliense* and *A. ceylanicum*. Trans R Soc Trop Med Hyg 57(1):76. https://doi.org/10.1016/0035-9203(63)90015-4
Hellmann K, Knoppe T, Radeloff I, Heine J (2003) The anthelmintic efficacy and the safety of a combination of imidacloprid and moxidectin spot-on in cats and dogs under field conditions in Europe. Parasitol Res 90(Suppl 3):S142–S143. https://doi.org/10.1007/s00436-003-0919-1
Herrick CA (1928) A quantitative study of infections with *Ancylostoma caninum* in dogs. Am J Epidemiol 8(2):125–157. https://doi.org/10.1093/oxfordjournals.aje.a120994
Herrmann A, Christoph T, Sebastian G (2004) Larva migrans "saxoniae": larva migrans infection in Saxony (Larva migrans "saxoniae": Auftreten einer Infektion mit Larva migrans in Sachsen). J Dtsch Dermatol Ges 2(1):46–48. https://doi.org/10.1046/j.1439-0353.2004.03641.x
Hess LB, Millward LM, Rudinsky A, Vincent E, Marsh A (2019) Combination anthelmintic treatment for persistent *Ancylostoma caninum* ova shedding in greyhounds. J Am Anim Hosp Assoc 55(3):160–166. https://doi.org/10.5326/JAAHA-MS-6904
Heukelbach J, Feldmeier H (2008) Epidemiological and clinical characteristics of hookworm-related cutaneous larva migrans. Lancet Infect Dis 8(5):302–309. https://doi.org/10.1016/S1473-3099(08)70098-7
Heukelbach J, Wilcke T, Feldmeier H (2004) Cutaneous larva migrans (creeping eruption) in an urban slum in Brazil. Int J Dermatol 43(7):511–515. https://doi.org/10.1111/j.1365-4632.2004.02152.x
Heukelbach J, Jackson A, Ariza L, Feldmeier H (2008) Prevalence and risk factors of hookworm-related cutaneous larva migrans in a rural community in Brazil. Ann Trop Med Parasitol 102(1):53–61. https://doi.org/10.1179/136485908X252205
Hochedez P, Caumes E (2007) Hookworm-related cutaneous larva migrans. J Travel Med 14(5):326–333. https://doi.org/10.1111/j.1708-8305.2007.00148.x
Hodgkinson JE, Kaplan RM, Kenyon F, Morgan ER, Park AW, Paterson S, Babayan SA, Beesley NJ, Britton C, Chaudhry U, Doyle SR, Ezenwa VO, Fenton A, Howell SB, Laing R, Mable BK, Matthews L, McIntyre J, Milne CE, Morrison TA, Prentice JC, Sargison ND, Williams DJL, Wolstenholme AJ, Devaney E (2019) Refugia and anthelmintic resistance: concepts and challenges. Int J Parasitol Drugs Drug Resist 10:51–57. https://doi.org/10.1016/j.ijpddr.2019.05.001
Hopkins T, Gyr P (1991) Synergism of a combination of febantel and pyrantel embonate against *Ancylostoma caninum* on dogs. Vet Med Rev 61:3–9

Hopkins T, Gyr P, Schimmel A (1998) The effect of pyrantel embonate with oxantel embonate praziquantel, pyrantel embonate with febantel-praziquantel and milbemycin oxime on natural infestations of *Ancylostoma caninum* in dogs. Aust Vet Pract 28(2):53–56
Horton SH Jr (1949) Creeping eruption; report of a case with Loeffler's syndrome. US Nav Med Bull 49(4):703–706
Hotez PJ, Hawdon JM, Schad GA (1993) Hookworm larval infectivity, arrest, and amphiparatenesis: the *Caenorhabditis elegans* daf-c paradigm. Parasitol Today 9:23–26
Hughes J, Macdonald DW (2013) A review of the interactions between free-roaming domestic dogs and wildlife. Biol Conserv 157:341–351. https://doi.org/10.1016/j.biocon.2012.07.005
Inpankaew T, Schär F, Dalsgaard A, Khieu V, Chimnoi W, Chhoun C, Sok D, Marti H, Muth S, Odermatt P, Traub RJ (2014) High prevalence of *Ancylostoma ceylanicum* hookworm infections in humans, Cambodia, 2012. Emerg Infect Dis 20(6):976–982. https://doi.org/10.3201/eid2006.131770
Irwin PJ (2002) Companion animal parasitology: a clinical perspective. Int J Parasitol 32. https://doi.org/10.1016/s0020-7519(01)00361-7
Jackson R, Lance D, Townsend K, Stewart K (1987) Isolation of anthelmintic resistant *Ancylostoma caninum*. N Z Vet J 35(12):215–216. https://doi.org/10.1080/00480169./1987.35456
Jelinek T, Maiwald H, Nothdurft HD, Loscher T (1994) Cutaneous larva migrans in travelers: synopsis of histories, symptoms, and treatment of 98 patients. Clin Infect Dis 19(6):1062–1066. https://doi.org/10.1093/clinids/19.6.1062
Jiang D, Zhan B, Mayor RS, Gillespie P, Keegan B, Bottazzi ME, Hotez P (2011) Ac-AP-12, a novel factor Xa anticoagulant peptide from the esophageal glands of adult *Ancylostoma caninum*. Mol Biochem Parasitol 177(1):42–48. https://doi.org/10.1016/j.molbiopara.2011.01.008
Jimenez Castro PD, Howell SB, Schaefer JJ, Avramenko RW, Gilleard JS, Kaplan RM (2019) Multiple drug resistance in the canine hookworm *Ancylostoma caninum*: an emerging threat? Parasit Vectors 12(1):576. https://doi.org/10.1186/s13071-019-3828-6
Jiraanankul V, Aphijirawat W, Mungthin M, Khositnithikul R, Rangsin R, Traub RJ, Piyaraj P, Naaglor T, Taamasri P, Leelayoova S (2011) Incidence and risk factors of hookworm infection in a rural community of central Thailand. Am J Trop Med Hyg 84(4):594–598. https://doi.org/10.4269/ajtmh.2011.10-0189
Jones WB 2nd (1993) Cutaneous larva migrans. South Med J 86(11):1311–1313. https://doi.org/10.1097/00007611-199311000-00032
Jung B-K, Lee J-Y, Chang T, Song H, Chai J-Y (2020) Rare case of enteric *Ancylostoma caninum* hookworm infection, South Korea. Emerg Infect Dis. https://doi.org/10.3201/eid2601.191335
Kalmon EH (1954) Creeping eruption associated with transient pulmonary infiltrations. Radiology 62(2):222–226. https://doi.org/10.1148/62.2.222
Kang HM, Lee CS (2015) Diffuse unilateral subacute neuroretinitis in a healthy Korean male: the first case report in Korea. J Korean Med Sci 30(3):346–349. https://doi.org/10.3346/jkms.2015.30.3.346
Kaplan RM (2004) Drug resistance in nematodes of veterinary importance: a status report. Trends Parasitol 20(10):477–481
Kaplan RM (2020) Biology, epidemiology, diagnosis, and management of anthelmintic resistance in gastrointestinal nematodes of livestock. Vet Clin North Am Food Anim Pract 36(1):17–30. https://doi.org/10.1016/j.cvfa.2019.12.001
Kaplan RM, Vidyashankar AN, Howell SB, Neiss JM, Williamson LH, Terrill TH (2007) A novel approach for combining the use of in vitro and in vivo data to measure and detect emerging moxidectin resistance in gastrointestinal nematodes of goats. Int J Parasitol 37(7):795–804. https://doi.org/10.1016/j.ijpara.2007.01.001
Katz R, Ziegler J, Blank H (1965) The natural course of creeping eruption and treatment with thiabendazole. Arch Dermatol 91:420–424. https://doi.org/10.1001/archderm.1965.01600110006002

Kaur S, Jindal N, Sahu P, Jairath V, Jain V (2015) Creeping eruption on the move: a case series from Northern India. Indian J Dermatol 60(4):422–422. https://doi.org/10.4103/0019-5154.160530

Khoshoo V, Schantz P, Craver R, Stern GM, Loukas A, Prociv P (1994) Dog hookworm: a cause of eosinophilic enterocolitis in humans. J Pediatr Gastroenterol Nutr 19(4):448–452

Khoshoo V, Craver R, Schantz P, Loukas A, Prociv P (1995) Abdominal pain, pan-gut eosinophilia, and dog hookworm infection. J Pediatr Gastroenterol Nutr 21(4):481. https://doi.org/10.1097/00005176-199511000-00022

Kienast A, Bialek R, Hoeger PH (2007) Cutaneous larva migrans in northern Germany. Eur J Pediatr 166(11):1183–1185. https://doi.org/10.1007/s00431-006-0364-0

Kim TH, Lee BS, Sohn WM (2006) Three clinical cases of cutaneous larva migrans. Korean J Parasitol 44(2):145–149. https://doi.org/10.3347/kjp.2006.44.2.145

Kirby-Smith JL, Dove WE, White GF (1926) Creeping eruption. Arch Derm Syphilol 13 (2):137–175. https://doi.org/10.1001/archderm.1926.02370140003001

Kitchen S, Ratnappan R, Han S, Leasure C, Grill E, Iqbal Z, Granger O, O'Halloran DM, Hawdon JM (2019) Isolation and characterization of a naturally occurring multidrug-resistant strain of the canine hookworm, *Ancylostoma caninum*. Int J Parasitol 49:397–406. https://doi.org/10.1016/j.ijpara.2018.12.004

Klose C, Mravak S, Geb M, Bienzle U, Meyer CG (1996) Autochthonous cutaneous larva migrans in Germany. Trop Med Int Health 1(4):503–504. https://doi.org/10.1046/j.1365-3156.1996.d01-86.x

Kopp SR, Kotze AC, McCarthy JS, Coleman GT (2007) High-level pyrantel resistance in the hookworm *Ancylostoma caninum*. Vet Parasitol 143(3–4):299–304

Kopp SR, Kotze AC, McCarthy JS, Traub RJ, Coleman GT (2008a) Pyrantel in small animal medicine: 30 years on. Vet J 178(2):177–184. https://doi.org/10.1016/j.tvjl.2007.06.021

Kopp SR, Coleman GT, McCarthy JS, Kotze AC (2008b) Phenotypic characterization of two *Ancylostoma caninum* isolates with different susceptibilities to the anthelmintic pyrantel. Antimicrob Agents Chemother 52(11):3980–3986. https://doi.org/10.1128/AAC.00523-08

Kopp SR, Coleman GT, McCarthy JS, Kotze AC (2008c) Application of in vitro anthelmintic sensitivity assays to canine parasitology: detecting resistance to pyrantel in *Ancylostoma caninum*. Vet Parasitol 152(3-4):284–293

Kopp SR, Coleman GT, Traub RJ, McCarthy JS, Kotze AC (2009) Acetylcholine receptor subunit genes from *Ancylostoma caninum*: altered transcription patterns associated with pyrantel resistance. Int J Parasitol 39(4):435–441. https://doi.org/10.1016/j.ijpara.2008.08.005

Kwa MS, Veenstra JG, Roos MH (1994) Benzimidazole resistance in *Haemonchus contortus* is correlated with a conserved mutation at amino acid 200 in beta-tubulin isotype 1. Mol Biochem Parasitol 63(2):299–303

Kwiatkowska-Kawecka Z (1973) *Ancyclostoma caninum* (Ercolani) larva in eye's anterior segment (Larwa Ancylostomatis canini (Ercolani) w przednim odcinku oka). Klin Oczna 43(3):329–330

Kwon I-H, Kim H-S, Lee J-H, Choi M-H, Chai J-Y, Nakamura-Uchiyama F, Nawa Y, Cho K-H (2003) A serologically diagnosed human case of cutaneous larva migrans caused by *Ancylostoma caninum*. Korean J Parasitol 41(4):233–237. https://doi.org/10.3347/kjp.2003.41.4.233

Labarthe N, Serrão ML, Ferreira AMR, Almeida NKO, Guerrero J (2004) A survey of gastrointestinal helminths in cats of the metropolitan region of Rio de Janeiro, Brazil. Vet Parasitol 123 (1):133–139. https://doi.org/10.1016/j.vetpar.2004.06.002

Lacey E (1988) The role of the cytoskeletal protein, tubulin, in the mode of action and mechanism of drug resistance to benzimidazoles. Inter J Parasitol 18(7):885–936

Laing R, Gillan V, Devaney E (2017) Ivermectin – old drug, new tricks? Trends Parasitol 33 (6):463–472. https://doi.org/10.1016/j.pt.2017.02.004

Landmann JK, Prociv P (2003) Experimental human infection with the dog hookworm, *Ancylostoma caninum*. Med J Aust 178(2):69–71

Lane HC (1913) *Ancylostoma ceylanicum*, a new human parasite. Indian Med Gaz 48:216–218

Le Joncour A, Lacour SA, Lecso G, Regnier S, Guillot J, Caumes E (2012) Molecular characterization of *Ancylostoma braziliense* larvae in a patient with hookworm-related cutaneous larva migrans. Am J Trop Med Hyg 86(5):843–845. https://doi.org/10.4269/ajtmh.2012.11-0734

Lee KT, Little MD, Beaver PC (1975) Intracellular (muscle-fiber) habitat of *Ancylostoma caninum* in some mammalian hosts. J Parasitol 61(4):589–598

Li D, He Q, Kang T, Yin H, Jin X, Li H, Gan W, Yang C, Hu J, Wu Y, Peng L (2010) Identification of an anticoagulant peptide that inhibits both fXIa and fVIIa/tissue factor from the blood-feeding nematode *Ancylostoma caninum*. Biochem Biophys Res Commun 392(2):155–159. https://doi.org/10.1016/j.bbrc.2009.12.177

Liotta JL, Youn H, Aksel S, Bienhoff SE, Bowman DD (2012) Prevalence of *Ancylostoma braziliense* in dogs from Alachua and Marion Counties, Florida, United States. J Parasitol 98(5):1039–1040. https://doi.org/10.1645/GE-2927.1

Little MD (1961) Observations on the possible role of insects as paratenic hosts for *Ancylostoma caninum*. J Parasitol 47(2):263–267

Little MD, Halsey NA, Cline BL, Katz SP (1983) *Ancylostoma* larva in a muscle fiber of man following cutaneous larva migrans. Am J Trop Med Hyg 32(6):1285–1288

Little SE, Johnson EM, Lewis D, Jaklitsch RP, Payton ME, Blagburn BL, Bowman DD, Moroff S, Tams T, Rich L, Aucoin D (2009) Prevalence of intestinal parasites in pet dogs in the United States. Vet Parasitol 166(1):144–152. https://doi.org/10.1016/j.vetpar.2009.07.044

Liu Y, Zheng G, Alsarakibi M, Zhang X, Hu W, Lu P, Lin L, Tan L, Luo Q, Li G (2013) Molecular identification of *Ancylostoma caninum* isolated from cats in southern China based on complete ITS sequence. Biomed Res Int 2013:868050. https://doi.org/10.1155/2013/868050

Liu YJ, Zheng GC, Zhang P, Alsarakibi M, Zhang XH, Li YW, Liu T, Ren SN, Chen ZX, Liu YL, Li SJ, Li GQ (2015) Molecular identification of hookworms in stray and shelter dogs from Guangzhou city, China using ITS sequences. J Helminthol 89(2):196–202. https://doi.org/10.1017/S0022149X13000783

Lockmann A, Seitz CS, Schön MP, Mößner R (2018) Creeping eruption and eosinophilic folliculitis: atypical cutaneous larva migrans. J Dtsch Dermatol Ges 16(2):202–204. https://doi.org/10.1111/ddg.13414

Looss A (1911) The anatomy and life history of *Agchlostoma duodenale*, Dubini. Records of the School of Medicine. Egypt Min Educ Cairo 4:531–536

Lopes MA, Zaia AA, de Almeida OP, Scully C (1994) Larva migrans that affect the mouth. Oral Surg Oral Med Oral Pathol 77(4):362–367. https://doi.org/10.1016/0030-4220(94)90198-8

Loukas A, Croese J, Opdebeeck J, Prociv P (1992) Detection of antibodies to secretions of *Ancylostoma caninum* in human eosinophilic enteritis. Trans R Soc Trop Med Hyg 86(6):650–653. https://doi.org/10.1016/0035-9203(92)90175-c

Loukas A, Opdebeeck J, Croese J, Prociv P (1994) Immunologic incrimination of *Ancylostoma caninum* as a human enteric pathogen. Am J Trop Med Hyg 50(1):69–77. https://doi.org/10.4269/ajtmh.1994.50.69

Loukas A, Opdebeeck J, Croese J, Prociv P (1996) Immunoglobulin G subclass antibodies against excretory/secretory antigens of *Ancylostoma caninum* in human enteric infections. Am J Trop Med Hyg 54(6):672–676

Loukas A, Hotez PJ, Diemert D, Yazdanbakhsh M, McCarthy JS, Correa-Oliveira R, Croese J, Bethony JM (2016) Hookworm infection. Nat Rev Dis Primers 2:16088. https://doi.org/10.1038/nrdp.2016.88

Mackenstedt U, Jenkins D, Romig T (2015) The role of wildlife in the transmission of parasitic zoonoses in peri-urban and urban areas. Int J Parasitol Parasit Wildl 4(1):71–79. https://doi.org/10.1016/j.ijppaw.2015.01.006

Magri F, Chello C, Pranteda G, Pranteda G (2019) Complete resolution of cutaneous larva migrans with topical ivermectin: a case report. Dermatol Ther 32(3):e12845–e12845. https://doi.org/10.1111/dth.12845

Malgor R, Oku Y, Gallardo R, Yarzabal I (1996) High prevalence of *Ancylostoma* spp. infection in dogs, associated with endemic focus of human cutaneous larva migrans, in Tacuarembo, Uruguay. Parasite 3(2):131–134. https://doi.org/10.1051/parasite/1996032131

Malvy D, Ezzedine K, Pistone T, Receveur MC, Longy-Boursier M (2006) Extensive cutaneous larva migrans with folliculitis mimicking multimetameric herpes zoster presentation in an adult traveler returning from Thailand. J Travel Med 13(4):244–247. https://doi.org/10.1111/j.1708-8305.2006.00040.x

Manning L, Chambers S, Paltridge G, Maurice P (2006) Cutaneous larva migrans (hookworm) acquired in Christchurch, New Zealand. N Z Med J 119(1231):U1910

Margono SS, Koesharjono C, Kosin E (1979) Hookworm in dogs and cats in the area of Jakarta. Trop Geogr Med 31(2):257–261

Martin LK (1972) Hookworm in Georgia. II. Survey of intestinal helminth infections in members of rural households of southeastern Georgia. Am J Trop Med Hyg 21(4):930–943

Martin RJ, Clark CL, Trailovic SM, Robertson AP (2004) Oxantel is an N-type (methyridine and nicotine) agonist not an L-type (levamisole and pyrantel) agonist: classification of cholinergic anthelmintics in *Ascaris*. Int J Parasitol 34(9):1083–1090. https://doi.org/10.1016/j.ijpara.2004.04.014

Martin RJ, Robertson AP, Buxton SK, Beech RN, Charvet CL, Neveu C (2012) Levamisole receptors: a second awakening. Trends Parasitol 28(7):289–296. https://doi.org/10.1016/j.pt.2012.04.003

Matsusaki G (1951) Studies on the life history of the hookworm. Part VII: on the development of *Ancylostoma caninum* in the abnormal host. Yok Med Bull 2(3):154–160

Mazzeo T, Dos Santos Motta MM, Curi ALL (2019) Diffuse unilateral subacute neuroretinitis: review article. J Ophthalmic Inflamm Infect 9(1):23. https://doi.org/10.1186/s12348-019-0191-x

McCoy OR (1931) The egg production of two physiological strains of the dog hookworm, *Ancylostoma caninum*. Am J Epidemiol 14(1):194–202. https://doi.org/10.1093/oxfordjournals.aje.a117756

McKenna ML, McAtee S, Bryan PE, Jeun R, Ward T, Kraus J, Bottazzi ME, Hotez PJ, Flowers CC, Mejia R (2017) Human intestinal parasite burden and poor sanitation in rural Alabama. Am J Trop Med Hyg 97(5):1623–1628. https://doi.org/10.4269/ajtmh.17-0396

Merino-Tejedor A, Nejsum P, Mkupasi EM, Johansen MV, Olsen A (2019) Molecular identification of zoonotic hookworm species in dog faeces from Tanzania. J Helminthol 93(3):313–318. https://doi.org/10.1017/S0022149X18000263

Mieszczanek J, Harrison LM, Vlasuk GP, Cappello M (2004) Anticoagulant peptides from *Ancylostoma caninum* are immunologically distinct and localize to separate structures within the adult hookworm. Mol Biochem Parasitol 133(2):319–323

Miller TA (1966) Blood loss during hookworm infection, determined by erythrocyte labeling with radioactive ^{51}Chromium. II. Pathogenesis of *Ancylostoma braziliense* infection in dogs and cats. J Parasitol 52(5):856–865. https://doi.org/10.2307/3276523

Miller TA (1968) Pathogenesis and immunity in hookworm infection. Trans R Soc Trop Med Hyg 62(4):473–485

Miller TA (1970) Potential transport hosts in the life cycle of canine and feline hookworms. J Parasitol 56(Suppl 4):238

Miller AC, Walker J, Jaworski R, de Launey W, Paver R (1991) Hookworm folliculitis. Arch Dermatol 127(4):547–549

Minnaar WN, Krecek RC (2001) Helminths in dogs belonging to people in a resource-limited urban community in Gauteng, South Africa. Onderstepoort J Vet Res 68(2):111–117

Minnaar WN, Krecek RC, Fourie LJ (2002) Helminths in dogs from a peri-urban resource-limited community in Free State Province, South Africa. Vet Parasitol 107(4):343–349. https://doi.org/10.1016/S0304-4017(02)00155-3

Miro G, Mateo M, Montoya A, Vela E, Calonge R (2007) Survey of intestinal parasites in stray dogs in the Madrid area and comparison of the efficacy of three anthelmintics in naturally infected dogs. Parasitol Res 100(2):317–320. https://doi.org/10.1007/s00436-006-0258-0

Moraes LR, Cialdini AP, Avila MP, Elsner AE (2002) Identifying live nematodes in diffuse unilateral subacute neuroretinitis by using the scanning laser ophthalmoscope. Arch Ophthalmol 120(2):135–138. https://doi.org/10.1001/archopht.120.2.135

Muhleisen JP (1953) Demonstration of pulmonary migration of the causative organism of creeping eruption. Ann Intern Med 38(3):595–600. https://doi.org/10.7326/0003-4819-38-3-595

Mulinge E, Njenga SM, Odongo D, Magambo J, Zeyhle E, Mbae C, Kagendo D, Kanyi H, Traub RJ, Wassermann M, Kern P, Romig T (2020) Molecular identification of zoonotic hookworms in dogs from four counties of Kenya. J Helminthol 94:e43. https://doi.org/10.1017/S0022149X1900018X

Müller-Stöver I, Richter J, Häussinger D (2010) Cutaneous larva migrans (creeping eruption) acquired in Germany (In Deutschland erworbene Larva migrans cutanea). Dtsch Med Wochenschr 135(17):859–861. https://doi.org/10.1055/s-0030-1253669

Munjal A, Al-Sabban A, Bull-Henry K (2017) Eosinophilic enteritis: a delayed diagnosis. J Investig Med High Impact Case Rep 5(4):2324709617734246. https://doi.org/10.1177/2324709617734246

Murphy MD, Spickler AR (2013) Zoonotic hookworms. http://www.cfsph.iastate.edu/DiseaseInfo/factsheets.php

Nadbath RP, Lawlor PP (1965) Nematode (*Ancylostoma*) in the cornea; a case report. Am J Ophthalmol 59:486–490. https://doi.org/10.1016/0002-9394(65)93754-2

Nawalinski TA, Schad GA (1974) Arrested development in *Ancylostoma duodenale:* course of a self-induced infection in man. Am J Trop Med Hyg 23(5):895–898. https://doi.org/10.4269/ajtmh.1974.23.895

Neveu C, Charvet CL, Fauvin A, Cortet J, Beech RN, Cabaret J (2010) Genetic diversity of levamisole receptor subunits in parasitic nematode species and abbreviated transcripts associated with resistance. Pharmacogenet Genomics 20(7):414–425. https://doi.org/10.1097/FPC.0b013e328338ac8c

Ngcamphalala PI, Lamb J, Mukaratirwa S (2019) Molecular identification of hookworm isolates from stray dogs, humans and selected wildlife from South Africa. J Helminthol 94:e39. https://doi.org/10.1017/S0022149X19000130

Ng-Nguyen D, Hii SF, Nguyen VA, Van Nguyen T, Van Nguyen D, Traub RJ (2015) Re-evaluation of the species of hookworms infecting dogs in Central Vietnam. Parasit Vectors 8:401. https://doi.org/10.1186/s13071-015-1015-y

Ngui R, Lim YA, Traub R, Mahmud R, Mistam MS (2012) Epidemiological and genetic data supporting the transmission of *Ancylostoma ceylanicum* among human and domestic animals. PLoS Negl Trop Dis 6(2):e1522. https://doi.org/10.1371/journal.pntd.0001522

Nichols RL (1956) The etiology of visceral larva migrans. II. Comparative larval morphology of *Ascaris lumbricoides, Necator americanus, Strongyloides stercoralis* and *Ancylostoma caninum.* J Parasitol 42(4 Section 1):363–399

Okulewicz A (2017) The impact of global climate change on the spread of parasitic nematodes. Ann Parasitol 63(1):15–20. https://doi.org/10.17420/ap6301.79

Oliveira-Arbex AP, David EB, Oliveira-Sequeira TC, Katagiri S, Coradi ST, Guimaraes S (2017) Molecular identification of *Ancylostoma* species from dogs and an assessment of zoonotic risk in low-income households, Sao Paulo State, Brazil. J Helminthol 91(1):14–19. https://doi.org/10.1017/S0022149X15001145

Oosthuizen WT, Erasmus JB (1993) Efficacy of moxidectin against a strain of *Haemonchus contortus* resistant to ivermectin, a benzimidazole and a salicylanilide. J S Afr Vet Assoc 64(1):9–12

Otranto D, Dantas-Torres F, Mihalca AD, Traub RJ, Lappin M, Baneth G (2017) Zoonotic parasites of sheltered and stray dogs in the era of the global economic and political crisis. Trends Parasitol 33(10):813–825. https://doi.org/10.1016/j.pt.2017.05.013

Pa Pa Aung W, Htoon TT, Tin HH, Sanpool O, Jongthawin J, Sadaow L, Phosuk I, Ropai R, Intapan PM, Maleewong W (2017) First molecular identifications of *Necator americanus* and *Ancylostoma ceylanicum* infecting rural communities in lower Myanmar. Am J Trop Med Hyg 96(1):214–216. https://doi.org/10.4269/ajtmh.16-0610

Panés-Rodríguez A, Piera-Tuneu L, López-Pestaña A, Ormaetxea-Pérez N, Gutiérrez-Támara P, Ibarbia-Oruezabal S, Tuneu-Valls A (2016) Autochthonous cutaneous larva migrans infection in Guipúzcoa (Larva migrans cutánea de origen autóctono en Guipúzcoa). Actas Dermosifiliogr 107(5):407–413. https://doi.org/10.1016/j.ad.2016.01.002

Pankavich JA, Berger H, Simkins KL (1992) Efficacy of moxidectin, nemadectin and ivermectin against an ivermectin-resistant strain of *Haemonchus contortus* in sheep. Vet Rec 130 (12):241–243. https://doi.org/10.1136/vr.130.12.241

Pasricha J (2011) A rash and cough in a traveler. Clin Infect Dis 53(2):167–206. https://doi.org/10.1093/cid/cir341

Pasyk K (1978) Dermatitis rhabditidosa in an 11-year-old girl: a new cutaneous parasitic disease of man. Br J Dermatol 98(1):107–112. https://doi.org/10.1111/j.1365-2133.1978.tb07340.x

Patterson CRS, Kersey PJW (2003) Cutaneous larva migrans acquired in England. Clin Exp Dermatol 28(6):671–672. https://doi.org/10.1046/j.1365-2230.2003.01347.x

Podder I, Chandra S, Gharami RC (2016) Loeffler's syndrome following cutaneous larva migrans: an uncommon sequel. Indian J Dermatol 61(2):190–192. https://doi.org/10.4103/0019-5154.177753

Poppert S, Heideking M, Agostini H, Fritzenwanker M, Wuppenhorst N, Muntau B, Henneke P, Kern W, Krucken J, Junker B, Hufnagel M (2017) Diffuse unilateral subacute neuroretinitis caused by *Ancylostoma* Hookworm. Emerg Infect Dis 23(2):343–344. https://doi.org/10.3201/eid2302.142064

Prichard RK (1990) Anthelmintic resistance in nematodes: extent, recent understanding and future directions for control and research. Int J Parasitol 20(4):515–523. https://doi.org/10.1016/0020-7519(90)90199-W

Prickett KA, Ferringer TC (2015) What's eating you? Cutaneous larva migrans. Cutis 95 (3):126–128

Prieto Fernández R, Villegas Zulueta A, Orta Huerta T (1978) First finding of the *Ancylostoma braziliense* nematode (Gómez de Faria, 1910) in *Canis familiaris* in Cuba (Primer hallazgo del nematodo Ancylostoma braziliense (Gómez de Faria, 1910) en Canis familiaris de Cuba). Rev Cubana Med Trop 30(1):39–44

Prociv P, Croese J (1990) Human eosinophilic enteritis caused by dog hookworm *Ancylostoma caninum*. Lancet 335. https://doi.org/10.1016/0140-6736(90)91186-e

Prociv P, Croese J (1996) Human enteric infection with *Ancylostoma caninum*: hookworms reappraised in the light of a "new" zoonosis. Acta Trop 62. https://doi.org/10.1016/s0001-706x(96)00016-2

Prociv P, Collyer R, Lim K, Tang R (1994) Brisbane dogs as a reservoir of zoonotic hookworms. Med J Aust 161(2):172–173

Purdy KS, Langley RG, Webb AN, Walsh N, Haldane D (2011) Cutaneous larva migrans. Lancet 377(9781):1948. https://doi.org/10.1016/S0140-6736(10)61149-X

Ray DK, Bhopale KK (1972) Complete development of *Ancylostoma ceylanicum* (Looss, 1911) in golden hamsters, *Mesocricetus auratus*. Experientia 28:359–360

Ray DK, Bhopale KK, Shrivastava VB (1972) Migration and growth of *Ancylostoma ceylanicum* in golden hamsters *Mesocricetus auratus*. J Helminth 46(4):357–362

Ray DK, Bhopale KK, Shrivastava VB (1975) Development of *Ancylostoma ceylanicum* Looss, 1911 (hamster strain) in the albino mouse, *Mus musculus*, with and without cortisone. Parasitology 71(2):193–197

Reichert F, Pilger D, Schuster A, Lesshafft H, Guedes de Oliveira S, Ignatius R, Feldmeier H (2018) Epidemiology and morbidity of hookworm-related cutaneous larva migrans (HrCLM): results of a cohort study over a period of six months in a resource-poor community in Manaus, Brazil. PLoS Negl Trop Dis 12(7):e0006662. https://doi.org/10.1371/journal.pntd.0006662

Rep BH (1963) On the polyxenia of *Ancylostomidae* and the validity of the characters used for their differentiation (II). Trop Geogr Med 12:271–316

Rep BH (1975) Intestinal helminths in dogs and cats on the Antillian Islands Aruba, Curaçao and Bonaire. Trop Geogr Med 27(3):317–323

Rep BH (1980) Pathogenicity of hookworms. The significance of population regression for the pathogenicity of hookworms. Trop Geogr Med 32(3):251–255
Rep BH, Heinemann DW (1976) Changes in hookworm distribution in Surinam. Trop Geogr Med 28(2):104–110
Rep BH, Vetter JC, Eijsker M (1968) Cross breeding experiments in *Ancylostoma braziliense* De Faria, 1910 and *A. ceylanicum* Looss, 1911. Trop Geogr Med 20(4):367–378
Richey TK, Gentry RH, Fitzpatrick JE, Morgan AM (1996) Persistent cutaneous larva migrans due to *Ancylostoma* species. South Med J 89(6):609–611. https://doi.org/10.1097/00007611-199606000-00010
Ridley RK, Dryden MW, Gabbert NH, Schoning P (1994) Epidemiology and control of helminth parasites in Greyhound breeding farms. Comp Cont Educ Pract Vet 16(5):585–599
Rivera-Roig V, Sanchez JL, Hillyer GV (2008) Hookworm folliculitis. Int J Dermatol 47 (3):246–248. https://doi.org/10.1111/j.1365-4632.2008.03469.x
Robert MG, Faisant A, Cognet O, Rabodonirina M, Peyron F, Piquemal M, Mazet R, Pelloux H, Brenier-Pinchart MP (2019) Autochthonous and persistent cutaneous larva migrans in an infant successfully treated by topic albendazole ointment. J Eur Acad Dermatol Venereol 33(4):e163–e164. https://doi.org/10.1111/jdv.15356
Robertson SJ, Pennington AJ, Evans AM, Martin RJ (1994) The action of pyrantel as an agonist and an open channel blocker at acetylcholine receptors in isolated *Ascaris suum* muscle vesicles. Eur J Pharmacol 271(2–3):273–282. https://doi.org/10.1016/0014-2999(94)90784-6
Robertson AP, Bjorn HE, Martin RJ (1999) Resistance to levamisole resolved at the single-channel level. FASEB J 13(6):749–760. https://doi.org/10.1096/fasebj.13.6.749
Robertson AP, Bjorn HE, Martin RJ (2000) Pyrantel resistance alters nematode nicotinic acetylcholine receptor single-channel properties. Eur J Pharmacol 394(1):1–8. https://doi.org/10.1016/s0014-2999(00)00135-7
Roche M (1966) Influence of male and female *Ancylostoma caninum* on each other's distribution in the intestine of the dog. Exp Parasitol 19(3):327–331. https://doi.org/10.1016/0014-4894(66)90083-x
Roche M, Martinez-Torres C, Macpherson L (1962) Electroesophagogram of individual hookworm (*Ancylostoma caninum*). Science 136(3511):148–150. https://doi.org/10.1126/science.136.3511.148
Roest MA, Ratnavel R (2001) Cutaneous larva migrans contracted in England: a reminder. Clin Exp Dermatol 26(5):389–390. https://doi.org/10.1046/j.1365-2230.2001.00841.x
Rosh A, McStay CM (2012) Cutaneous larva migrans. J Emerg Med 43(2):e135–e136. https://doi.org/10.1016/j.jemermed.2010.11.019
Sandground JH (1939) Creeping eruption in the Netherland East Indies caused by the invasion of the larva of *Ancylostoma braziliense*. Ned Tijdschr Geneeskd 79(13):805–810
Sato M, Sanguankiat S, Yoonuan T, Pongvongsa T, Keomoungkhoun M, Phimmayoi I, Boupa B, Moji K, Waikagul J (2010) Copro-molecular identification of infections with hookworm eggs in rural Lao PDR. Trans R Soc Trop Med Hyg 104(9):617–622. https://doi.org/10.1016/j.trstmh.2010.06.006
Schad GA (1979) *Ancylostoma duodenale*: maintenance through six generations in helminth-naive pups. Exp Parasitol 47(2):246–253
Schad GA (1982) Arrested development of *Ancylostoma caninum* in dogs: influence of photoperiod and temperature on induction of a potential to arrest. In: Meerovitch E (ed) Aspects of parasitology: a festschrift dedicated to the fiftieth anniversary of the Institute of Parasitology of McGill University. McGill University, Montreal, pp 361–391
Schad GA (1990) Hypobiosis and related phenomena in hookworm infection. In: Schad GA, Warren KS (eds) Hookworm disease. Current status and new directions. Taylor and Francis, London, pp 71–88
Schad GA, Page MR (1982) *Ancylostoma caninu* : adult worm removal, corticosteroid treatment, and resumed development of arrested larvae in dogs. Exp Parasitol 54:303–309

Schad GA, Chowdhury AB, Dean CG, Kochar VK, Nawalinski TA, Thomas J, Tonascia JA (1973) Arrested development in human hookworm infections: an adaptation to a seasonally unfavorable external environment. Science 180(4085):502–504. https://doi.org/10.1126/science.180.4085.502

Schad GA, Murrell KD, Fayer R, El Naggar HMS, Page MR, Parrish PK, Stewart TB (1984) Paratenesis in *Ancylostoma duodenale* suggests possible meat-borne human infection. Trans R Soc Trop Med Hyg 78:203–204

Schimmel A, Altreuther G, Schroeder I, Charles S, Cruthers L, Ketzis J, Kok DJ, Kraemer F, McCall JW, Krieger KJ (2009) Efficacy of emodepside plus praziquantel tablets (Profender tablets for dogs) against mature and immature adult *Ancylostoma caninum* and *Uncinaria stenocephala* infections in dogs. Parasitol Res 105. https://doi.org/10.1007/s00436-009-1490-1

Schmid K, Rohdich N, Zschiesche E, Kok DJ, Allan MJ (2010) Efficacy, safety and palatability of a new broad-spectrum anthelmintic formulation in dogs. Vet Rec 167(17):647–651. https://doi.org/10.1136/vr.c4661

Schwenkenbecher JM, Albonico M, Bickle Q, Kaplan RM (2007) Characterization of beta-tubulin genes in hookworms and investigation of resistance-associated mutations using real-time PCR. Mol Biochem Parasitol 156(2):167–174

Setasuban P, Vajrasthira S, Muennoo C (1976) Prevalence and zoonotic potential of *Ancylostoma ceylanicum* in cats in Thailand. Southeast Asian J Trop Med Public Health 7(4):534–539

Shelmire B (1928) Experimental creeping eruption from a cat and dog hookworm (*A. braziliense*). J Am Med Assoc 91(13):938–944. https://doi.org/10.1001/jama.1928.02700130016005

Shoop WL (1991) Vertical transmission of helminths: hypobiosis and amphiparatenesis. Parasitol Today 7(2):51–54

Siddalingappa K, Murthy SC, Herakal K, Kusuma MR (2015) Cutaneous larva migrans in early infancy. Indian J Dermatol 60(5):522–522. https://doi.org/10.4103/0019-5154.164436

Silvestre A, Cabaret J (2002) Mutation in position 167 of isotype 1 beta-tubulin gene of Trichostrongylid nematodes: role in benzimidazole resistance? Mol Biochem Parasitol 120 (2):297–300

Smout FA, Thompson RC, Skerratt LF (2013) First report of *Ancylostoma ceylanicum* in wild canids. Int J Parasitol Parasit Wildl 2:173–177. https://doi.org/10.1016/j.ijppaw.2013.04.003

Smout FA, Skerratt LF, Butler JRA, Johnson CN, Congdon BC, Thompson RCA (2017) The hookworm *Ancylostoma ceylanicum*: an emerging public health risk in Australian tropical rainforests and Indigenous communities. One Health 3:66–69. https://doi.org/10.1016/j.onehlt.2017.04.002

Soriano LF, Piansay-Soriano ME (2017) Treatment of cutaneous larva migrans with a single session of carbon dioxide laser: a study of ten cases in the Philippines. J Cosmet Dermatol 16(1):91–94. https://doi.org/10.1111/jocd.12296

Souza EC, Casella AM, Nakashima Y, Monteiro ML (2005) Clinical features and outcomes of patients with diffuse unilateral subacute neuroretinitis treated with oral albendazole. Am J Ophthalmol 140(3):437–445. https://doi.org/10.1016/j.ajo.2005.03.065

Sow D, Soro F, Javelle E, Simon F, Parola P, Gautret P (2017) Epidemiological profile of cutaneous larva migrans in travelers returning to France between 2003 and 2015. Travel Med Infect Dis 20:61–64. https://doi.org/10.1016/j.tmaid.2017.06.003

Sowemimo OA, Asaolu SO (2008) The daily egg production of *Ancylostoma caninum* and the distribution of the worm along the digestive tract of the dog. Res J Parasitol 3(3):92–97. https://doi.org/10.3923/jp.2008.92.97

Starr MC, Montgomery SP (2011) Soil-transmitted Helminthiasis in the United States: a systematic review – 1940-2010. Am J Trop Med Hyg 85(4):680–684. https://doi.org/10.4269/ajtmh.2011.11-0214

Stassens P, Bergum PW, Gansemans Y, Jespers L, Laroche Y, Huang S, Maki S, Messens J, Lauwereys M, Cappello M, Hotez PJ, Lasters I, Vlasuk GP (1996) Anticoagulant repertoire of the hookworm *Ancylostoma caninum*. Proc Natl Acad Sci USA 93(5):2149–2154. https://doi.org/10.1073/pnas.93.5.2149

Stone WM, Girardeau MH (1960) *Ancylostoma caninum* larvae present in the colostrum of a bitch. Vet Rec 79(24):773–774
Stone WM, Peckham JC (1970) Infectivity of *Ancylostoma caninum* larvae from canine milk. Am J Vet Res 31(9):1693–1694
Stoye N (1973) Untersuchungen uber die moglichkeit pranataler und galaktogener infektionen mit *Ancylostoma caninum* Ercolani 1859 (Ancylostomidae) beim hund. Zbl Vet Med B 20:1–29
Stoye M, Krause J (1976) Versuche zur reaktivierung inhibierter larven von *Ancylostoma caninum*: die wirkung von oestradiol und progesteron. Zbl Vet Med B 23:822–839
Talley NJ, Shorter RG, Phillips SF, Zinsmeister AR (1990) Eosinophilic gastroenteritis: a clinicopathological study of patients with disease of the mucosa, muscle layer, and subserosal tissues. Gut 31(1):54–58. https://doi.org/10.1136/gut.31.1.54
Tamminga N, Bierman WFW, de Vries PJ (2009) Cutaneous larva migrans acquired in Brittany, France. Emerg Infect Dis 15(11):1856–1858. https://doi.org/10.3201/eid1511.090261
Tan SK, Liu TT (2010) Cutaneous larva migrans complicated by Löffler syndrome. Arch Dermatol 146(2):210–212. https://doi.org/10.1001/archdermatol.2009.392
Te Booij M, de Jong EMGJ, Bovenschen HJ (2010) Löffler syndrome caused by extensive cutaneous larva migrans: a case report and review of the literature. Dermatol Online J 16 (10):2–2
Thompson RC, Roberts MG (2001) Does pet helminth prophylaxis increase the rate of selection for drug resistance? Trends Parasitol 17(12):576–578. https://doi.org/10.1016/s1471-4922(01) 02084-0
Traub RJ (2013) *Ancylostoma ceylanicum*, a re-emerging but neglected parasitic zoonosis. Int J Parasitol 43(12–13):1009–1015. https://doi.org/10.1016/j.ijpara.2013.07.006
Traub RJ, Robertson ID, Irwin P, Mencke N, Thompson RCA (2004) Application of a species-specific PCR-RFLP to identify *Ancylostoma* eggs directly from canine faeces. Vet Parasitol 123 (3–4):245–255
Traub RJ, Robertson ID, Irwin PJ, Mencke N, Thompson RC (2005) Canine gastrointestinal parasitic zoonoses in India. Trends Parasitol 21(1):42–48. https://doi.org/10.1016/j.pt.2004.10.011
Traub RJ, Inpankaew T, Sutthikornchai C, Sukthana Y, Thompson RC (2008) PCR-based coprodiagnostic tools reveal dogs as reservoirs of zoonotic ancylostomiasis caused by *Ancylostoma ceylanicum* in temple communities in Bangkok. Vet Parasitol 155(1–2):67–73
Traversa D (2012) Pet roundworms and hookworms: a continuing need for global worming. Parasit Vectors 5(1):91. https://doi.org/10.1186/1756-3305-5-91
Tremblay A, MacLean JD, Gyorkos T, Macpherson DW (2000) Outbreak of cutaneous larva migrans in a group of travellers. Trop Med Int Health 5(5):330–334. https://doi.org/10.1046/j.1365-3156.2000.00557.x
Triantafillidis J, Parasi A, Sklavaina M (2002) Eosinophilic gastroenteritis: current aspects on etiology, pathogenesis, diagnosis and treatment. Ann Gastroent 15(2):106–115
Uenishi T, Sakata C, Tanaka S, Yamamoto T, Shuto T, Hirohashi K, Kubo S, Kinoshita H (2003) Eosinophilic enteritis presenting as acute intestinal obstruction: a case report and review of the literature. Dig Surg 20(4):326–329. https://doi.org/10.1159/000071759
Vanhaecke C, Perignon A, Monsel G, Regnier S, Paris L, Caumes E (2014a) Aetiologies of creeping eruption: 78 cases. Br J Dermatol 170(5):1166–1169. https://doi.org/10.1111/bjd.12637
Vanhaecke C, Perignon A, Monsel G, Regnier S, Bricaire F, Caumes E (2014b) The efficacy of single dose ivermectin in the treatment of hookworm related cutaneous larva migrans varies depending on the clinical presentation. J Eur Acad Dermatol Venereol 28(5):655–657. https://doi.org/10.1111/jdv.12097
Vaughan TK, English JC 3rd (1998) Cutaneous larva migrans complicated by erythema multiforme. Cutis 62(1):33–35
Velasquez CC, Cabrera BC (1968) *Ancylostoma ceylanicum* (Looss, 1911) in a Filipino woman. J Parasitol 54(3):430–431

Venkatesh P, Sarkar S, Garg S (2005) Diffuse unilateral subacute neuroretinitis: report of a case from the Indian subcontinent and the importance of immediate photocoagulation. Int Ophthalmol 26(6):251–254. https://doi.org/10.1007/s10792-007-9045-x

Veraldi S, Carrera C (1999) Clinical presentations of cutaneous larva migrans. J Eur Acad Sermatol Venereol 12(Suppl. 2):260

Veraldi S, Bottini S, Carrera C, Gianotti R (2005) Cutaneous larva migrans with folliculitis: a new clinical presentation of this infestation. J Eur Acad Dermatol Venereol 19(5):628–630. https://doi.org/10.1111/j.1468-3083.2005.01235.x

Veraldi S, Bottini S, Rizzitelli G, Persico MC (2012) One-week therapy with oral albendazole in hookworm-related cutaneous larva migrans: a retrospective study on 78 patients. J Dermatol Treat 23(3):189–191. https://doi.org/10.3109/09546634.2010.544707

Veraldi S, Persico MC, Francia C, Schianchi R (2013a) Chronic hookworm-related cutaneous larva migrans. Int J Infect Dis 17(4):e277–e279. https://doi.org/10.1016/j.ijid.2012.11.002

Veraldi S, Persico MC, Francia C, Nazzaro G, Gianotti R (2013b) Follicular cutaneous larva migrans: a report of three cases and review of the literature. Int J Dermatol 52(3):327–330. https://doi.org/10.1111/j.1365-4632.2012.05723.x

Veraldi S, Çuka E, Pontini P, Vaira F (2017) Bullous cutaneous larva migrans: case series and review of atypical clinical presentations. G Ital Dermatol Venereol 152(5):516–519. https://doi.org/10.23736/S0392-0488.16.04832-X

Vercruysse J, Albonico M, Behnke JM, Kotze AC, Prichard RK, McCarthy JS, Montresor A, Levecke B (2011) Is anthelmintic resistance a concern for the control of human soil-transmitted helminths? Int J Parasitol Drugs Drug Resist 1(1):14–27. https://doi.org/10.1016/j.ijpddr.2011.09.002

Vetter JCM, Leegwater-v.d. Linden ME (1977) Skin Penetration of infective hookworm larvae. III. Comparative studies on *Ancylostoma braziliense*, *Ancylostoma ceylanicum*, and *Ancylostoma caninum*. Z Parasitenk 53:155–158

Walker NI, Croese J, Clouston AD, Parry M, Loukas A, Prociv P (1995) Eosinophilic enteritis in northeastern Australia. Pathology, association with *Ancylostoma caninum*, and implications. Am J Surg Pathol 19(3):328–337. https://doi.org/10.1097/00000478-199503000-00011

Wang S, Xu W, Li LF (2017) Cutaneous larva migrans associated with Loffler's Syndrome in a 6-year-old boy. Pediatr Infect Dis J 36(9):912–914. https://doi.org/10.1097/INF.0000000000001593

Wang MW, Yan XX, Hang JX, Shi XL, Fu YQ, Zhang P, Yang F, Pan WD, Li GQ (2019) Molecular differentiation of three canine and feline hookworms in South China through HRM analysis. J Helminthol 93(2):159–165. https://doi.org/10.1017/S0022149X18000068

Watson TG, Nettles VF, Davidson WR (1981) Endoparasites and selected infectious agents in bobcats (*Felis rufus*) from West Virginia and Georgia. J Wildl Dis 17(4):547–554. https://doi.org/10.7589/0090-3558-17.4.547

Weaver HJ, Hawdon JM, Hoberg EP (2010) Soil-transmitted helminthiases: implications of climate change and human behavior. Trends Parasitol 26(12):574–581

Wells HS (1931) Observations on the blood sucking activities of the hookworm, *Ancylostoma caninum*. J Parasitol 17(4):167–182. https://doi.org/10.2307/3271452

White GF, Dove WE (1929) A dermatitis caused by larvae of *Ancylostoma caninum*. Arch Derm Syphilol 20(2):191–200. https://doi.org/10.1001/archderm.1929.01440020043009

Wijers DJ, Smit AM (1966) Early symptoms after experimental infection of man with *Ancylostoma braziliense* var. *ceylanicum*. Trop Geogr Med 18(1):48–52

Williamson SM, Walsh TK, Wolstenholme AJ (2007) The cys-loop ligand-gated ion channel gene family of *Brugia malayi* and *Trichinella spiralis*: a comparison with *Caenorhabditis elegans*. Invert Neurosci 7(4):219–226. https://doi.org/10.1007/s10158-007-0056-0

Wolstenholme AJ, Fairweather I, Prichard R, von Samson-Himmelstjerna G, Sangster NC (2004) Drug resistance in veterinary helminths. Trends Parasitol 20(10):469–476

Wong-Waldamez A, Silva-Lizama E (1995) Bullous larva migrans accompanied by Loeffler's syndrome. Int J Dermatol 34(8):570–571. https://doi.org/10.1111/j.1365-4362.1995.tb02957.x

Wright DO, Gold EM (1945) Loeffler's syndrome associated with creeping eruption (cutaneous helminthiasis). J Am Med Assoc 128(15):1082–1083. https://doi.org/10.1001/jama.1945.02860320024006

Wright DO, Gold EM (1946) Loeffler's syndrome associated with creeping eruption (cutaneous helminthiasis): report of twenty-six Cases. Arch Int Med 78(3):303–312. https://doi.org/10.1001/archinte.1946.00220030050003

Yoshida Y (1971) Comparative studies on *Ancylostoma braziliense* and *Ancylostoma ceylanicum*. I. The adult stage. J Parasitol 57(5):983–989. https://doi.org/10.2307/3277850

Yoshida Y, Okamoto K, Chiu JK (1968) *Ancylostoma ceylanicum* infection in dogs, cats, and man in Taiwan. Am J Trop Med Hyg 17(3):378–381. https://doi.org/10.4269/ajtmh.1968.17.378

Yoshida Y, Okamoto K, Matsuo K, Kwo EH, Retnasabapathy A (1973) The occurrence of *Ancylostoma braziliense* (de Faria, 1910) and *Ancylostoma ceylanicum* (Looss, 1911) in Malaysia. Southeast Asian J Trop Med Public Health 4(4):498–503

Yoshida Y, Kondo K, Kurimoto H, Fukutome S, Shirasaka S (1974) Comparative studies on *Ancylostoma braziliense* and *Ancylostoma ceylanicum*. III. Life history in the definitive host. J Parasitol 60(4):636–641

Yun MY, Cho YU, Park IS, Choi SK, Kim SJ, Shin SH, Kim KR (2007) Eosinophilic gastroenteritis presenting as small bowel obstruction: a case report and review of the literature. World J Gastroenterol 13(11):1758–1760. https://doi.org/10.3748/wjg.v13.i11.1758

Chapter 10
Strongyloidiasis: Really a Zoonosis?

Adrian Streit

Abstract More than 600 million people are estimated to be infected with the nematode *Strongyloides stercoralis*, which is often overlooked during routine parasitological diagnostics. Some of the young worms mature to infective larvae within the host, enabling this parasite to establish long-lived, self-sustaining infections. While most of these infections are mild, they can self-enhance, resulting in complicated strongyloidisas, which if not treated in time, is normally lethal.

While it is undisputed that dogs are susceptible to experimental infection with human-derived *S. stercoralis*, it has long been debated if *Strongyloides* sp. naturally found in dogs are human infective *S. stercoralis* or belong to a different species, *S. canis*.

Based on recent studies, I argue that dogs naturally carry *S. stercoralis*, in addition to at least one other species, for which we have no indication that it infects people. I argue that for all practical purposes of *S. stercoralis* treatment and prevention, dogs should be seriously considered as possible sources for human *S. stercoralis*. However, from a rigorous scientific point of view, we must admit that currently we have no idea about the relative importance of zoonotic transmission compared with human to human transmission.

Keywords Strongyloidiasis · Neglected tropical disease · *Strongyloides stercoralis* · *Strongyloides canis* · *Strongyloides fuelleborni* · Zoonosis

10.1 Introduction

This paragraph is summarized from Grove (1989).

A. Streit (✉)
Department for Integrative Evolutionary Biology, Max Planck Institute for Developmental Biology, Tübingen, Germany
e-mail: adrian.streit@tuebingen.mpg.de

C. Strube, H. Mehlhorn (eds.), *Dog Parasites Endangering Human Health*, Parasitology Research Monographs 13,
https://doi.org/10.1007/978-3-030-53230-7_10

In 1876 Louis Normand, a physician at the naval hospital in Toulon and Arthur Bavay, a professor of pharmacy in the French navy, reported the isolation of two different new species of worms isolated from French soldiers who had been repatriated from Cochinchina (Vietnam). These soldiers suffered and sometimes died from a disease called Cochinchina diarrhea. By 1877, they had concluded that these worms represented two new species of *Anguillula*, which they named *A. stercoralis* and *A. intestinalis*. *A. stercoralis*, they found as larvae in the patients' intestines and in freshly deposited stool and they described that these larvae developed into free-living males and females within a few days. *A. intestinalis* was found as parasitic adults in the patients' intestines and these adults produced progeny, which, after leaving the host, developed into larvae with an esophagus that extended almost to the middle of the body and a truncated, not pointy, tail. Several other authors studied these worms and confirmed the findings by Normand and Bavay. But confusion soon arose, when Grassi and Parona discovered that the small larvae produced by *A. intestinalis* looked exactly like the *A. stercoralis* larvae Normand and Bavay had described. In their hands, however, these larvae did not develop to free-living adults but to *A. intestinalis* larvae. Grassi then erected the new genus *Strongyloides* and called the worms he observed *Strongyloide*s *intestinalis*. To further add to the confusion, Perroncito succeeded in culturing *A. stercoralis* free-living adults, which he called *Pseudorhabditis stercoralis*, and noticed that the larvae they produced also looked like the *A. intestinalis* larvae described by Normand and Bavay. But it was not until 1883 that Leukart, who called the species *Rhabdonema strongyloides*, realized that all these larvae and adults represented different stages of a complex heterogonic life cycle, i.e., this parasite alternated between parasitic and free-living generations. Finally, the confusion about the name of this worm was resolved by the Commission on Zoological Nomenclature in 1915, designating the species as *Strongyloides stercoralis*.

In English, this worm is sometimes referred to as "threadworm", however, I will avoid this name because "threadworm" it is also commonly used for *Enterobius* spp., otherwise also known as "pinworm."

The interesting life cycle of *Strongyloides* spp. caught the attention of numerous researchers since its discovery, and many aspects of this life cycle were eventually elucidated (Streit 2008, 2017). During this time, numerous other species of *Strongyloides* that are parasites of a wide variety of vertebrates were discovered (Speare 1989). But, *S. stercoralis*, despite the fact that it was from its discovery on always associated with a potentially fatal disease (Cochinchina diarrhea, Grove 1989), received comparatively little attention as a pathogen. Fortunately, this perspective appears to be changing. Strongyloidiasis is now formally recognized by the World Health Organisation (WHO) as one of the Neglected Tropical Diseases (NTDs), but it is still often overlooked in comparison with other Soil-Transmitted Helminthiases (STHs), and has therefore been considered (one of) the most neglected NTDs by various authors (Albonico et al. 2016; Bisoffi et al. 2013; Olsen et al. 2009). Many physicians I talked to had never heard about *S. stercoralis* and it may be symptomatic that this chapter is written by a biologist and not a medical professional. Below, I summarize our knowledge of this disease and its causative agent from the perspective of a modern biologist, with a special focus on the possible role of dogs as sources for zoonotic human *S. stercoralis* infections. For the more general sections at the beginning, I used predominantly recent and historical review articles as sources. Some of them are listed in Box 10.1 as recommended entry points to the field of *S. stercoralis* and strongyloidiasis. The discussion of the role and importance of dogs, which is the main focus of this book, is overwhelmingly based on original primary literature.

Box 10.1 Recommended Review Articles as Entry Points for *Strongyloides stercoralis* and Strongyloidiasis

Life cycle and basic biology

Schad GA (1989) Morphology and life history of *Strongyloides stercoralis*. In: Grove DI (ed) Strongyloidiasis: a major roundworm infection of man. Taylor & Francis, London, pp 85–104

Streit A (2008) Reproduction in *Strongyloides* (Nematoda): a life between sex and parthenogenesis. Parasitology 135(3):285–294

Streit A (2017) Genetics: modes of reproduction and genetic analysis. Parasitology 144:316–326. doi:https://doi.org/10.1017/S0031182016000342

Viney ME, Lok JB (2015) The biology of *Strongyloides* spp. (July 16, 2015). In: Community TCeR (ed) WormBook. 2015/07/18 edn. WormBook. doi:10.1895/wormbook.1.141.2. http://www.wormbook.org

Disease, diagnostics and treatment

Albonico M, Becker SL, Odermatt P, Angheben A, Anselmi M, Amor A, Barda B, Buonfrate D, Cooper P, Getaz L, Keiser J, Khieu V, Montresor A, Munoz J, Requena-Mendez A, Savioli L, Speare R, Steinmann P, van Lieshout L, Utzinger J, Bisoffi Z, StrongNet Working G (2016) StrongNet: an international network to improve diagnostics and access to treatment for strongyloidiasis control. PLoS Negl Trop Dis 10(9):e0004898

Bisoffi Z, Buonfrate D, Montresor A, Requena-Mendez A, Munoz J, Krolewiecki AJ, Gotuzzo E, Mena MA, Chiodini PL, Anselmi M, Moreira J, Albonico M (2013) *Strongyloides stercoralis*: a plea for action. PLoS Negl Trop Dis 7(5):e2214

Keiser PB, Nutman TB (2004) *Strongyloides stercoralis* in the immunocompromised population. Clin Microbiol Rev 17(1):208–217

Nutman TB (2017) Human infection with *Strongyloides stercoralis* and other related *Strongyloides* species. Parasitology 144(3):263–273

Diagnostic methods

Siddiqui AA, Berk SL (2001) Diagnosis of *Strongyloides stercoralis* infection. Clin Infect Dis 33(7):1040–1047

(continued)

Box 10.1 (continued)

Watts MR, Robertson G, Bradbury RS (2016) The laboratory diagnosis of *Strongyloides stercoralis*. Microbiol Aust 37(1). https://doi.org/10.1071/MA16003

Case reports

Tamarozzi F, Martello E, Giorli G, Fittipaldo A, Staffolani S, Montresor A, Bisoffi Z, Buonfrate D (2019) Morbidity associated with chronic *Strongyloides stercoralis* infection: a systematic review and meta-analysis. Am J Trop Med Hyg 100(6):1305–1311

Buonfrate D, Requena-Mendez A, Angheben A, Munoz J, Gobbi F, Van Den Ende J, Bisoffi Z (2013) Severe strongyloidiasis: a systematic review of case reports. BMC Infect Dis 13:78

Barroso M, Salvador F, Sanchez-Montalva A, Bosch-Nicolau P, Molina I (2019) *Strongyloides stercoralis* infection: a systematic review of endemic cases in Spain. PLoS Negl Trop Dis 13(3):e0007230

Prevalence, epidemiology and risk factors

Beknazarova M, Whiley H, Ross K (2016) Strongyloidiasis: a disease of socioeconomic disadvantage. Int J Environ Res Public Health 13(5). doi: https://doi.org/10.3390/ijerph13050517

Buonfrate D, Bisanzio D, Giorli G, Odermatt P, Furst T, Greenaway C, French M, Reithinger R, Gobbi F, Montresor A, Bisoffi Z (2020) The global prevalence of strongyloides stercoralis infection. Pathogens 9(6). https://doi.org/10.3390/pathogens9060468

Puthiyakunnon S, Boddu S, Li Y, Zhou X, Wang C, Li J, Chen X (2014) Strongyloidiasis—an insight into its global prevalence and management. PLoS Negl Trop Dis 8(8):e3018

Schär F, Trostdorf U, Giardina F, Khieu V, Muth S, Marti H, Vounatsou P, Odermatt P (2013) *Strongyloides stercoralis*: global distribution and risk factors. PLoS Negl Trop Dis 7(7):e2288

Collections of review articles

Grove DI (ed) (1989) Strongyloidiasis: A major roundworm infection of man. Taylor & Francis, London. ISBN: 0-85066-732-1

Special issue of Parasitology: volume 144, issue 3, March 2017

Remark In addition to its medical importance, *S. stercoralis*, together with some of its close relatives, have become important research models in translational and basic research. This includes the study of host–parasite interactions, immunology, parasite genomics, life cycle switches, or the evolution of parasitism. This aspect of *Strongyloides* spp. is not the focus of this chapter and the reader is referred to the following publications and references therein as examples and entry points to the different subjects (Hunt et al. 2016; Lok 2007, 2018; Streit 2014, 2017; Viney 2017; Viney and Kikuchi 2017; Viney and Lok 2015; Dulovic and Streit 2019; Gang et al. 2017; Jaleta and Lok 2019; Maeda et al. 2019; Breloer and Abraham 2017).

10.2 The Pathogen: *Strongyloides stercoralis*

In this chapter, I will concentrate on *S. stercoralis* as the vast majority of human *Strongyloides* infections are with this species (Nutman 2017; Albonico et al. 2016) and because it is the human infective *Strongyloides* species that is also associated with dogs (see below). However, two other species of *Strongyloides*, namely *S. fuelleborni fuelleborni* and *S. fuelleborni kellyi* have also been reported to infect humans (Nutman 2017). These two worms had originally been described as two subspecies of the species *S. fuelleborni* (Viney et al. 1991), but molecular phylogeny suggests that they are in fact separate species (Dorris et al. 2002). While *S. fuelleborni fuelleborni* has been reported from multiple continents in nonhuman primates with occasional zoonotic human infections, *S. fuelleborni kellyi* appears to be restricted to Papua New Guinea and, so far, no animal host has been identified (Viney et al. 1991, #477; Nutman 2017, #1517).

10.2.1 Phylogenetic Position

While classical taxonomy based upon morphology, had placed the genus *Strongyloides* within the order of Rhabditidae, relatively close to *Caenorhabditis elegans*, a nonparasitic species widely used for research (Nigon and Felix 2017), later taxonomical studies based upon DNA sequences showed that *Strongyloides* spp. actually belongs to an entirely different group of nematodes (Blaxter et al. 1998; Holterman et al. 2006). According to Blaxter et al. (1998), who defined five major clades of nematodes, *Strongyloides* spp. belongs to clade IV and is now known to be phylogenetically remote from all the other well-known human parasitic nematodes. This includes the hookworms (*Ancylostoma* spp. and *Necator* spp.), which (along with *C. elegans*) belong to clade V, the filarial nematodes, pinworms (*Enterobius*) and the giant roundworms (*Ascaris*), which belong to clade III or the whipworms (*Trichuris*) and *Trichinella*, both of which belong to clade I. Additionally, a later classification of nematode phylogeny, which proposed a phylogeny with 12 main groups (Holterman et al. 2006), was still in agreement with Blaxter et al. (1998) that *Strongyloides* spp. is very remote from most other human, or more general vertebrate, parasitic nematodes. *Strongyloides* spp. does, however, have very close facultative parasitic relatives of the genus *Parastrongyloides* spp. (found in marsupials) and nonparasitic relatives of the genus *Rhabditophanes* spp. (Grant et al. 2006; Mackerras 1959; Dorris et al. 2002; Hunt et al. 2016), such that one must propose that *Strongyloides* spp. has evolved its parasitic lifestyle independently of the other parasites mentioned above (Blaxter et al. 1998; Streit 2014).

10.2.2 *The Life Cycle of* Strongyloides stercoralis

10.2.2.1 Overview

S. stercoralis has a complex and rather unique life cycle, consisting of parasitic and free-living generations that have been repeatedly reviewed (e.g., Schad 1989; Streit 2017; Viney and Lok 2015). The following paragraph summarizes the lifecycle as it is generally described in the literature. This paragraph and the corresponding Fig. 10.1 were taken under the creative commons license from a recent article from my laboratory (Zhou et al. 2019b).

Infective third-stage larvae (iL3s), which are all females, invade a new host by skin penetration and eventually establish in the small intestine of the host. The parasitic adult females reproduce by parthenogenesis and their progeny have four developmental options:

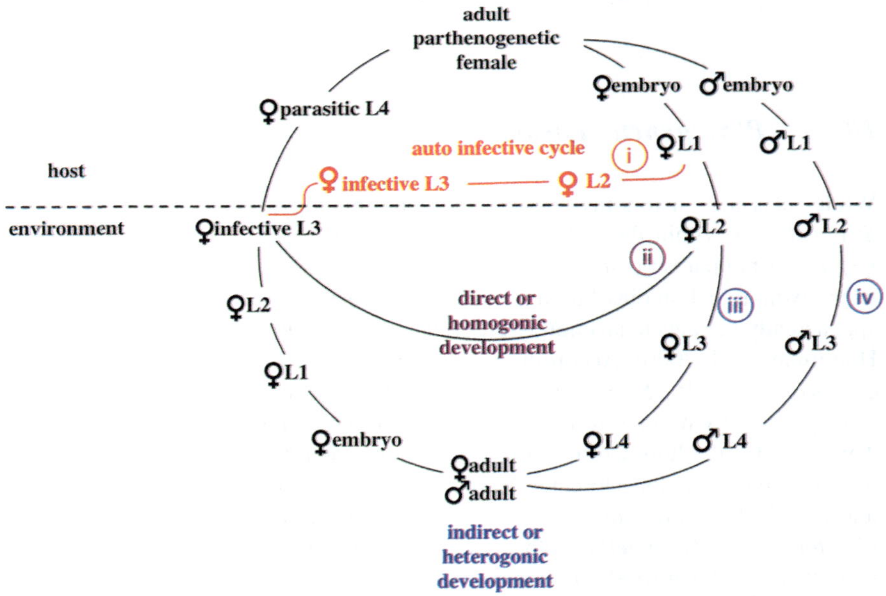

Fig. 10.1 The life cycle of *S. stercoralis*. The life cycle of *Strongyloides stercoralis*. (i)–(iv) refer to the corresponding descriptions of the developmental options in the text. Copyright: Creative Commons Attribution 4.0 License (https://creativecommons.org/licenses/by/4.0/). Citation: Zhou et al. (2019) From the feces to the genome: a guideline for the isolation and preservation of *Strongyloides stercoralis* in the field for genetic and genomic analysis of individual worms. *Parasites & Vectors,* 12, 496

i. They may become female, and develop into infective third-stage larvae (iL3) within the host and reinfect the same host individual (autoinfective cycle).
ii. They may become female, leave the host as first-stage larvae and develop into iL3 in the environment and search for a new host (direct/homogonic development).
iii. They may become female, leave the host as first-stage larvae, and develop into free-living, noninfective third-stage larvae and subsequently into adult females (indirect/heterogonic development).
iv. They may become male and develop into free-living adult males (indirect/heterogonic cycle).

The free-living adults mate and reproduce sexually in the environment and all their progeny are females and develop to iL3s.

As complex as this life cycle already is, it should nevertheless be noted that multiple authors, in particular Schad and colleagues (e.g., Mansfield et al. 1996; Schad 1989), have repeatedly pointed out that there are observations described in the literature that cannot be explained with the proposed life cycles alone. In particular, they stress that the possibility of the existence of alternative routes through the host's body, bypassing the lungs (see also Page et al. 2018), and the occasional, possibly strainspecific, occurrence of parasitic males should not be excluded.

10.2.2.2 Description of Selected Life Cycle Stages

Below I describe selected features of some of the developmental stages. They were chosen for their importance for the following paragraphs of this chapter. All size information was taken from Little (1966), Speare (1989), and Schad (1989) who also provide precise measurements and detailed drawings. Descriptions and developmental time indications are based on the same three references and Zhou et al. (2019b). The description is limited to features that are observable with rather low magnification using a dissecting microscope.

Parasitic Adults All parasitic individuals are females. The parasitic worms are usually described as "filariform" They are about 2.5-mm long and very slender with a diameter of about 0.035 mm. With this they are slightly smaller than *S. fuelleborni* (approx. 3.5 mm long and 0.05 mm in diameter). The shape of the ovary is straight, which is a feature that distinguishes it from *S. fuelleborni fuelleborni*, which has a spiral-shaped ovary. Both these species belong to the "medium sized" (2–5 mm in length) class of *Strongyloides* spp. as defined by Speare (1989). Compared with adults of other common small intestinal nematodes of humans, such as hookworms or Ascardis (Anderson 2000), they are much smaller hence their German trivial name "Zwergfadenwurm" (dwarf roundworm). *Trichinella spiralis* is comparable in size, but the two worms are easily morphologically distinguishable (compare, for example, the images in Eckert et al. (2005, p. 332) and Schad (1989, p. 87); e.g., the vulva is located anteriorly, close to the end of the pharynx in *Trichinella* but in the posterior third in *S. stercoralis*).

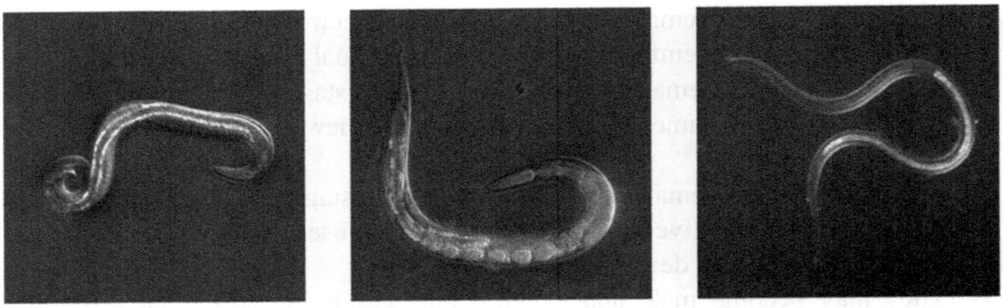

Fig. 10.2 Free-living adults and infective L3. Differential interference contrast (DIC) images of a free-living male, a free-living female and an infective third-stage larva (left to right). Male and female are at the same magnification (the female is about 1-mm long), the iL3 is at higher magnification (total body length is about 0.6 mm). The worms shown are *S. papillosus* but at the given magnification and resolution they are morphologically indistinguishable from *S. stercoralis*. Copyright: Creative Commons Attribution 4.0 License (https://creativecommons.org/licenses/by/4.0/). Citation: Zhou et al. (2019) From the feces to the genome: a guideline for the isolation and preservation of *Strongyloides stercoralis* in the field for genetic and genomic analysis of individual worms. *Parasites & Vectors,* 12, 496

First-Stage Larvae (L1) The L1 is very small (about 0.25 mm in length and 0.017 mm in diameter at the time when they are shed with the feces). They are an important diagnostic stage because *S. stercoralis* is the only common human parasitic nematode that sheds larvae and not eggs. Morphologically, L1 and also the later larval stages other than infective larvae are difficult to tell apart from many other parasitic and nonparasitic nematode species. They may be differentiated from L1 larvae of the hookworms by their short buccal cavity and prominent genital primordium.

Infective Third-Stage Larvae (iL3) With a diameter of up to only about 0.02 mm, the iL3s are very slender, about 0.6 mm in length and highly motile. The tail of this larval stage ends in a bifurcate point. The typical feature is their long pharynx that is very clear in appearance and comprises between 40 and 50% of the body length (Fig. 10.2). This makes them the developmental stage that is most easily unambiguously identifiable as *Strongyloides* spp. (the species cannot be determined), even by unexperienced observers and even if there is the possibility that the sample was contaminated with soil-dwelling nematodes. The only other genus of nematodes that has essentially indistinguishable iL3s is *Parastrongyloides* (Grant et al. 2006; Mackerras 1959), which is, however, restricted to marsupial hosts.

Free-Living Adult Females The free-living females (Fig. 10.2) are described as "rhabditiform." They measure about 1–1.5 mm in length and about 0.085 mm in diameter. Their pharynx (esophagus) consists of three clearly distinguishable parts (procorpus, isthmus, and bulb, from anterior to posterior). The females have a centrally located vulva and a didelphic gonad with the two arms extending anteriorly and posteriorly, respectively. Typically, they carry several developing embryos in their uteri on either side of the vulva. If contamination by soil born nematodes can be excluded, *Strongyloides* spp. are the only adult nematode females of their size class to be expected in a human fecal culture. The time required for the development of the adult is about one and a half to 2 days, depending on the conditions (e.g., temperature).

Free-Living Adult Males The free-living males (Fig. 10.2) are slightly smaller than the females. Their pharynx is comparable in terms of size and structure to that of females. They have a monodelphic gonad terminating in a posterior copulatory structure. Again, if contamination by soil born nematodes can be excluded, *Strongyloides* spp. are the only adult nematode males of their size class to be expected in a human fecal culture and will develop at the same time as the females.

10.2.2.3 Modes of Reproduction

Both sexual and asexual reproduction have their advantages but also costs associated with them (Butlin 2002; Engelstadter 2008; Beukeboom and Perrin 2014). Sexual reproduction means that in the progeny genetic material from a mother and a father are combined to generate new combinations of genetic material (genes). Thanks to meiotic recombination, new combinations of alleles can arise even on a particular chromosome. Frequently mentioned putative advantages of this are that the process allows for the combination of different independently arisen advantageous mutations in one individual or on one chromosome and at the same time also can separate beneficial and deleterious mutations that arose in the same individual or on the same chromosome. Further, as sexual reproduction generates genetically diverse progeny, it increases the chances that at least some of them are suited for changing environments. On the downside, sexual reproduction also breaks up favorable combinations of genetic materials that might have been perfectly adapted to the current environment, such that at times where conditions are stable over extended periods of time, asexual reproduction might be an advantage. In those cases where the gametes of the two sexes are produced in different individuals (meaning that separate males and females exist) there is a cost associated with the production of "unproductive" males, which is known as the "twofold cost of sex" (References for the entire paragraph Butlin 2002; Engelstadter 2008; Beukeboom and Perrin 2014).

Therefore, *S. stercoralis* might be able to get the best of both strategies with minimal repercussions as its complex life cycle enables it to alternate between asexual reproduction in the parasitic generation and sexual reproduction in the free-living generation (Streit 2017).

The following two paragraphs are mainly based on Streit (2008, 2017).

Reproduction in Parasitic Adults There is a wide agreement that all parasitic *S. stercoralis* individuals are females. Except for two reports from the 1930s from the same laboratory describing the same material (Faust 1933; Kreis 1932), no parasitic males have ever been reported in *S. stercoralis* or any other species of *Strongyloides* and these two reports have been disregarded by later authors for various reasons. For several other species of *Strongyloides*, but never to my knowledge in *S. stercoralis*, it has been demonstrated cytologically and by using molecular genetic methods that the reproduction in the parasitic generation is by mitotic parthenogenesis. In this process, the chromosomes never undergo pairing as would be typical for the first meiotic division. Instead, the single maturation division is mitotic (without reduction of chromosome number) resulting in an oocyte and later one-cell embryo, which is genetically identical with its mother. While it seems safe to conclude that the parasitic females of *S. stercoralis* normally reproduce parthenogenetically, Schad (1989) stressed that one should not exclude that under certain circumstances and possibly only in certain populations, parasitic males might occasionally occur. It is also noteworthy that in the closely related genus *Parastrongyloides* parasitic females and males exist (Grant et al. 2006; Mackerras 1959) and the reproduction between them is sexual (Kulkarni et al. 2013). Therefore, the parthenogenetic mode of reproduction must have arisen in *Strongyloides* after its evolutionary separation from *Parastrongyloides* (Streit 2014). A very important consequence of the capacity of *S. stercoralis* females to self-reproduce is that it makes this species an excellent colonizer. In other species of intestinal parasites with males and females, a particular host individual needs to be infected with at least two individuals (one female and one male) and these two individuals need to find each other in the intestine of the host. This is not the case for *Strongyloides* spp. and a single female may be sufficient to start a new population. For the closely related *S. ratti*, experimental infection attempts with single infective larvae were successful in roughly half the cases (Graham 1936; Viney et al. 1992).

Reproduction in Free-Living Adults Most authors agree that in the free-living generations of all species of *Strongyloides* studied so far, males are required for successful reproduction. Until the 1990s, several authors, studying various species of *Strongyloides* (including *S. stercoralis*, Hammond and Robinson 1994), had proposed that based on cytological observations, the reproduction is nevertheless parthenogenic with the sperm only being required for triggering development and probably providing the centriole. Such a process is known as sperm-dependent parthenogenesis or pseudogamy. However, molecular genetic studies showed that *S. stercoralis* males did contribute genetically to the progeny both, in experimental crosses in the laboratory and in field samples and inheritance appeared to follow the Mendelian rules (Jaleta et al. 2017), demonstrating that the free-living generations of *S. stercoralis* reproduces sexually, as it had been shown to be the case in other species of *Strongyloides* (Viney et al. 1993; Eberhardt et al. 2007; Kulkarni et al. 2013; Streit 2017). In *S. stercoralis* and most but not all other species of *Strongyloides* that have been studied (Yamada et al. 1991), all progeny generated in this sexual event are females and invariably develop into infective larvae, meaning consecutive free-living generations are not possible. This is again different in *Parastrongyloides* spp., where a high, if not unlimited number of consecutive bisexual free-living generations are possible (Grant et al. 2006).

10.2.2.4 Some Thoughts About the Different Life Cycles

The Direct or Homogonic Cycle This cycle is a purely asexual cycle resulting in a population of genetically uniform individuals. With the exception of newly arising mutations, all progeny of a particular individual derived through this cycle over many generations are genetically identical to each other. Therefore, under constant conditions, this cycle might present the opportunity for a perfectly adapted genotype to spread rapidly. However, if conditions change, all individuals will also be equally maladapted to the new environment.

The Indirect or Heterogonic Cycle Through this sexual cycle, *S. stercoralis* is capable of generating new genetic diversity as explained above, which may facilitate the adaptation to changing environments. Additionally, this host independent reproductive step considerably increases the number of infective larvae a parasitic female gives rise to. While we do not know how many progeny free-living females produce under natural circumstances, from laboratory experiments on *S. ratti*, we can estimate that it might be in the order of a few dozen progeny produced over its short reproductive life during this indirect cycle (Gardner et al. 2004, 2006; Dulovic et al. 2016). This production of progeny undergoing the indirect life cycle may explain the evolutionary maintenance of the free-living cycle independent of the advantages of sexual reproduction as it helps to generate sufficient numbers of infective larvae (Fenton et al. 2004). As explained above, this investment in males is, however, not without a cost, as males do not produce eggs and furthermore, recombination can cause the breaking up of particularly favorable gene combinations reducing overall fitness.

The Auto Infective Cycle Like the direct cycle, of which it is a variant, this is a fully asexual cycle. It makes perfect sense for a female that is well adapted to the particular host individual to produce genetically identical progeny that also inhabit the very same host. Therefore, this cycle allows the parasite to maintain, normally very mild infections, for much longer than the life expectancy of a parasitic individual. Indeed, while parasitic females live in the order of a few weeks to months, (Nolan et al. 1993) *S. stercoralis* chronic infections have been found that have been going on for decades (e.g., Rothe et al. 2020; Nordheim et al. 2018). Further, as parasitic females can produce progeny of the homogonic and the heterogonic cycle at the same time as auto infective larvae, propagation to new host individuals is still possible. However, under certain circumstances (see below), this cycle gets out of control and can lead to a worm burden that kills the host. While at first this may appear to not be in the parasites best interest and may be considered an accident, it can also be speculated that it may be beneficial for the parasite to rapidly reproduce if it senses that its host has deteriorating health and might be likely to die soon anyway from other causes.

10.2.2.5 The Life History Switches

This paragraph is mainly based on Streit (2008, 2017).

Sex Determination It is remarkable that the parasitic female in a parthenogenetic reproductive event produces males and females that differ in their chromosome number. Females have two pairs of autosomes and one pair of X chromosomes, while males have only one X chromosome along with the two autosomes (Hammond and Robinson 1994). This is called an XX/XO chromosomal sex determination system and is very common but by no means universal among nematodes (Pires-daSilva 2007). In spite of this difference in chromosome number, sex determination is actually influenced by the environment, namely the immune status of the host. A stronger immune response to *Strongyloides* spp. generally appears to lead to the generation of more males (notice that most studies to this topic were performed in animal parasitic species of *Strongyloides*). The genetic background of the parasites is also an important factor, resulting in some isolates that have a strong tendency to produce males (and free-living females), while others produce only very few sexual animals (Viney et al. 1992; Zhou et al. 2019a). The mechanism by which the one X chromosome is eliminated in males is currently unknown. Based on studies in other species of *Strongyloides* (Harvey and Viney 2001; Albertson et al. 1979; Nemetschke et al. 2010), we can assume that the chromosome reduction in *S. stercoralis* also occurs during the single maturation division that leads to the diploid oocyte/one-cell embryo.

The Direct–Indirect Switch This female only switch occurs at the L1 or L2 stage and is arguably the best-studied developmental biological processes in *Strongyloides* spp. in general and in *S. stercoralis* in particular. As with sex determination, the genetic background and the host's immune status (stronger immune response causes more heterogonic development) are important factors and predispose the females toward one or the other developmental route. However, other than with sex, which appears to be irreversibly determined at the one-cell stage, female *Strongyloides* spp. (including *S. stercoralis*, Nolan et al. 2004) can still switch between the two developmental routes after leaving the host in response to environmental influence. The main factors influencing this switch are temperature, food availability, and population density. The generalizing conclusion by Schad (1989) that "favorable conditions within the host and adverse conditions outside the host favor direct larval development to infectivity, whereas unfavorable internal conditions and favorable external conditions favor indirect development" still holds true today in light of the recent literature.

The genetic control involved in the formation of infective larvae is the developmental regulatory process of which we have the best (although compared with other organisms still very limited) molecular genetic understanding in *Strongyloides* spp. (Dulovic and Streit 2019; Ogawa et al. 2009; Patton et al. 2017; Stoltzfus et al. 2014; Wang et al. 2009, 2015). The basic molecular biology of *Strongyloides* spp. is not the topic of this chapter and the interested reader is referred to the references listed above and the references listed at the end of the introduction. However, I would like to point out, that at the heart of this process is a nuclear hormone receptor called DAF-12, which can be inhibited by the application of Delta7-dafachronic acid (Δ7-DA). Δ7-DA, which is the natural ligand of DAF-12 in the free-living nematode *Caenorhabditis elegans* and may or may not be the natural ligand in *Strongyloides* spp. has been found to prevent the formation of infective larvae in two different animal parasitic species of *Strongyloides* (Dulovic and Streit 2019; Ogawa et al. 2009) and in *S. stercoralis* (Wang et al. 2009). DAF-12 has therefore been proposed as a possible target for therapeutic intervention (Ogawa et al. 2009; Wang et al. 2009). While we know very little about the induction of auto-infective iL3s, there is an encouraging report that Δ7-DA might indeed suppress the auto-infective cycle as well as the homogonic cycle (Patton et al. 2017).

10.3 The Disease: Strongyloidiasis

This section is mainly based on Nutman (2017) and Albonico et al. (2016).

10.3.1 Acute and Chronic Strongyloides stercoralis *Infections Without Complications*

Infection occurs percutaneously upon contact with contaminated fecal material or soil. Laboratory experiments with the closely related *S. ratti* in rats showed that infection with a single larva led to a patent infection in more than half of the attempts (Graham 1936; Viney et al. 1992), suggesting that only very few larvae are necessary to establish an infection. The infection event may lead to skin irritation at the site of entry and to visible tracks of the larvae migrating through the skin. However, in most cases, the infection goes unnoticed and the infective larvae develop, establish in the small intestine and start releasing progeny in a matter of days to a few weeks (Freedman (1991), reviewing the very few controlled experimental infections conducted in humans, lists pre-patient periods between 3 and 28 days with most cases between 23 and 28 days). This stage is known as acute strongyloidiasis. *S. stercoralis* infection triggers a Th2 immune response (Breloer and Abraham 2017), which may lead to clearance of the infection. The exact time an individual parasitic female can survive and reproduce is probably variable but, based on laboratory experiments in surrogate hosts (Nolan et al. 1993) and experiments in other species of *Strongyloides* (Gardner et al. 2006), lies in the range of several weeks to months. However, it is possible that the infection becomes chronic through the auto infective cycle (see Fig. 10.1). I am not aware of any studies addressing what portion of *S. stercoralis* infections indeed become chronic. The ability to form self-sustaining infections distinguishes *S. stercoralis* (and possibly a few not yet well-studied other species of *Strongyloides*) from all other mammalian parasitic nematodes I am aware of. Chronically infected, otherwise healthy, individuals usually have little to no *S. stercoralis*-associated clinical symptoms and these infections are rather unlikely to be detected.

10.3.2 Hyperinfection Syndrome and Disseminated Strongyloidiasis

If a chronically infected patient fails to control the *S. stercoralis* infection at a low level, this may lead to a dramatic increase of the worm burden and a self-enhancing progression known as "hyperinfection syndrome." This leads to a very high number of migrating larvae that then tend to appear almost everywhere in body, causing a variety of rather unspecific symptoms due to damage to various organs. This "disseminated strongyloidiasis" is associated with very high (>80%) mortality if not treated in time.

10.3.2.1 Factors Promoting Hyperinfection Syndrome and Dissemination

A variety of conditions, which mainly but not exclusively lead to a weakening of the immune system may lead to the loss of control of chronic but clinically inapparent *S. stercoralis* infections. Nutman (2017) provides a fairly comprehensive list of drugs, treatments diseases, and syndromes that have been described to promote *S. stercoralis* hyperinfection and dissemination. This list contains Immunosuppressives (most importantly corticosteroids), antineoplastic agents, biologics, medical treatments like whole-body irradiation or solid organ transplantation, malnutrition, noninfectious diseases, for example, certain blood cancers and infectious diseases, among them very importantly HTLV-1 (human T-lymphotropic virus 1) infections. Other than originally suspected (Keiser and Nutman 2004), HIV infection appears not to be a major promotor of strongyloidiasis (Nutman 2017). Individual case studies, listing the respective suspected hyperinfection promoting conditions and treatments are summarized in Barroso et al. (2019), Buonfrate et al. (2013), and Tamarozzi et al. (2019).

10.4 Diagnosis and Treatment

10.4.1 Symptoms

This section is mainly based on Nutman (2017) and Albonico et al. (2016).

As mentioned above, most *S. stercoralis* infections progress with little to no clinical manifestations. During the acute phase of infection, the first symptoms that may occur are associated with the skin penetration and larval migration to the small intestine. Patients may experience irritation at the site of larval skin penetration, sometimes followed by local urticaria or edema. Within about a week postinfection, a dry cough and tracheal irritation may be observed. Once the parasite has established in the small intestine, gastrointestinal symptoms like diarrhea, constipation, abdominal pain, and a lack of appetite are sometimes observed. Chronic strongyloidiasis is most often not associated with clinical symptoms although some gastrointestinal and dermatological (e.g., "larva currens", subcutaneously migrating auto-infective larvae) problems may occur. Up to 75% of chronically *S. stecoralis* infected individuals exhibit peripheral eosinophilia or elevated IgE levels. Therefore, Nutman (2017) suggests that strongyloidiasis should be considered in cases of severe or persistent eosinophilia in patients who did spend time in endemic regions. However, the value of eosinophilia as a predictor for *S. stercoralis* infection has also been contested (Baaten et al. 2011; Naidu et al. 2013). Although eosinophilia may also be observed in patients with hyper infection syndrome, more often such patients show reduced eosinophil counts. Even once *S. stercoralis* infections become severely pathogenic, the symptoms are variable and rather unspecific. In addition to general conditions, like fatigue, weakness, or total body pain Nutman (2017) lists more than 30 gastrointestinal, cardiopulmonary, dermatological, and central nervous system manifestations of *S. stercoralis* hyperinfection/dissemination. All of these symptoms have more common causes other than strongyloidiasis (for examples of specific cases with symptom descriptions for individual patients see Barroso et al. 2019; Buonfrate et al. 2013; Tamarozzi et al. 2019).

10.4.2 Detection of Strongyloides stercoralis

As outlined above, a reliable diagnosis of an *S. stercoralis* infection based purely upon symptoms is not possible. Various immunological and molecular diagnostic tools are available and have been improved upon over the last few years (Nutman 2017). They rely on the detection of either antibodies against *S. stercoralis* in the serum of patients or of *S. stercoralis* antigens or DNA in fecal samples. However, issues remain with sensitivity and or specificity with these indirect methods and the definitive diagnosis relies on the direct detection of *Strongyloides* worms in stool (Nutman 2017). In cases of hyperinfection syndrome, the detection of *S. stercoralis* is fairly straight forward, due to the large number of larvae that can be found in the stool or are detectable within the body for example upon duodenal aspiration (Nutman 2017). Nevertheless, specific testing for *S. stercoralis* is required and there have been multiple reports of fatal strongyloidiasis in well-developed countries mainly because *S. stercoralis* was only diagnosed postmortem or too late for successful therapeutic intervention (Barroso et al. 2019; Buonfrate et al. 2013; Tamarozzi et al. 2019). Detection of chronic, clinically inapparent *S. stercoralis* infection is much more difficult. *S. stercoralis* infections are frequently not detected by routine parasitological testing because of the usually very low worm burden and correspondingly low and discontinuous larval output a and because tests frequently used in medical laboratories detect helminth eggs, but have poor sensitivity for the detection of larvae in the stool (Page and Speare 2016; Watts et al. 2016). As mentioned previously, *S. stercoralis* does not normally pass eggs but already hatched larvae. This is different in the two other human infective species of *Strongyloides*, *S. fuelleborni fuelleborni* and *S. fuelleborni kellyi* (Viney et al. 1991). Therefore, asymptomatic *S. stercoralis* must be specifically looked for if it is to be diagnosed with satisfactory sensitivity (Page and Speare 2016). To this end, serological tests to detect the presence of antibodies against *S. stercoralis* have been improved over the last few years and are increasingly used for diagnosis and screening (Nutman 2017; Kalantari et al. 2020). However, these methods still show variable specificity and, in particular in immunosuppressed patients, insufficient sensitivity with an inability to distinguish between past and present infection (for a recent comparison of different serological assays for *S. stercoralis* detection see Kalantari et al. 2020). With respect to specificity, the direct observation of larvae from stool samples is arguably the most accurate method. As mentioned above, the presence of young already hatched larvae in fresh stool is a reliable indication for *Strongyloides*. The identification of *Strongyloides* can be facilitated by culturing the stool samples for between 2 and 4 days under aerobic conditions, such that free-living adult males and females, and infective third-stage larvae develop, all of which are easily recognizable morphologically (Fig. 10.2, Zhou et al. 2019b). There are various methods for how to isolate the larvae from the fecal sample for observation, including direct smearing of feces in saline–Lugol iodine stain, agar plate culture (APC, e.g., Koga agar plate), the Harada/Mori filter paper culture, the formalin ethyl acetate concentration technique and the Baermann technique (for a description and comparison of these techniques see Siddiqui and Berk 2001; Watts et al. 2016). The drawback of the coprological methods is their limited sensitivity. Only through analyzing multiple samples from consecutive days and combining coprological and serological or PCR based methods can the sensitivity be increased to a satisfactory level (Siddiqui and Berk 2001; Albonico et al. 2016; Nutman 2017; Watts et al. 2016). APC and Baermann are currently the most frequently used techniques. APC, essentially placing a fecal sample on an agar plate and observing the worms that emerge within 2–4 days is possibly the easiest coprological method to perform in the context of large-scale testing (Watts et al. 2016) and some authors found it to be slightly more sensitive than the Baermann technique when performed with comparable quantities of fecal sample (Khieu et al. 2013; Steinmann et al. 2007). The Baermann technique, illustrated in Fig. 10.3, is somewhat more laborious and requires more equipment that cannot be transported while the assay is running. However, this technique allows for analyzing larger sample volumes. Because unlike APC, Baermann funnels cannot be set up in the field, the transportation of the samples is a crucial step. It is important to keep the samples aerated, for example, by mixing the feces with sawdust and we found refrigeration during transport to be highly detrimental for later Baermann analysis (Zhou et al. 2019b).

a b

Fig. 10.3 The Baermann technique. (**a**) Setting up a simple Baermann funnel: place a piece of rubber tubing at the bottom of a glass funnel, close the end of the tube with a clamp and fill the funnel with tap water. Wrap the fecal sample or fecal culture in tissue paper and place it into the funnel. Incubate at ambient temperature. For a more sophisticated setup, see Lok (2007). (**b**) Harvesting the worms: After the worms have crawled out of the feces and accumulated at the bottom of the funnel (allow at least 2 h), place a second clamp 1–2 cm above the first clamp, remove the lower clamp and collect the worms into a watch glass or small Petri dish. Copyright: Creative Commons Attribution 4.0 License (https://creativecommons.org/licenses/by/4.0/). Citation: Zhou et al. (2019) From the feces to the genome: a guideline for the isolation and preservation of *Strongyloides stercoralis* in the field for genetic and genomic analysis of individual worms. *Parasites & Vectors*, 12, 496

10.4.3 Treatment

This section is mainly based on Nutman (2017) and Albonico et al. (2016).

The recommended treatment for strongyloidiasis is Ivermectin because it targets both adult worms as well as larvae. For uncomplicated *S. stercoralis* infections, oral dose of 200 μg/kg for 2 days is recommended. Unfortunately, access to ivermectin is limited in many countries and under certain circumstances (e.g., coinfections with *Loa loa*) this drug is contraindicated. As an alternative therapy, 400-μg albendazole twice a day for 3–7 days can be used, although slightly less effective when compared with ivermectin. If hyperinfection syndrome is suspected, this should be considered a medical emergency and treatment with ivermectin should be initiated immediately. Although no controlled trials have been conducted, daily treatment for at least 2 weeks has emerged as the therapy. In cases where patients cannot take oral medication, because of severe systemic illness or intestinal paralysis, rectal application can be tried as an alternative. There are no injectable ivermectin formulations approved for human use but in extreme cases of emergency, doctors do turn to subcutaneous administration of injectable veterinary formulations of ivermectin as a last resort.

10.5 Epidemiology and Risk Factors

Estimates of the numbers of people currently infected with *S. stercoralis* in the world range from "30–100 million" (Bethony et al. 2006; Nutman 2017) more than 600 million (Buonfrate et al. 2020). Given the difficulties with diagnosis of uncomplicated infections (as discussed above), the true number may even be considerably higher. Locally, the prevalence of *S. stercoralis* can be very high in some tropical and subtropical areas (Puthiyakunnon et al. 2014; Schär et al. 2013 list examples of >70%). However, for most countries only very few studies were conducted, if any (Schär et al. 2013). Further, the prevalence may differ considerably even between neighboring villages (Forrer et al. 2019; Khieu et al. 2014a; Steinmann et al. 2015; Zhou et al. 2019a) such that prevalence estimates for entire countries by extrapolation from a few local studies are not reliable, making global estimates even more difficult to calculate.

As an NTD, strongyloidiasis is frequently regarded as a tropical disease. While climatic factors such high moisture and high temperature, may indeed contribute to high prevalence of *S. stercoralis*, poor living and health conditions such as inadequate sanitary infrastructure or limited access to health care are more important risk factors for *S. stercoralis* infections (Beknazarova et al. 2016; Schär et al. 2013). High prevalence disease may occur in marginalized communities with poor sanitation in temperate countries (Strkolcova et al. 2017). Accordingly, Beknazarova et al. (2016) strongly suggested that strongyloidiasis should not be primarily regarded as a tropical disease but rather as a disease of the socioeconomically disadvantaged.

Although the vast majority of *S. stercoralis* infected people are poor people living in underdeveloped tropical and subtropical regions, *S. stercoralis* is a cosmopolitan parasite and is neither limited to this geographic region nor this group of people (Puthiyakunnon et al. 2014; Schär et al. 2013). The presence of *S. stercoralis* and fatal cases of strongyloidiasis have also been reported from well-developed regions with temperate climates such as the European Union and North America (Page and Speare 2016; Strkolcova et al. 2017; Agarwala et al. 2014; Buonfrate et al. 2013, 2016; Choksi et al. 2016; Jones et al. 2016; Roseman et al. 2013; Rothe et al. 2020; Barroso et al. 2019). As mentioned previously, patients receiving immunosuppressive treatments are particularly at risk (Keiser and Nutman 2004; Barroso et al. 2019; Buonfrate et al. 2013; Tamarozzi et al. 2019). *S. stercoralis* is therefore an important complication of organ transplantations and cancer chemotherapies, in particular in regions where it is highly prevalent (Getaz et al. 2019; Miglioli-Galvao et al. 2020). In addition to cases where organ recipients were infected naturally with *S. stercoralis* before or after the transplantation, there are several well-documented cases where *S. stercoralis* was transferred to organ recipients from the donor along with the organ (e.g., Kim et al. 2016; Elzein et al. 2020; Nordheim et al. 2018; Hasan et al. 2013; Camargo et al. 2019; Roseman et al. 2013). Therefore, transplantation centers recommend *Strongyloides* screening for organ donors and recipients (Elzein et al. 2020; Malinis et al. 2019; Winnicki et al. 2018; Camargo et al. 2019) even in clearly nontropical places like Scandinavia (Nordheim et al. 2018).

10.6 *Strongyloides* spp. in Dogs and Humans: Are They the Same or Are They Different?

Strongyloides spp. infections are generally not considered a major health problem for dogs. Nevertheless, such infections are rather common in certain regions and they have also been reported from highly developed countries with nontropical climates. Furthermore, infections with severe clinical complications do occur (Basso et al. 2019; Thamsborg et al. 2017; Strkolcova et al. 2017; White et al. 2019). The role of *Strongyloides* spp. for dog health is not part of this chapter and the reader is referred to Basso et al. (2019) and Thamsborg et al. (2017) as entry points to this topic.

10.6.1 The Pre-molecular Era

While parasitic adults and infective larvae are fairly easy to recognize as *Strongyloides* spp., it is generally very difficult to safely determine the species of a *Strongyloides* worm based on morphological criteria alone. Although some morphological features that allow the distinction of at least some species from each other have been described, frequently the host species that the sample was found in was, and still is, used to infer the species (Speare 1986, 1989). Very early after its discovery, Fölleborn (1914) noticed that dogs in East Asia carried a *Strongyloides* sp. that was very similar to the human one, and that dogs were susceptible to experimental infection with human-derived *S. stercoralis*, although sometimes with difficulty. Fülleborn (1914) also described the *Strongyloides* of dogs to undergo only the heterogonic cycle, while the *S. stercoralis* in humans showed a mixture of heterogonic and homogonic development. Nevertheless, Fulleborn (1914) concluded that the worms he found in dogs were a variety of *S. stercoralis*. Brumpt (1922) summarized the early findings but concluded from the same data that he "prefer to consider the dog parasitic species studied by Fülleborn as a distinct species (*Strongyloides canis*, n. sp.)" (translated from the French original by the author). However, Brumpt did not provide any further description of this new species. Later some authors (Augustine 1940; Augustine and Davey 1939) strongly supported the view of Brumpt (1922) based on the study of naturally infected dogs in the United States. To further support the notion that *Strongyloides* in humans and dogs are different, Augustine (1940) also reported the only attempt I am aware of, to experimentally infect a human volunteer with a dog-derived *Strongyloides*, which failed. Although the name *S. canis* stayed alive in the literature, most authors did not adopt it and Speare (1986) considered it a "nomen dubium" also pointing out that there are numerous reports showing that dogs are susceptible occasional hosts for various described species of *Strongyloides* other than *S. stercoralis*. However, in order to evaluate a possible role of dogs as sources for zoonotic human strongyloidiasis, the crucial question is not if all *Strongyloides* spp. in dogs are *S. stercoralis* or how many different species of *Strongyloides* dogs can carry or if they can be infected with human-derived *S. stercoralis* in the laboratory, but rather if dogs in a natural setting carry and transmit human pathogenic *S. stercoralis*? To address this question, molecular techniques have proved pivotal.

10.6.2 First Molecular Studies

Ramachandran et al. (1997) employed PCR amplification and restriction digestion of a portion of the ribosomal RNA encoding locus to compare different species of *Strongyloides* and *S. stercoralis* from different hosts and noticed differences between human and dog derived *S. stercoralis*. Later, protocols were established for the molecular taxonomy of nematodes based upon PCR amplification and sequencing of regions of the coding unit for the 18S ribosomal RNA, also known as Small ribosomal SubUnit (*SSU*) locus (Blaxter et al. 1998; Floyd et al. 2002; Eyualem and Blaxter 2003). These regions tend to be fairly invariable within particular nematode species, but in most cases differ between species (Herrmann et al. 2006; Floyd et al. 2002; Eyualem and Blaxter 2003) but see Box 10.2. Dorris et al. (2002) employed this method to propose a phylogeny of selected species of *Strongyloides* and close relatives.

Box 10.2
It is important to notice that sequence differentiation is an indirect measure of time of phylogenetic separation and not a test for species affiliation. However, as normally the time that passed since their last common ancestor is shorter for two individuals of the same species than for individuals of different species, the chance that a sequence change occurred in at least one of the lineages is larger for the more distantly related individuals of different species. However, neither is having the same sequence a proof that the carriers belong to the same species nor is having a different sequence proof for different species. Nevertheless, sequence similarities or differences do in general correlate with relatedness and therefore do have a certain predictive value with respect to species.

Later, Hasegawa and colleagues optimized molecular taxonomy for *Strongyloides* spp. by systematically evaluating the degree of sequence conservation in the four "hyper variable regions" in the *SSU* within and between species of *Strongyloides* (Hasegawa et al. 2009). Further, they complemented the analysis of the nuclear *SSU* by simultaneously analyzing a portion of the gene encoding the Cytochrome C oxidase subunit I (*coxI*) in the organellar genome of the mitochondrium (Hasegawa et al. 2010). Using these selected sequences, Hasegawa and colleagues analyzed the human infective *Strongyloides* spp., i.e., *S. stercoralis* and *S. fuelleborni* from humans, other primates, and dogs (Hasegawa et al. 2009,

2010, 2016). The authors noticed that, in agreement with Ramachandran et al. (1997), *S. stercoralis* tended to fall into phylogenetic groups according to host (primate or dogs) rather than geographic location i.e., based upon *cox1* sequence, four human-derived *S. stercoralis* from Japan were more similar to a worm from a human in Tanzania, Africa than to two *S. stercoralis* isolated from dogs in Japan (Hasegawa et al. 2010). From this, Hasegawa et al. (2010) concluded that *S. stercoralis* in dogs and humans belonged to different sub-clades of *S. stercoralis*, which made zoonotic transmission appear rather unlikely. However, while these studies represented a wide geographical range, they only included small numbers of individual worms from any given place and any given host.

10.6.3 The Picture Today

The picture changed when Jaleta et al. (2017) and Nagayasu et al. (2017) conducted studies on larger numbers of *S. stercoralis* worms isolated from humans and dogs in much more restricted geographical areas. While Nagayasu et al. (2017) analyzed worms from humans and dogs from Myanmar and Japan, Jaleta et al. (2017) sampled specifically in two villages in northern Cambodia and made sure that the dog samples were from dogs where at least one member of the owner's family was *S. stercoralis* positive. Both studies included more sequence information than just the *SSU* and *cox1*, with Nagayasu et al. (2017) using additional genes and Jaleta et al. (2017) including whole genome information. Both studies identified one clade of worms that were represented in humans and in dogs and at least one (probably two) clade that was found only in dogs. Based on the molecular information, these dog restricted *Strongyloides* spp. are more closely related to *S. stercoralis* than any other species for which the corresponding information is known, and their free-living females and males are morphometrically indistinguishable from *S. stercoralis* (Nagayasu et al. 2017). The presence of the same two types (dog and human infecting and dog restricted) of *Strongyloides* spp. was later also observed in aboriginal communities in Australia (Beknazarova et al. 2019). *S. stercoralis* that based upon their *SSU* and *cox1* sequences, belonged to the "human type" were also found in dogs in Thailand (Sanpool et al. 2019). Basso et al. (2019) molecularly analyzed *S. stercorals* from dogs that originated from European countries and were presented to the university small animal clinic at Berne, Switzerland. All *S. stercoralis* tested were molecularly of the "human type." Recently, Barratt et al. (2019) summarized the published molecular taxonomic studies in *S. stercoralis* and *S. fuelleborni* and added additional samples. As part of this publication the authors calculated a dendrogram for the different isolates based on the mitochondrial *cox1* sequence, which is reproduced here as Fig. 10.4 under the creative commons license. In this overview, two dog restricted clades, which are well separated from the human and dog infective clade and from each other, are easily visible. In the human infective clade, there is no obvious subdivision according to either host species or geography. Notice, the tree as shown, is a dendrogram based on very limited

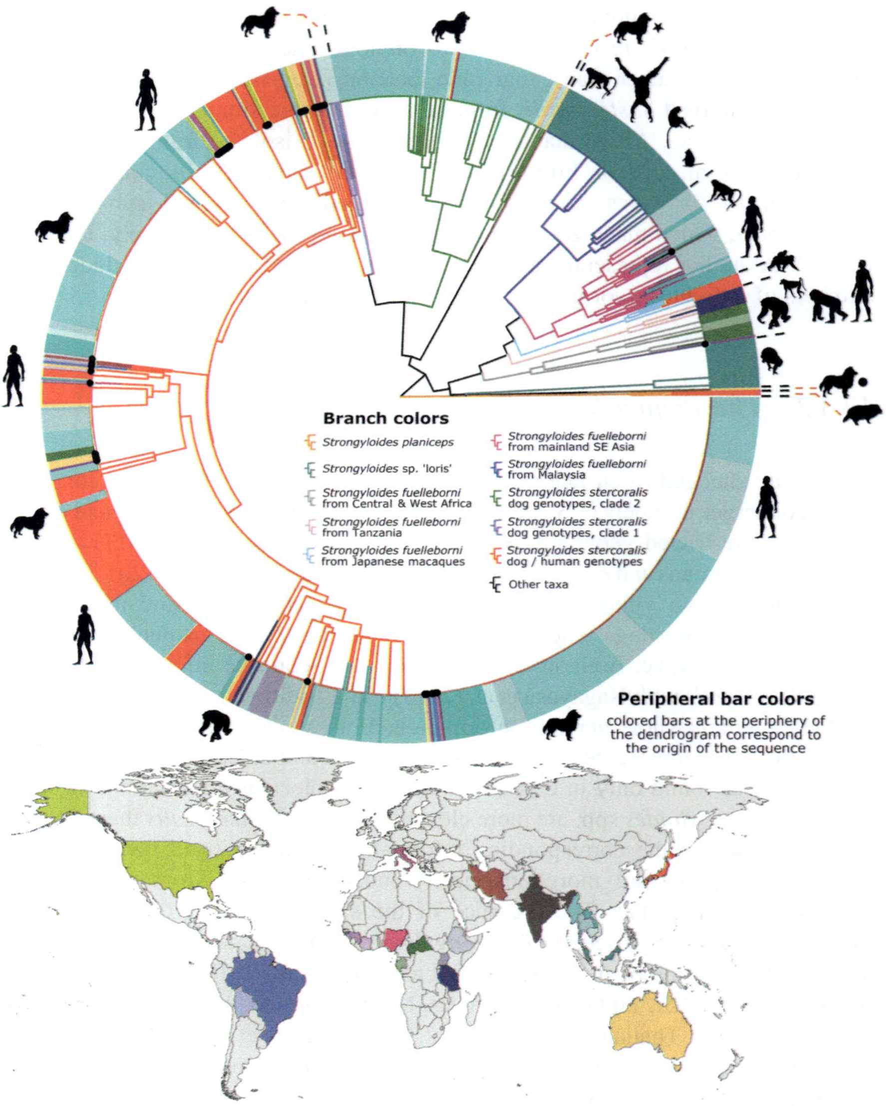

Fig. 10.4 Dendrogram of *cox1* sequences. This figure was reproduced from the addendum of Barratt et al. (2019) (corrected version of the figure as submitted to the journal by the authors, for copy right statement, see below). The legend was taken from the original reference and has only been modified to the extent necessary to achieve that the references adhere to the style of this article and refer now to the reference list of this article. This dendrogram represents 787 *cox1* sequences, including those generated by Barratt et al. (2019) (branches tipped in a black dot) and all published *cox1* sequences from GB (at the time this figure was originally prepared) that overlap completely with the 217 base *cox1* amplicon used by Barratt et al. (2019) (to these authors knowledge). Peripheral bars are colored according to their site of origin, which corresponds to the colored countries on the map. Branches are color-coded separately, according to their identity; either a species assignment, a genus, or their *S. stercoralis* genotype. The dog image with a black star indicates a sequence from an Australian dog generated by the same authors previously

sequence information. While it is suitable to show groups of closely related individuals, it should not be interpreted as a reliable reconstruction of the exact phylogenetic relationships. In particular, it should not be concluded that one of the two dog restricted clades is more closely related to the "human" clade than to the other "dog" clade. In Jaleta et al. (2017), the two dog restricted clades appear more closely related to each other than to the "human" clade, although with very low statistical support. Therefore, the phylogenetic relationship of these three clades has currently to be considered unknown. However, whole genome sequence analysis strongly suggested that the human (and dog) infective clade is genetically isolated from both dog-specific clades which suggest that the worms in the "dog" clades belong to different species than the ones in the "human" clade (Jaleta et al. 2017).

10.7 Conclusions and Final Remarks

So far, I have tried to summarize the relevant literature. In this last section, I will discuss and interpret these findings from my own point of view, which naturally is more subjective and has to be considered in part an opinion rather than fact.

At least in South East Asia and in Australia, dogs are infected with one or two other species of *Strongyloides* that are very closely related to *S. stercoralis*. They are unlikely to belong to other described species of *Strongyloides* with widely accepted names. Therefore, the name *S. canis* could be resurrected, although a clear species description is missing and it would still not be clear for which one of the probably two non human infective species in dogs it stands. In the past, various authors have discussed the importance or lack thereof of dogs as sources of human strongyloidiasis (Augustine 1940; Basso et al. 2019; Brumpt 1922; Thamsborg et al. 2017) with no convincing conclusion. In the few epidemiological studies where dogs were

Fig. 10.4 (continued) (Beknazarova et al. 2019), that is, distinct from other *Strongyloides* spp. and clusters between the *S. stercoralis* and *S. fuelleborni* groups. The dog image with a black circle highlights a published sequence (Beknazarova et al. 2019) that clusters close to, yet is distinct from *Strongyloides* spp. detected previously in lorises (Frias et al. 2018). Animal images reflect the mammalian hosts that the sequences were associated with. Two sequences of *Strongyloides planiceps* (orange branches) from Japanese raccoon dogs serve as an outgroup. The identity of each sequence is provided in S1 Fig of Barratt et al. (2019), which is a searchable PDF of the same dendrogram with all GB accession numbers, the countries of origin, and host species provided. The GB accession numbers for sequences in this dendrogram that were generated newly by Barratt et al. (2019) (branches tipped in a black dot) are provided in S1 File of Barratt et al. (2019). The sequences used to construct this dendrogram are provided in S2 File of Barratt et al. (2019). Copyright: This figure is free of copyright and was made available by the original authors under the Creative Commons Public Domain Dedication (https://creativecommons.org/publicdomain/zero/1.0/). Citation: Barratt et al. (2019) A global genotyping survey of *Strongyloides stercoralis* and *Strongyloides fuelleborni* using deep amplicon sequencing. *PLoS Negl Trop Dis*, 13, e0007609. The reproduced Figure is a corrected version that the authors submitted to the journal and made available to me at the same time

specifically evaluated as possible risk factors for *S. stercoralis*, they did not appear as such (Steinmann et al. 2007; Khieu et al. 2014b). However, as dogs tend to roam freely in these study areas, exposure to dog fecal contamination was probably not much different for the dog owner and his neighbor. The recent studies summarized above showed that in dogs *Strongyloides* spp. can be found that are neither morphologically nor molecularly distinguishable from human infective *S. stercoralis*. In my opinion, this leaves little doubt that dogs are susceptible to human infective *S. stercoralis* and do carry such worms under natural conditions. Furthermore, as we know that dogs do shed larvae of this species, which develop to the infective stage (Jaleta et al. 2017), it seems highly unlikely to me that these larvae would be unable to infect a human being who gets in contact with them. In fact, a few experimental infections of humans with *S. stercoralis* raised in dogs that had been experimentally infected with *S. stercoralis* of human origin were successful (Freedman 1991), demonstrating that human infective *S. stercoralis* can reproduce and give rise to infection competent iL3s in dogs. Therefore, in my opinion, for all practical purposes of strongyloidiasis control and prevention, dogs should be treated as a putative source of *S. stercoralis*. Dogs should be treated along with the humans they live with, examples for which have been reported (Wilson and Fearon 2018). Dog fecal contamination should be a factor when devising hygienic improvement, not only because of *S. stercoralis* but also because of numerous other diseases as described in other chapters of this book.

However, a conclusion in the interest of the prudent and careful handling of a putatively fatal pathogen is one thing, but a reliable scientific fact is another. For the first, it is sufficient that the danger appears likely enough that assuming that it does not exist is unacceptable. For the second a rigorous scientific evaluation of all the evidence needs to be done. From such a scientific stand point, we must admit that all the field data presented could also be explained with a one-way human to dog transfer of *S. stercoralis* alone. I am not aware of any case where it was clearly demonstrated that an *S. stercoralis* infection in a human was from larvae that were derived from a naturally infected dog. For obvious reasons, researchers were reluctant to test this experimentally and the only attempt to do so I am aware of, failed (Augustine 1940). However, knowing what we know now, chances are high that this experiment was done with a dog-specific type of *Strongloides* spp. Unambiguously demonstrating that a person was naturally infected by an *S. stercoralis* whose last parasitic ancestor (mother or grandmother, depending on the life cycle it went through) lived in a naturally infected dog will be challenging.

Nevertheless, the circumstantial evidence presented above is probably sufficient to conclude that, dog to human transfer is extremely likely to occur, even if only sporadically. However, we must admit, that currently, we have no idea what the relative importance of the zoonotic transmission of *S. stercoralis* is for human strongyloidiasis compared with human-to-human transmission. As a possible patient and a father, I would like to see dogs seriously treated as putative sources for zoonotic *S. stercoralis* (as for other diseases described in this book), while as a scientist, I would argue that we need more genetic and epidemiological research to further our understanding.

Acknowledgments I thank Alex Dulovic, James Lightfoot and Richard Bradbury for critically reading the manuscript and language editing. I thank Richard Bradbury for providing the corrected version of Fig. 10.2 prior to publication. I apologize to all my colleagues whose work was not cited or was cited only indirectly through review articles. The choice of references was in many places arbitrary. The work in my laboratory was funded by the Max Planck Society.

References

Agarwala R, Wasielewski J, Biman B (2014) Pulmonary strongyloidiasis following renal transplantation without travel to an endemic area. Oxf Med Case Rep 2014(4):83–85. https://doi.org/10.1093/omcr/omu032

Albertson DG, Nwaorgu OC, Sulston JE (1979) Chromatin diminution and a chromosomal mechanism of sexual differentiation in *Strongyloides papillosus*. Chromosoma 75(1):75–87

Albonico M, Becker SL, Odermatt P, Angheben A, Anselmi M, Amor A, Barda B, Buonfrate D, Cooper P, Getaz L, Keiser J, Khieu V, Montresor A, Munoz J, Requena-Mendez A, Savioli L, Speare R, Steinmann P, van Lieshout L, Utzinger J, Bisoffi Z, StrongNet Working G (2016) StrongNet: an international network to improve diagnostics and access to treatment for strongyloidiasis control. PLoS Negl Trop Dis 10(9):e0004898. https://doi.org/10.1371/journal.pntd.0004898

Anderson RC (2000) Nematode parasites of vertebrates. Their development ad transmission, 2nd edn. CAB International, Oxon

Augustine DL (1940) Experimental studies on the validity of species in the genus *Strongyloides*. Am J Hyg 32(Sect. D):24–32

Augustine DL, Davey DG (1939) Observations on a natural infection with *Strongyloides* in the dog. J Parasitol 25(2):117–119. https://doi.org/10.2307/3272350

Baaten GG, Sonder GJ, van Gool T, Kint JA, van den Hoek A (2011) Travel-related schistosomiasis, strongyloidiasis, filariasis, and toxocariasis: the risk of infection and the diagnostic relevance of blood eosinophilia. BMC Infect Dis 11:84

Barratt JLN, Lane M, Talundzic E, Richins T, Robertson G, Formenti F, Pritt B, Verocai G, Nascimento de Souza J, Mato Soares N, Traub R, Buonfrate D, Bradbury RS (2019) A global genotyping survey of *Strongyloides stercoralis* and *Strongyloides fuelleborni* using deep amplicon sequencing. PLoS Negl Trop Dis 13(9):e0007609. https://doi.org/10.1371/journal.pntd.0007609

Barroso M, Salvador F, Sanchez-Montalva A, Bosch-Nicolau P, Molina I (2019) *Strongyloides stercoralis* infection: a systematic review of endemic cases in Spain. PLoS Negl Trop Dis 13(3):e0007230. https://doi.org/10.1371/journal.pntd.0007230

Basso W, Grandt LM, Magnenat AL, Gottstein B, Campos M (2019) *Strongyloides stercoralis* infection in imported and local dogs in Switzerland: from clinics to molecular genetics. Parasitol Res 118(1):255–266. https://doi.org/10.1007/s00436-018-6173-3

Beknazarova M, Whiley H, Ross K (2016) Strongyloidiasis: a disease of socioeconomic disadvantage. Int J Environ Res Public Health 13(5). https://doi.org/10.3390/ijerph13050517

Beknazarova M, Barratt JLN, Bradbury RS, Lane M, Whiley H, Ross K (2019) Detection of classic and cryptic *Strongyloides* genotypes by deep amplicon sequencing: a preliminary survey of dog and human specimens collected from remote Australian communities. PLoS Negl Trop Dis 13(8):e0007241. https://doi.org/10.1371/journal.pntd.0007241

Bethony J, Brooker S, Albonico M, Geiger SM, Loukas A, Diemert D, Hotez PJ (2006) Soil-transmitted helminth infections: ascariasis, trichuriasis, and hookworm. Lancet 367(9521):1521–1532. https://doi.org/10.1016/S0140-6736(06)68653-4

Beukeboom L, Perrin N (2014) The evolution of sex determination. Oxford University Press, New York

Bisoffi Z, Buonfrate D, Montresor A, Requena-Mendez A, Munoz J, Krolewiecki AJ, Gotuzzo E, Mena MA, Chiodini PL, Anselmi M, Moreira J, Albonico M (2013) *Strongyloides stercoralis*: a plea for action. PLoS Negl Trop Dis 7(5):e2214. https://doi.org/10.1371/journal.pntd.0002214

Blaxter ML, De Ley P, Garey JR, Liu LX, Scheldeman P, Vierstraete A, Vanfleteren JR, Mackey LY, Dorris M, Frisse LM, Vida JT, Thomas WK (1998) A molecular evolutionary framework for the phylum Nematoda. Nature 392(6671):71–75

Breloer M, Abraham D (2017) *Strongyloides* infection in rodents: immune response and immune regulation. Parasitology 144(3):295–315. https://doi.org/10.1017/S0031182016000111

Brumpt E (1922) *Strongyloides stercoralis* (Bavay, 1877) [in French]. In: Brumpt E (ed) Précis de Parasitologie. Collection de Précis médicaux, 3ème edn. Mason et Cie, Paris, pp 691–697

Buonfrate D, Requena-Mendez A, Angheben A, Munoz J, Gobbi F, Van Den Ende J, Bisoffi Z (2013) Severe strongyloidiasis: a systematic review of case reports. BMC Infect Dis 13:78. https://doi.org/10.1186/1471-2334-13-78

Buonfrate D, Baldissera M, Abrescia F, Bassetti M, Caramaschi G, Giobbia M, Mascarello M, Rodari P, Scattolo N, Napoletano G, Bisoffi Z, Group CCMSS (2016) Epidemiology of *Strongyloides stercoralis* in northern Italy: results of a multicentre case-control study, February 2013 to July 2014. Euro Surveill 21(31):30310. https://doi.org/10.2807/1560-7917.ES.2016.21.31.30310

Buonfrate D, Bisanzio D, Giorli G, Odermatt P, Furst T, Greenaway C, French M, Reithinger R, Gobbi F, Montresor A, Bisoffi Z (2020) The global prevalence of *Strongyloides stercoralis* infection. Pathogens 9(6). https://doi.org/10.3390/pathogens9060468

Butlin R (2002) Evolution of sex: the costs and benefits of sex: new insights from old asexual lineages. Nat Rev Genet 3(4):311–317. https://doi.org/10.1038/nrg749

Camargo JF, Simkins J, Anjan S, Guerra G, Vianna R, Salama S, Albright C, Shipman E, Montoya J, Morris MI, Abbo LM (2019) Implementation of a *Strongyloides* screening strategy in solid organ transplant donors and recipients. Clin Transpl 33(4):e13497. https://doi.org/10.1111/ctr.13497

Choksi TT, Madison G, Dar T, Asif M, Fleming K, Clarke L, Danilewitz M, Hennawy R (2016) Multiorgan dysfunction syndrome from *Strongyloides stercoralis* hyperinfection in a patient with human T-cell lymphotropic virus-1 coinfection after initiation of ivermectin treatment. Am J Trop Med Hyg 95(4):864–867. https://doi.org/10.4269/ajtmh.16-0259

Dorris M, Viney ME, Blaxter ML (2002) Molecular phylogenetic analysis of the genus *Strongyloides* and related nematodes. Int J Parasitol 32(12):1507–1517

Dulovic A, Streit A (2019) RNAi-mediated knockdown of *daf-12* in the model parasitic nematode *Strongyloides ratti*. PLoS Pathog 15(3):e1007705. https://doi.org/10.1371/journal.ppat.1007705

Dulovic A, Puller V, Streit A (2016) Optimizing culture conditions for free-living stages of the nematode parasite *Strongyloides ratti*. Exp Parasitol 168:25–30. https://doi.org/10.1016/j.exppara.2016.06.005

Eberhardt AG, Mayer WE, Streit A (2007) The free-living generation of the nematode *Strongyloides papillosus* undergoes sexual reproduction. Int J Parasitol 37:989–1000

Eckert J, Friedhoff KT, Zahner H, Deplazes P (2005) Lehrbuch der Parasitologie für die Tiermedizin. Enke Verlag, Stuttgart

Elzein F, Albahili H, Bahloul A, Alonazi T, Alghamdi A, Alsufyani E, Musa A, Alsaeed M (2020) Transplant-related strongyloidiasis in solid organ transplant recipients in Saudi Arabia and the Gulf cooperation council countries. Int J Infect Dis 93:133–138. https://doi.org/10.1016/j.ijid.2020.01.032

Engelstadter J (2008) Constraints on the evolution of asexual reproduction. BioEssays 30 (11–12):1138–1150

Eyualem A, Blaxter M (2003) Comparison of biological, molecular, and morphological methods of species identification in a set of cultured Panagrolaimus isolates. J Nematol 35(1):119–128

Faust EC (1933) Experimental studies on human and primate species of *Strongyloides*. II. The development of *Strongyloides* in the experimental host. Am J Hyg 18:114–132

Fenton A, Paterson S, Viney ME, Gardner MP (2004) Determining the optimal developmental route of *Strongyloides ratti*: an evolutionarily stable strategy approach. Evolution 58(5):989–1000

Floyd R, Abebe E, Papert A, Blaxter M (2002) Molecular barcodes for soil nematode identification. Mol Ecol 11(4):839–850

Forrer A, Khieu V, Vounatsou P, Sithithaworn P, Ruantip S, Huy R, Muth S, Odermatt P (2019) *Strongyloides stercoralis*: spatial distribution of a highly prevalent and ubiquitous soil-transmitted helminth in Cambodia. PLoS Negl Trop Dis 13(6):e0006943. https://doi.org/10.1371/journal.pntd.0006943

Freedman DO (1991) Experimental infection of human subject with *Strongyloides* species. Rev Infect Dis 13(6):1221–1226. https://doi.org/10.1093/clinids/13.6.1221

Frias L, Stark DJ, Lynn MS, Nathan SK, Goossens B, Okamoto M, MacIntosh AJJ (2018) Lurking in the dark: cryptic *Strongyloides* in a Bornean slow loris. Int J Parasitol Parasites Wildl 7 (2):141–146. https://doi.org/10.1016/j.ijppaw.2018.03.003

Fulleborn F (1914) Untersuchungen über den Infektionsweg bei *Strongyloides* und *Ancylostomum* und die Biologie dieser Parasiten. Arch Schiff Tropenhyg 18:182–236

Gang SS, Castelletto ML, Bryant AS, Yang E, Mancuso N, Lopez JB, Pellegrini M, Hallem EA (2017) Targeted mutagenesis in a human-parasitic nematode. PLoS Pathog 13(10):e1006675. https://doi.org/10.1371/journal.ppat.1006675

Gardner MP, Gems D, Viney ME (2004) Aging in a very short-lived nematode. Exp Gerontol 39 (9):1267–1276

Gardner MP, Gems D, Viney ME (2006) Extraordinary plasticity in aging in *Strongyloides ratti* implies a gene-regulatory mechanism of lifespan evolution. Aging Cell 5(4):315–323

Getaz L, Castro R, Zamora P, Kramer M, Gareca N, Torrico-Espinoza MDC, Macias J, Lisarazu-Velasquez S, Rodriguez G, Valencia-Rivero C, Perneger T, Chappuis F (2019) Epidemiology of *Strongyloides stercoralis* infection in Bolivian patients at high risk of complications. PLoS Negl Trop Dis 13(1):e0007028. https://doi.org/10.1371/journal.pntd.0007028

Graham GL (1936) Studies on *Strongyloides*. I. *S. ratti* in parasitic series, each generation in the rat established with a single homogonic larva. Am J Hyg 24:71–87

Grant WN, Stasiuk S, Newton-Howes J, Ralston M, Bisset SA, Heath DD, Shoemaker CB (2006) *Parastrongyloides trichosuri*, a nematode parasite of mammals that is uniquely suited to genetic analysis. Int J Parasitol 36(4):453–466

Grove DI (1989) Historical introduction. In: Grove DI (ed) Strongyloidiasis: a major roundworm infection. Taylor & Francis, London, pp 1–9

Hammond MP, Robinson RD (1994) Chromosome complement, gametogenesis, and development of *Strongyloides stercoralis*. J Parasitol 80(5):689–695

Harvey SC, Viney ME (2001) Sex determination in the parasitic nematode *Strongyloides ratti*. Genetics 158(4):1527–1533

Hasan A, Le M, Pasko J, Ravin KA, Clauss H, Hasz R, Hunt E, Bosserman E, McAuliffe I, Montgomery SP, Kuehnert MJ, Hocevar SN, Abanyie F (2013) Transmission of *Strongyloides stercoralis* through transplantation of solid organs—Pennsylvania, 2012. Centers for Disease Control and Prevention. Morb Mortal Wkly Rep 62(14):264–266

Hasegawa H, Hayashida S, Ikeda Y, Sato H (2009) Hyper-variable regions in 18S rDNA of *Strongyloides* spp. as markers for species-specific diagnosis. Parasitol Res 104(4):869–874

Hasegawa H, Sato H, Fujita S, Nguema PP, Nobusue K, Miyagi K, Kooriyama T, Takenoshita Y, Noda S, Sato A, Morimoto A, Ikeda Y, Nishida T (2010) Molecular identification of the causative agent of human strongyloidiasis acquired in Tanzania: dispersal and diversity of *Strongyloides* spp. and their hosts. Parasitol Int 59(3):407–413. https://doi.org/10.1016/j.parint.2010.05.007

Hasegawa H, Kalousova B, McLennan MR, Modry D, Profousova-Psenkova I, Shutt-Phillips KA, Todd A, Huffman MA, Petrzelkova KJ (2016) *Strongyloides* infections of humans and great apes in Dzanga-Sangha protected areas, Central African Republic and in degraded forest fragments in Bulindi, Uganda. Parasitol Int 65(5 Pt A):367–370. https://doi.org/10.1016/j.parint.2016.05.004

Herrmann M, Mayer WE, Sommer RJ (2006) Nematodes of the genus *Pristionchus* are closely associated with scarab beetles and the Colorado potato beetle in Western Europe. Zoology (Jena) 109(2):96–108

Holterman M, van der Wurff A, van den Elsen S, van Megen H, Bongers T, Holovachov O, Bakker J, Helder J (2006) Phylum-wide analysis of SSU rDNA reveals deep phylogenetic relationships among nematodes and accelerated evolution toward crown clades. Mol Biol Evol 23(9):1792–1800

Hunt VL, Tsai IJ, Coghlan A, Reid AJ, Holroyd N, Foth BJ, Tracey A, Cotton JA, Stanley EJ, Beasley H, Bennett HM, Brooks K, Harsha B, Kajitani R, Kulkarni A, Harbecke D, Nagayasu E, Nichol S, Ogura Y, Quail MA, Randle N, Xia D, Brattig NW, Soblik H, Ribeiro DM, Sanchez-Flores A, Hayashi T, Itoh T, Denver DR, Grant W, Stoltzfus JD, Lok JB, Murayama H, Wastling J, Streit A, Kikuchi T, Viney M, Berriman M (2016) The genomic basis of parasitism in the *Strongyloides* clade of nematodes. Nat Genet 48(3):299–307. https://doi.org/10.1038/ng.3495

Jaleta TG, Lok JB (2019) Advances in the molecular and cellular biology of *Strongyloides* spp. Curr Trop Med Rep 6(4):161–178. https://doi.org/10.1007/s40475-019-00186-x

Jaleta TG, Zhou S, Bemm FM, Schar F, Khieu V, Muth S, Odermatt P, Lok JB, Streit A (2017) Different but overlapping populations of *Strongyloides stercoralis* in dogs and humans-dogs as a possible source for zoonotic strongyloidiasis. PLoS Negl Trop Dis 11(8):e0005752. https://doi.org/10.1371/journal.pntd.0005752

Jones JM, Hill C, Briggs G, Gray E, Handali S, McAuliffe I, Montgomery S, Komatsu K, Adams L (2016) Notes from the field: strongyloidiasis at a long-term-care facility for the developmentally disabled – Arizona, 2015. Morb Mortal Wkly Rep 65(23):608–609. https://doi.org/10.15585/mmwr.mm6523a5

Kalantari N, Chehrazi M, Ghaffari S, Gorgani-Firouzjaee T (2020) Serological assays for the diagnosis of *Strongyloides stercoralis* infection: a systematic review and meta-analysis of diagnostic test accuracy. Trans R Soc Trop Med Hyg 114:459. https://doi.org/10.1093/trstmh/trz135

Keiser PB, Nutman TB (2004) *Strongyloides stercoralis* in the immunocompromised population. Clin Microbiol Rev 17(1):208–217

Khieu V, Schar F, Marti H, Sayasone S, Duong S, Muth S, Odermatt P (2013) Diagnosis, treatment and risk factors of *Strongyloides stercoralis* in schoolchildren in Cambodia. PLoS Negl Trop Dis 7(2):e2035. https://doi.org/10.1371/journal.pntd.0002035

Khieu V, Schar F, Forrer A, Hattendorf J, Marti H, Duong S, Vounatsou P, Muth S, Odermatt P (2014a) High prevalence and spatial distribution of *Strongyloides stercoralis* in rural Cambodia. PLoS Negl Trop Dis 8(6):e2854. https://doi.org/10.1371/journal.pntd.0002854

Khieu V, Schar F, Marti H, Bless PJ, Char MC, Muth S, Odermatt P (2014b) Prevalence and risk factors of *Strongyloides stercoralis* in Takeo Province, Cambodia. Parasit Vectors 7:221. https://doi.org/10.1186/1756-3305-7-221

Kim JH, Kim DS, Yoon YK, Sohn JW, Kim MJ (2016) Donor-derived strongyloidiasis infection in solid organ transplant recipients: a review and pooled analysis. Transplant Proc 48 (7):2442–2449. https://doi.org/10.1016/j.transproceed.2015.11.045

Kreis HA (1932) Studies on the genus *Strongyloides* (nematodes). Am J Hyg 16:450–491

Kulkarni A, Dyka A, Nemetschke L, Grant WN, Streit A (2013) *Parastrongyloides trichosuri* suggests that XX/XO sex determination is ancestral in Strongyloididae (Nematoda). Parasitology 140(14):1822–1830. https://doi.org/10.1017/S0031182013001315

Little MD (1966) Comparative morphology of six species of *Strongyloides* (Nematoda) and redefinition of the genus. J Parasitol 52(1):69–84

Lok JB (2007) *Strongyloides stercoralis*: a model for translational research on parasitic nematode biology (February 17, 2007). In: Community TCeR (ed) WormBook. https://doi.org/10.1895/wormbook.1.134.1. http://www.wormbook.org

Lok JB (2018) Functional genomic study in *Strongyloides* and related parasitic nematodes by transgenesis and CRISPR/Cas9 mutagenesis. J Parasitol 104(6):586–589. https://doi.org/10.1645/18-134

Mackerras MJ (1959) *Strongyloides* and *Parastrongyloides* (Nematoda: Rhabdiasoidea) in Australian marsupials. Aust J Zool 7(2):87–104

Maeda Y, Palomares-Rius JE, Hino A, Afrin T, Mondal SI, Nakatake A, Maruyama H, Kikuchi T (2019) Secretome analysis of *Strongyloides venezuelensis* parasitic stages reveals that soluble and insoluble proteins are involved in its parasitism. Parasit Vectors 12(1):21. https://doi.org/10.1186/s13071-018-3266-x

Malinis M, Boucher HW, Practice ASTIDCo (2019) Screening of donor and candidate prior to solid organ transplantation-guidelines from the American Society of Transplantation Infectious Diseases Community of Practice. Clin Transpl 33(9):e13548. https://doi.org/10.1111/ctr.13548

Mansfield LS, Niamatali S, Bhopale V, Volk S, Smith G, Lok JB, Genta RM, Schad GA (1996) *Strongyloides stercoralis*: maintenance of exceedingly chronic infections. Am J Trop Med Hyg 55(6):617–624. https://doi.org/10.4269/ajtmh.1996.55.617

Miglioli-Galvao L, Pestana JOM, Lopes-Santoro G, Torres Goncalves R, Requiao Moura LR, Pacheco Silva A, Camera Pierrotti L, David Neto E, Santana Girao E, Costa de Oliveira CM, Saad Abboud C, Dias Franca JI, Devite Bittante C, Correa L, Aranha Camargo LF (2020) Severe *Strongyloides stercoralis* infection in kidney transplant recipients: a multicenter case-control study. PLoS Negl Trop Dis 14(1):e0007998. https://doi.org/10.1371/journal.pntd.0007998

Nagayasu E, Aung M, Hortiwakul T, Hino A, Tanaka T, Higashiarakawa M, Olia A, Taniguchi T, Win SMT, Ohashi I, Odongo-Aginya EI, Aye KM, Mon M, Win KK, Ota K, Torisu Y, Panthuwong S, Kimura E, Palacpac NMQ, Kikuchi T, Hirata T, Torisu S, Hisaeda H, Horii T, Fujita J, Htike WW, Maruyama H (2017) A possible origin population of pathogenic intestinal nematodes, *Strongyloides stercoralis*, unveiled by molecular phylogeny. Sci Rep 7(1):4844. https://doi.org/10.1038/s41598-017-05049-x

Naidu P, Yanow SK, Kowalewska-Grochowska KT (2013) Eosinophilia: a poor predictor of *Strongyloides* infection in refugees. Can J Infect Dis Med Microbiol 24(2):93–96

Nemetschke L, Eberhardt AG, Hertzberg H, Streit A (2010) Genetics, chromatin diminution, and sex chromosome evolution in the parasitic nematode genus *Strongyloides*. Curr Biol 20(19):1687–1696

Nigon VM, Felix MA (2017) History of research on *C. elegans* and other free-living nematodes as model organisms. WormBook, pp 1–91. https://doi.org/10.1895/wormbook.1.181.1

Nolan TJ, Megyeri Z, Bhopale VM, Schad GA (1993) *Strongyloides stercoralis*: the first rodent model for uncomplicated and hyperinfective strongyloidiasis, the Mongolian gerbil (*Meriones unguiculatus*). J Infect Dis 168(6):1479–1484

Nolan TJ, Brenes M, Ashton FT, Zhu X, Forbes WM, Boston R, Schad GA (2004) The amphidial neuron pair ALD controls the temperature-sensitive choice of alternative developmental pathways in the parasitic nematode, *Strongyloides stercoralis*. Parasitology 129(Pt 6):753–759

Nordheim E, Olafsson Storro M, Natvik AK, Birkeland Kro G, Midtvedt K, Varberg Reisaeter A, Hagness M, Fevang B, Pettersen FO (2018) Donor-derived strongyloidiasis after organ transplantation in Norway. Transpl Infect Dis 21:e13008. https://doi.org/10.1111/tid.13008

Nutman TB (2017) Human infection with *Strongyloides stercoralis* and other related *Strongyloides* species. Parasitology 144(3):263–273. https://doi.org/10.1017/S0031182016000834

Ogawa A, Streit A, Antebi A, Sommer RJ (2009) A conserved endocrine mechanism controls the formation of dauer and infective larvae in nematodes. Curr Biol 19(1):67–71

Olsen A, van Lieshout L, Marti H, Polderman T, Polman K, Steinmann P, Stothard R, Thybo S, Verweij JJ, Magnussen P (2009) Strongyloidiasis—the most neglected of the neglected tropical diseases? Trans R Soc Trop Med Hyg 103(10):967–972

Page W, Speare R (2016) Chronic strongyloidiasis – don't look and you won't find. Aust Fam Physician 45(1):40–44

Page W, Judd JA, Bradbury RS (2018) The unique life cycle of *Strongyloides stercoralis* and implications for public health action. Trop Med Infect Dis 3(2). https://doi.org/10.3390/tropicalmed3020053

Patton JB, Bonne-Annee S, Deckman J, Hess JA, Torigian A, Nolan TJ, Wang Z, Kliewer SA, Durham AC, Lee JJ, Eberhard ML, Mangelsdorf DJ, Lok JB, Abraham D (2017) Methylprednisolone acetate induces, and Delta7-dafachronic acid suppresses, *Strongyloides stercoralis* hyperinfection in NSG mice. Proc Natl Acad Sci USA 115:204. https://doi.org/10.1073/pnas.1712235114

Pires-daSilva A (2007) Evolution of the control of sexual identity in nematodes. Semin Cell Dev Biol 18(3):362–370

Puthiyakunnon S, Boddu S, Li Y, Zhou X, Wang C, Li J, Chen X (2014) Strongyloidiasis—an insight into its global prevalence and management. PLoS Negl Trop Dis 8(8):e3018. https://doi.org/10.1371/journal.pntd.0003018

Ramachandran S, Gam AA, Neva FA (1997) Molecular differences between several species of *Strongyloides* and comparison of selected isolates of *S. stercoralis* using a polymerase chain reaction-linked restriction fragment length polymorphism approach. Am J Trop Med Hyg 56 (1):61–65. https://doi.org/10.4269/ajtmh.1997.56.61

Roseman DA, Kabbani D, Kwah J, Bird D, Ingalls R, Gautam A, Nuhn M, Francis JM (2013) *Strongyloides stercoralis* transmission by kidney transplantation in two recipients from a common donor. Am J Transplant Off J Am Soc Transplant Am Soc Transplant Surg 13 (9):2483–2486. https://doi.org/10.1111/ajt.12390

Rothe K, Katchanov J, Schneider J, Spinner CD, Phillip V, Busch DH, Tappe D, Braren R, Schmid RM, Slotta-Huspenina J (2020) *Strongyloides stercoralis* hyperinfection syndrome presenting as mechanical ileus after short-course oral steroids for chronic obstructive pulmonary disease (COPD) exacerbation. Parasitol Int 76:102087. https://doi.org/10.1016/j.parint.2020.102087

Sanpool O, Intapan PM, Rodpai R, Laoraksawong P, Sadaow L, Tourtip S, Piratae S, Maleewong W, Thanchomnang T (2019) Dogs are reservoir hosts for possible transmission of human strongyloidiasis in Thailand: molecular identification and genetic diversity of causative parasite species. J Helminthol 94:e110. https://doi.org/10.1017/S0022149X1900107X

Schad GA (1989) Morphology and life history of *Strongyloides stercoralis*. In: Grove DI (ed) Strongyloidiasis: a major roundworm infection of man. Taylor & Francis, London, pp 85–104

Schär F, Trostdorf U, Giardina F, Khieu V, Muth S, Marti H, Vounatsou P, Odermatt P (2013) *Strongyloides stercoralis*: global distribution and risk factors. PLoS Negl Trop Dis 7(7):e2288. https://doi.org/10.1371/journal.pntd.0002288

Siddiqui AA, Berk SL (2001) Diagnosis of *Strongyloides stercoralis* infection. Clin Infect Dis 33 (7):1040–1047. https://doi.org/10.1086/322707

Speare R (1986) Studies on the taxonomy of *Strongyloides* (Nematoda; Strongyloididae). James Cook University of North Queensland

Speare R (1989) Identification of species of *Strongyloides*. In: Grove DI (ed) Strongyloidiasis: a major roundworm infection of man. Taylor & Francis, London, pp 11–83

Steinmann P, Zhou XN, Du ZW, Jiang JY, Wang LB, Wang XZ, Li LH, Marti H, Utzinger J (2007) Occurrence of *Strongyloides stercoralis* in Yunnan Province, China, and comparison of diagnostic methods. PLoS Negl Trop Dis 1(1):e75. https://doi.org/10.1371/journal.pntd.0000075

Steinmann P, Yap P, Utzinger J, Du ZW, Jiang JY, Chen R, Wu FW, Chen JX, Zhou H, Zhou XN (2015) Control of soil-transmitted helminthiasis in Yunnan province, People's Republic of China: experiences and lessons from a 5-year multi-intervention trial. Acta Trop 141 (Pt B):271–280. https://doi.org/10.1016/j.actatropica.2014.10.001

Stoltzfus JD, Bart SM, Lok JB (2014) cGMP and NHR signaling co-regulate expression of insulin-like peptides and developmental activation of infective larvae in *Strongyloides stercoralis*. PLoS Pathog 10(7):e1004235. https://doi.org/10.1371/journal.ppat.1004235

Streit A (2008) Reproduction in *Strongyloides* (Nematoda): a life between sex and parthenogenesis. Parasitology 135(3):285–294

Streit A (2014) How to become a parasite without sex chromosomes: a hypothesis for the evolution of *Strongyloides* spp. and related nematodes. Parasitology 141:1244–1254. https://doi.org/10.1017/S003118201400064X

Streit A (2017) Genetics: modes of reproduction and genetic analysis. Parasitology 144:316–326. https://doi.org/10.1017/S0031182016000342

Strkolcova G, Goldova M, Bockova E, Mojzisova J (2017) The roundworm *Strongyloides stercoralis* in children, dogs, and soil inside and outside a segregated settlement in Eastern Slovakia: frequent but hardly detectable parasite. Parasitol Res 116(3):891–900. https://doi.org/10.1007/s00436-016-5362-1

Tamarozzi F, Martello E, Giorli G, Fittipaldo A, Staffolani S, Montresor A, Bisoffi Z, Buonfrate D (2019) Morbidity associated with chronic *Strongyloides stercoralis* infection: a systematic review and meta-analysis. Am J Trop Med Hyg 100(6):1305–1311. https://doi.org/10.4269/ajtmh.18-0895

Thamsborg SM, Ketzis J, Horii Y, Matthews JB (2017) *Strongyloides* spp. infections of veterinary importance. Parasitology 144:274–284. https://doi.org/10.1017/S0031182016001116

Viney M (2017) *Strongyloides*. Parasitology 144(3):259–262. https://doi.org/10.1017/S0031182016001773

Viney M, Kikuchi T (2017) *Strongyloides ratti* and *S. venezuelensis* – rodent models of *Strongyloides* infection. Parasitology 144(3):285–294. https://doi.org/10.1017/S0031182016000020

Viney ME, Lok JB (2015) The biology of *Strongyloides* spp. (July 16, 2015). In: Community TCeR (ed) WormBook. 2015/07/18 edn. WormBook. https://doi.org/10.1895/wormbook.1.141.2s http://www.wormbook.org

Viney ME, Ashford RW, Barnish G (1991) A taxonomic study of *Strongyloides* Grassi, 1879 (Nematoda) with special reference to i von Linstow, 1905 in man in Papua New Guinea and the description of a new subspecies. Syst Parasitol 18:95–109

Viney ME, Matthews BE, Walliker D (1992) On the biological and biochemical nature of cloned populations of *Strongyloides ratti*. J Helminthol 66(1):45–52

Viney ME, Matthews BE, Walliker D (1993) Mating in the nematode parasite *Strongyloides ratti*: proof of genetic exchange. Proc R Soc – Biol Sci 254(1341):213–219

Wang Z, Zhou XE, Motola DL, Gao X, Suino-Powell K, Conneely A, Ogata C, Sharma KK, Auchus RJ, Lok JB, Hawdon JM, Kliewer SA, Xu HE, Mangelsdorf DJ (2009) Identification of the nuclear receptor DAF-12 as a therapeutic target in parasitic nematodes. Proc Natl Acad Sci USA 106(23):9138–9143

Wang Z, Stoltzfus J, You YJ, Ranjit N, Tang H, Xie Y, Lok JB, Mangelsdorf DJ, Kliewer SA (2015) The nuclear receptor DAF-12 regulates nutrient metabolism and reproductive growth in nematodes. PLoS Genet 11(3):e1005027. https://doi.org/10.1371/journal.pgen.1005027

Watts MR, Robertson G, Bradbury RS (2016) The laboratory diagnosis of *Strongyloides stercoralis*. Microbiol Aust 37(1). https://doi.org/10.1071/MA16003

White MAE, Whiley H, Ross KE (2019) A review of *Strongyloides* spp. environmental sources worldwide. Pathogens 8:91. https://doi.org/10.3390/pathogens8030091

Wilson A, Fearon D (2018) An unusual rash: *Strongyloides stercoralis* presenting as larva currens in a 12-year-old girl with Crohn's disease. J Paediatr Child Health 55:364. https://doi.org/10.1111/jpc.14265

Winnicki W, Eder M, Mazal P, Mayer FJ, Sengolge G, Wagner L (2018) Prevalence of *Strongyloides stercoralis* infection and hyperinfection syndrome among renal allograft recipients in Central Europe. Sci Rep 8(1):15406. https://doi.org/10.1038/s41598-018-33775-3

Yamada M, Matsuda S, Nakazawa M, Arizono N (1991) Species-specific differences in heterogonic development of serially transferred free-living generations of *Strongyloides planiceps* and *Strongyloides stercoralis*. J Parasitol 77(4):592–594

Zhou S, Fu X, Pei P, Kucka M, Liu J, Tang L, Zhan T, He S, Chan YF, Rödelsperger C, Liu D, Streit A (2019a) Characterization of a non-sexual population of *Strongyloides stercoralis* with hybrid 18S rDNA haplotypes in Guangxi, Southern China. PLoS Negl Trop Dis 13(5): e0007396. https://doi.org/10.1371/journal.pntd.0007396
Zhou S, Harbecke D, Streit A (2019b) From the feces to the genome: a guideline for the isolation and preservation of *Strongyloides stercoralis* in the field for genetic and genomic analysis of individual worms. Parasit Vectors 12(1):496. https://doi.org/10.1186/s13071-019-3748-5

Chapter 11
Dirofilaria spp. Do They Get Under Your Skin (Or into Your Heart)?

Alice Vismarra, Marco Genchi, Chiara Cattabiani, and Laura Kramer

Abstract The World Health Organization defines the "One Health" concept as being particularly relevant in the control of zoonoses. *Dirofilaria immitis* has a worldwide distribution, while *Dirofilaria repens* is found only in Europe, Asia, and Africa. Both parasites have zoonotic potential and wherever canine dirofilariosis exists, there is a risk of human infection. *D. immitis* causes heartworm disease in dogs and cats, while *D. repens* causes subcutaneous dirofilariosis, primarily in dogs. *D. immitis* is found in Europe, the United States, Africa, India, South America, and Australia. *D. repens*, on the other hand is endemic in many countries of the Old World but has not yet been found in the Americas. The geographical location of both parasites is changing and they are currently spreading into previously unaffected areas, due to movement of infected animals and climate change. Both parasites have zoonotic potential and wherever canine dirofilariaois exists, there is a risk of human infection. Human subcutaneous dirofilariosis caused by *D. repens* is currently considered an emerging disease in humans. *D. immitis* infections in humans are less frequent, but the disease is potentially severe.

This chapter describes the biology of the two nematodes, the disease they cause in dogs, the main animal host, and in humans.

Keywords *Dirofilaria immitis* · *Dirofilaria repens* · Human subcutaneous dirofilariosis · Prevention

A. Vismarra (✉) · M. Genchi · C. Cattabiani · L. Kramer
Department of Veterinary Science, University of Parma, Parma, Italy
e-mail: alice.vismarra@unipr.it

C. Strube, H. Mehlhorn (eds.), *Dog Parasites Endangering Human Health*, Parasitology Research Monographs 13,
https://doi.org/10.1007/978-3-030-53230-7_11

11.1 Life Cycle of *Dirofilaria immitis* (Fig. 11.1)

Dogs and cats with *Dirofilaria immitis* harbor adult parasites (females are approximately 25–31 cm long, males 12–20 cm) in the pulmonary arteries (Fig. 11.2). Small first-stage larvae, called "microfilariae" circulate in the blood and are ingested by mosquitoes. Approximately 15 days later, larvae become infective and are introduced into a new host when the mosquito bites. Following several months of tissue migration, parasites arrive in the pulmonary artery and begin to release microfilariae.

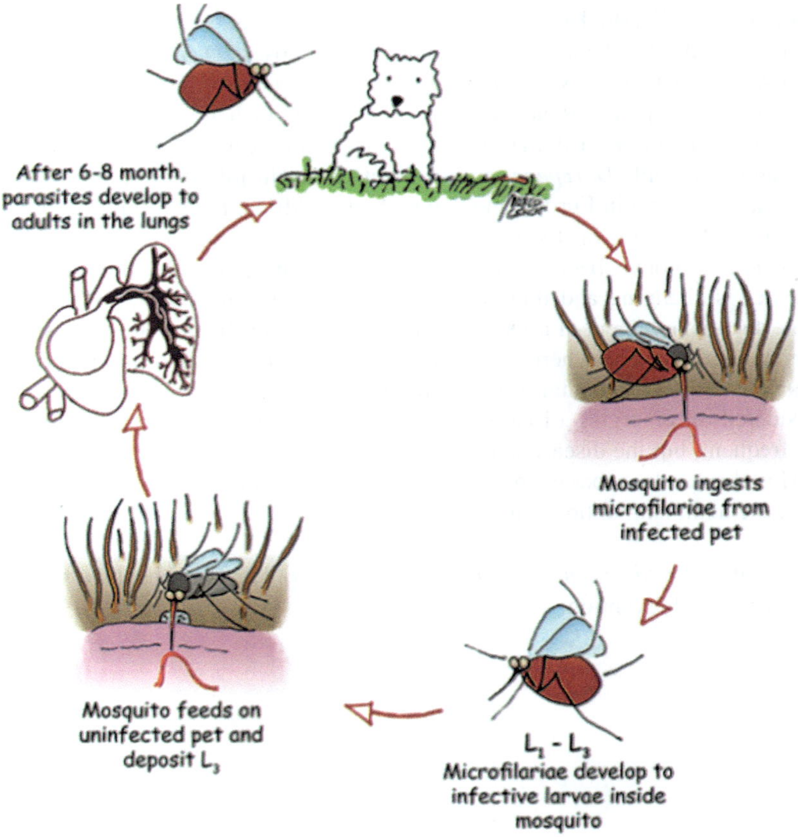

Fig. 11.1 *Dirofilaria immitis* life cycle

Fig. 11.2 Adult *Dirofilaria immitis* in the pulmonary artery of an infected dog (with kind permission of Dr. Luigi Venco)

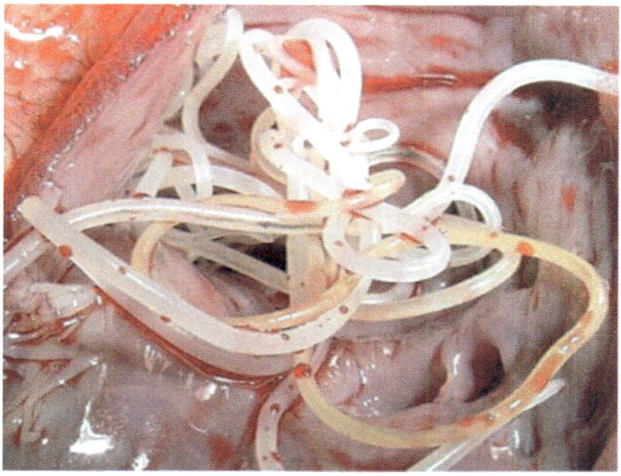

11.2 Life Cycle of *Dirofilaria repens* (Fig. 11.3)

The life cycle of *D. repens* is very similar to that of *D. immitis*. The same mosquito species are involved in transmission and infected animals have circulating microfilariae in the blood. The main difference is the location of the adult parasites that reside in the subcutaneous tissue. Normally, 5–8 months are necessary from the moment of infection to the moment that microfilariae are available for mosquitos to take up.

11.3 Canine Heartworm Disease (HWD; *Dirofilaria immitis*)

Box 11.1
The clinical presentations common in dogs with HWD include:

- Coughing, dyspnea due to inflammation
- "Fainting" (syncope) due to pulmonary hypertension
- Ascites (peritoneal effusion) due to right heart congestive failure
- "Caval Syndrome" due to a sudden rise in pulmonary pressure and the subsequent displacement of worms from the pulmonary artery into the right cardiac chambers. Breathing difficulty, loud heart murmur (right side of the thorax), and hemoglobinuria are pivotal clinical signs in this syndrome.

Dogs with HWD can appear healthy for quite a long time following infection. It takes several months for the parasites to reach the lungs and if the number of parasites is low and/or the dog is not an active dog (hunting, sports, agility, etc.), this "asymptomatic" phase can even last years. However, as time passes and the

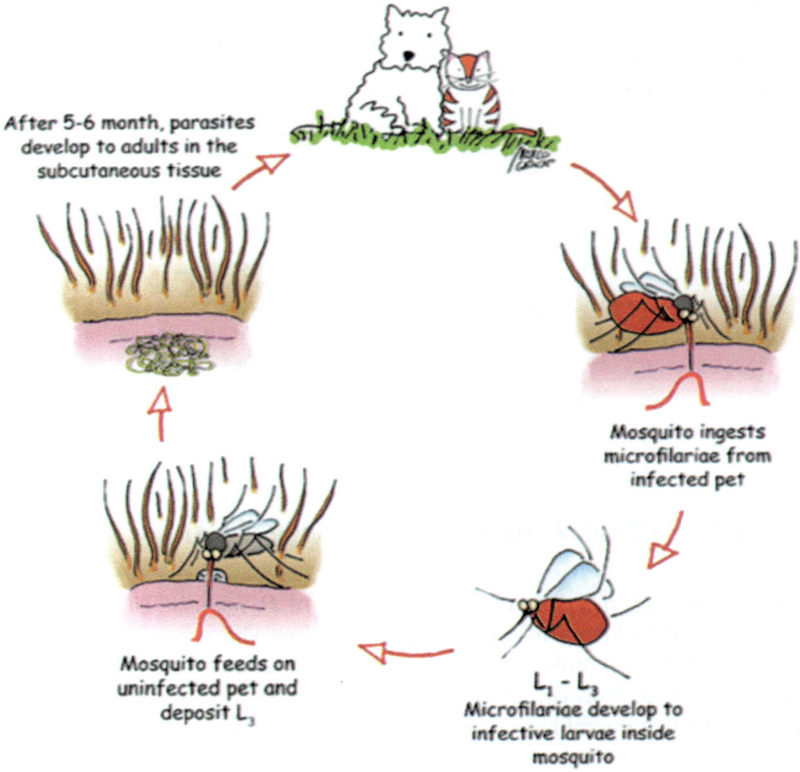

Fig. 11.3 *Dirofilaria repens* life cycle

parasites cause inflammation and hardening of the pulmonary arteries, different symptoms can appear (see Box 11.1). The disease involves primarily pulmonary arteries and lungs and the heart is involved only in the last stages of infection, which is characterized by right congestive heart failure. At times acute clinical signs can be observed in the late stage of the disease (pulmonary thromboembolism, Caval Syndrome).

Diagnosis of HWD in dogs requires a blood test for the identification of circulating microfilariae by microscopy and the parasite antigens through serology. Table 11.1 shows how to interpret test results in dogs.

Chest X-rays are also necessary to determine the severity of disease and the possible side effects of treatment. When adult worms die, they are flushed down into the smaller branches of the pulmonary arteries, where they can cause thromboembolism (a clot forms around the dead parasites and blocks the flow of blood into the lung). The only available drug for treatment is melarsomine di-hydrochloride. Several studies (Bazzocchi et al. 2008; Bendas et al. 2017; Grandi et al. 2010; Mavropoulou et al. 2014; Savadelis et al. 2017) have shown that a combination of

Table 11.1 Brief explanation on how to interpret diagnostic test results in dog and which actions need to be taken in each specific case

Knott test	Ag test	Interpretation	Action
Negative	Negative	False negatives for both tests include young parasites or young animals	Repeat test after 7 months
Positive	Positive	Positive	Treat
Positive	Negative	False negatives for Ag tests include low female worm burden	Treat
Negative	Positive	False negatives for mf include infections with only one sex (male or female), the use of drugs that kill mf or immune-mediated elimination	Treat

an anti-parasitic drug belonging to the macrocyclic lactones, together with the antibiotic doxycycline is also able to eliminate the parasite. Exercise limitations and support therapy with anti-inflammatory drugs are also necessary for several months following treatment to avoid posttreatment complications.

Prevention of HWD in dogs is based on the administration of larvicidal drugs that can eliminate the larvae that have been inoculated by infected mosquitos. In this way, the parasite will not reach adulthood and the disease is prevented. These include ivermectin, moxidectin, selamectin, and milbemycin oxima. Administration of these larvicidals is monthly or, in the case of injectable formulations, once a year. The timing of prevention is important and is based on the environmental conditions ideal for mosquito activity (Self et al. 2019). The current recommendation is to carry out prevention all year round; even in winter, urban heat islands allow mosquitoes to survive. It is also very important to prescribe preventives to dogs that will be travelling to areas that are endemic for the disease (https://www.esccap.org/travelling-pets-advice).

11.4 Canine Subcutaneous Dirofilariosis

The majority of *D. repens*-infected dogs do not show any clinical signs. When present, the primary clinical sign is the development of one or more skin nodules anywhere on the body. They can measure anywhere from 0.5 to 3 cm (Fig. 11.4). On histology, nodules are granulomatous to suppurative and may contain cross-sections of filarial nematodes and/or microfilariae. The infiltrate is mainly composed of lymphocytes, macrophages, plasma cells, neutrophils, and eosinophils in different proportions. Sporadic reports of erythema, papules, alopecia, and pruritus have also been described in dogs with natural *D. repens* infection (Hargis et al. 1999; Bourdeau and Roussel 2010; Albanese et al. 2013) (Fig. 11.4).

Diagnosis of canine subcutaneous dirofilariosis (SCD) relies on the morphological identification of *D. repens* microfilariae, that must be distinguished from *D. immitis* mf (Fig. 11.3; Box 11.2). There is currently no serological test available for diagnosing *D. repens* in dogs.

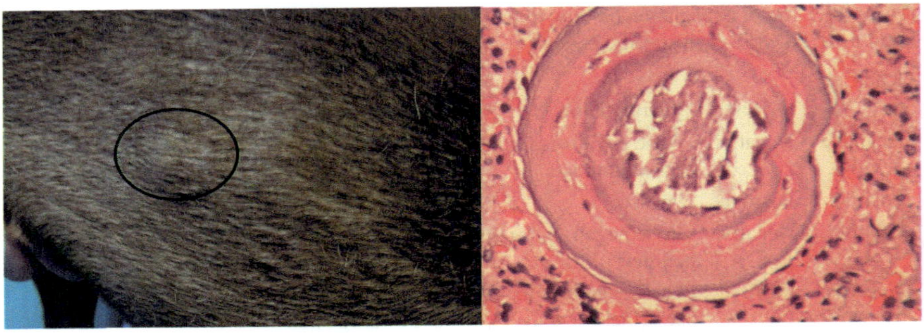

Fig. 11.4 Skin nodule caused by *Dirofilaria repens* in a dog

Treatment of infection requires surgical removal of nodules when present, and monthly administration of topical moxidectin (Petry et al. 2015). Prevention is the same as for *D. immitis*.

11.5 Human Dirofilariosis

Box 11.2 Differential Features of *D. immitis* vs. *D. repens* microfilariae

D. immitis:

Conical head, tail straight with the end pointed

Mean length 295.5–308.0 μm

D. repens: Blunt head, tail sharp and filiform often ending like an umbrella handle

Mean length 358.7–380.20 μm

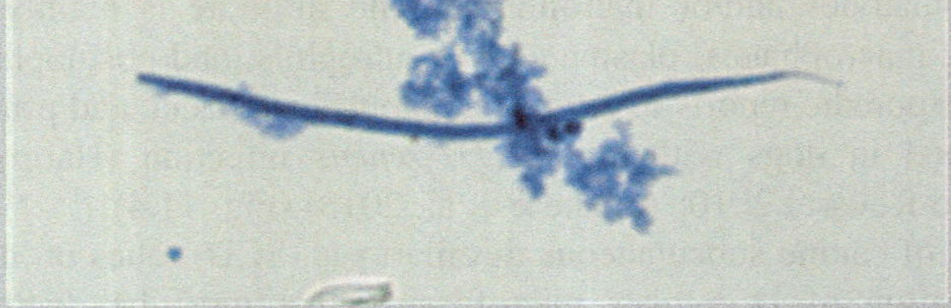

Microfilariae from *D. repens* (top) and *D. immitis* (bottom)

Human infections with *Dirofilaria* spp. have been reported from many countries where canine infection is prevalent. Humans can become infected with *Dirofilaria* anywhere infected dogs and competent mosquito vectors are present (Fig. 11.5). In most cases when humans are bitten by an infected mosquito, the immune system responds and the parasite does not develop. However, it can occur that the larvae begin to migrate and reach different areas of the body. There are several forms of human dirofilariosis.

1. Pulmonary dirofilariosis. This form of infection is caused by the migration of *D. immitis* or *D. repens* toward the pulmonary arteries. The immature worms die once they have reached the lungs and the inflammatory response from the patient leads to the development of a lung nodule. In the Americas, this form of human dirofilariosis is caused by *D. immitis*, while in the Old World, where both parasites are present, pulmonary dirofilariosis can be due to either nematode (Simón et al. 2012). Clinical signs are often absent and the infection may go unnoticed. When signs are present (coughing) and radiographs are taken, the nodule can be misdiagnosed as pulmonary neoplasia. Indeed, the greatest challenge for physicians in cases of pulmonary dirofilariosis is the differential diagnosis. The nodule should be removed surgically and histology carried out. While the infection is benign, the emotional consequences for the patient are many and surgery is invasive.
2. Subcutaneous and ocular dirofilariosis. This form of infection is caused mainly by *D. repens* and is currently considered an emerging and neglected disease in Europe (Genchi et al. 2009, 2011; Otranto et al. 2013; Penezic et al. 2014; Genchi and Kramer 2017; Capelli et al. 2018; Self et al. 2019). The parasite can be inoculated into different parts of the body and migrate to different sites while it develops. The most frequent localizations are around the eyes, eyelids, under the conjunctiva membrane (Fig. 11.6). Patients with ocular dirofilariosis report itching and burning of the eye, and the feeling of "something" in the eye. Swelling of the eyelids and the periorbital area has also been reported.

Subcutaneous dirofilariosis usually presents as a single, painless nodule that can be found on the chest, upper and lower limbs, and neck. In males, the scrotal area is often involved, while in females nodules in the subcutaneous tissue of the breast can be misdiagnosed as neoplasia (Genchi et al. 2011; Pampiglione and Rivasi 2007).

Human subcutaneous dirofilariosis is a diagnostic challenge for physicians. There is no reliable, sensitive and specific test for confirming infection. Human patients are not microfilariaemic and there is currently no commercially available serological test. Ultrasound examination of nodules has been reported as being able to identify live, motile parasites, thus differentiating them from neoplasia (Gopinath et al. 2013). However, in the vast majority of cases, surgical removal is necessary to make a definitive diagnosis. Worms can sometimes be seen macroscopically within the nodule. Other times, the inflammatory granuloma destroys much of the parasite and only histology, with the identification of worm debris, is diagnostic (Fig. 11.7).

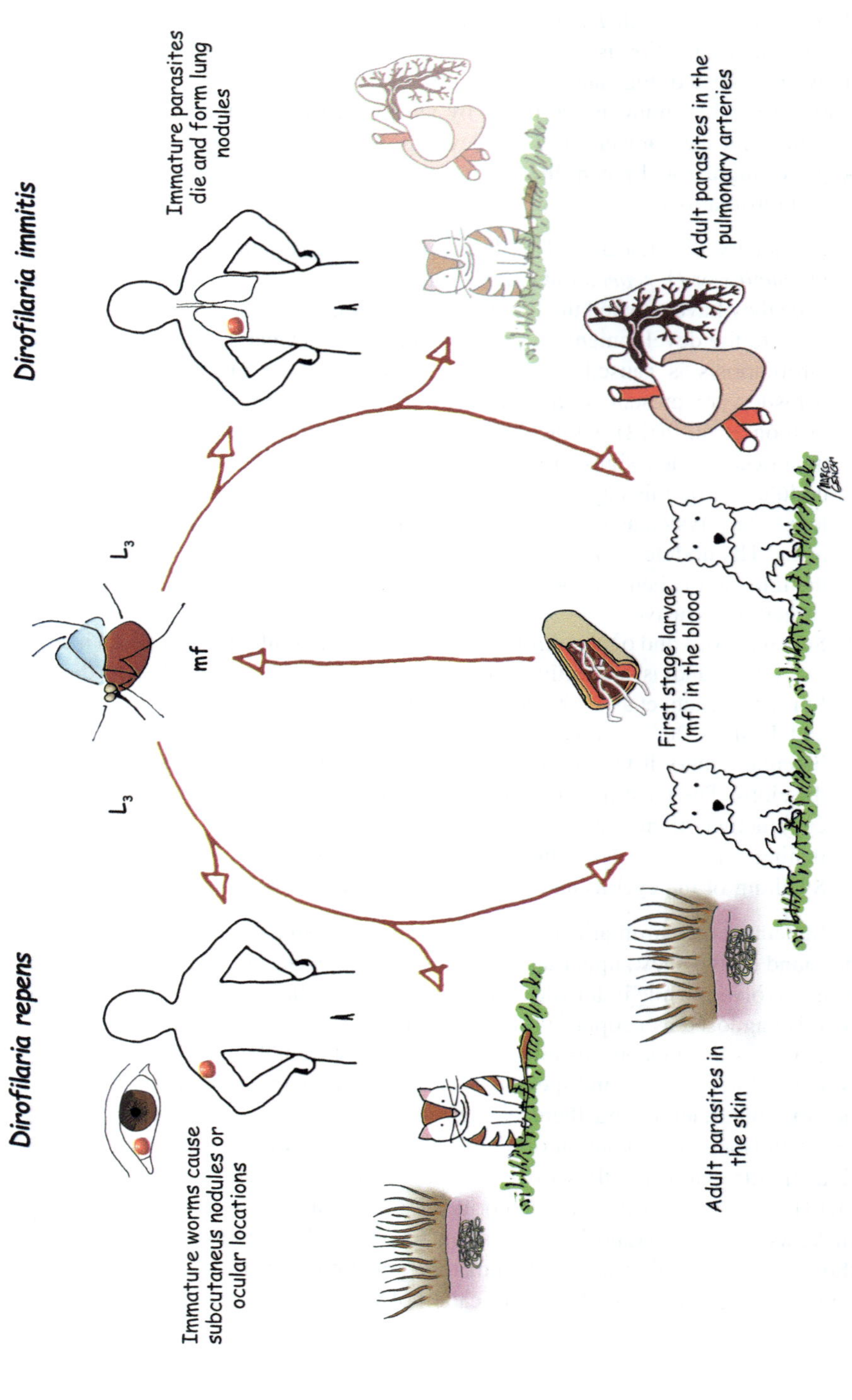

Fig. 11.5 Human infection by *Dirofilaria* spp.

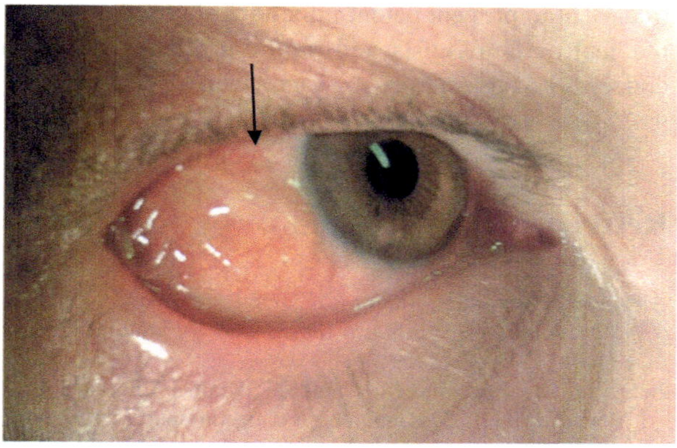

Fig. 11.6 *Dirofilaria repens* under the conjunctiva membrane

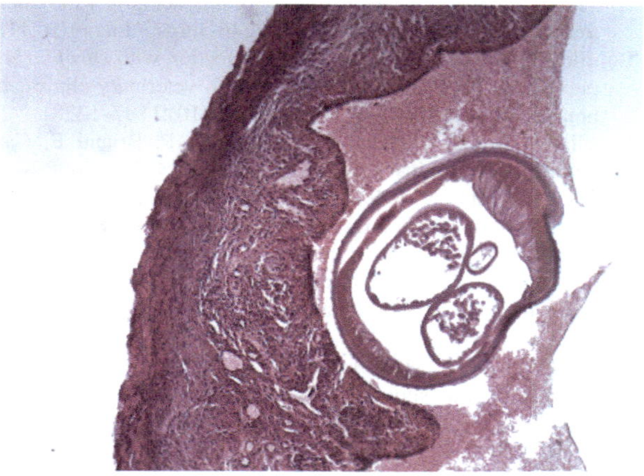

Fig. 11.7 Cross section of *Dirofilaria repens* in a nodule

11.6 Conclusions

For vector-borne diseases where an animal species serves as a reservoir, especially a pet, veterinarians play a significant role in prevention and should be more aware of their responsibility in reducing the impact of the zoonotic agents. The abundance of mosquitoes in the environment and the prevalence of microfilaraemic dogs in the human habitat are the main risk of *Dirofilaria* spp. infections for people. The direct contact between infected dogs and humans is not a risk factor for humans to become infected. Dogs living in areas where the parasite is present must receive preventives. In endemic areas, physicians should inform their clients that human dirofilariosis is an emerging infection. Physicians should suggest to their clients who own dogs to ask their veterinarians to examine the dog for circulating microfilariae. In case of positivity, dogs should be treated with a microfilaricidal drug and prophylactic treatment to prevent patent infections should be initiated. Finally, collaboration with medical entomologists and public health experts should be enhanced, under the concept (and the actions) of One Health.

References

Albanese F, Abramo F, Braglia C, Caporali C, Venco L, Vercelli A et al (2013) Nodular lesions due to infestation by *Dirofilaria repens* in dogs from Italy. Vet Dermatol 24:255–e56. https://doi.org/10.1111/vde.12009

Bazzocchi C, Mortarino M, Grandi G, Kramer LH, Genchi C, Bandi C, Genchi M, Sacchi L, McCall JW (2008) Combined ivermectin and doxycycline treatment has microfilaricidal and adulticidal activity against *Dirofilaria immitis* in experimentally infected dogs. Int J Parasitol 38 (12):1401–1410. https://doi.org/10.1016/j.ijpara.2008.03.002

Bendas AJR, Mendes-de-Almeida F, Von Simson C, Labarthe N (2017) Heat pretreatment of canine samples to evaluate efficacy of imidacloprid + moxidectin and doxycycline in heartworm treatment. Parasit Vectors 10(246):246. https://doi.org/10.1186/s13071-017-2189-2

Bourdeau P, Roussel A (2010) Canine infection with *Dirofilaria repens* in France: distribution and prevalence from a national survey with veterinary clinics. In: Proceedings ICOPA XII, Melbourne, August 2010 (Int J Parasitol 40(10):1107–1228)

Capelli G, Genchi C, Baneth G, Bourdeau P, Brianti E, Cardoso L, Danesi P, Fuehrer HP, Giannelli A, Ionică AM, Maia C, Modrý D, Montarsi F, Krücken J, Papadopoulos E, Petrić D, Pfeffer M, Savić S, Otranto D, Poppert S, Silaghi C (2018) Recent advances on *Dirofilaria repens* in dogs and humans in Europe. Parasit Vectors 11(1):663. https://doi.org/10.1186/s13071-018-3205-x

Genchi C, Kramer L (2017) Subcutaneous dirofilariosis (*Dirofilaria repens*): an infection spreading throughout the old world. Parasit Vectors 10(Suppl 2):517. https://doi.org/10.1186/s13071-017-2434-8

Genchi C, Rinaldi L, Mortarino M, Genchi M, Cringoli G (2009) Climate and dirofilaria infection in Europe. Vet Parasitol 163:286–292. https://doi.org/10.1016/j.vetpar.2009.03.026

Genchi C, Kramer LH, Rivasi F (2011) Dirofilarial infections in Europe. Vector Borne Zoonotic Dis 11(10):1307–1317. https://doi.org/10.1089/vbz.2010.0247

Gopinath TN, Lakshmi KP, Shaji PC, Rajalakshmi PC (2013) Periorbital dirofilariosis – clinical and imaging findings: live worm on ultrasound. Indian J Ophthalmol 61:298–300. https://doi.org/10.4103/0301-4738.114111

Grandi G, Quintavalla C, Mavropoulou A, Genchi M, Gnudi G, Bertoni G, Kramer L (2010) A combination of doxycycline and ivermectin is adulticidal in dogs with naturally acquired heartworm disease (*Dirofilaria immitis*). Vet Parasitol 169(3–4):347–351. https://doi.org/10.1016/j.vetpar.2010.01.025 (Erratum in: Vet Parasitol. 2011; 177(1–2), 196. Dosage error in published abstract)

Hargis AM, Lewis TP, Duclos DD, Loeffler DG, Rausch RI (1999) Dermatitis associated with microfilariae (Filarioidea) in 10 dogs. Vet Dermatol 10:95–107

Mavropoulou A, Gnudi G, Grandi G, Volta A, Kramer LH, Quintavalla C (2014) Clinical assessment of post-adulticide complications in *Dirofilaria immitis*-naturally infected dogs treated with doxycycline and ivermectin. Vet Parasitol 205(1–2):211–215. https://doi.org/10.1016/j.vetpar.2014.06.014

Otranto D, Dantas-Torres F, Brianti E, Traversa D, Petrić D, Genchi C, Capelli G (2013) Vector-borne helminths of dogs and humans in Europe. Parasit Vectors 16(6):16. https://doi.org/10.1186/1756-3305-6-16

Pampiglione S, Rivasi F (2007) Human dirofilariasas due to Dirofilaria (Nochtiella) repens: an update of world literature from 1995 to 2000. Parassitologia 42:231–254

Penezic A, Selakovic S, Pavlovic I, Cirovic D (2014) First findings and prevalence of adult heartworms (*Dirofilaria immitis*) in wild carnivores from Serbia. Parasitol Res 113:3281–3285. https://doi.org/10.1007/s00436-014-3991-9

Petry G, Genchi M, Schmidt H, Schaper R, Lawrenz B, Genchi C (2015) Evaluation of the adulticidal efficacy of Imidacloprid 10%/Moxidectin 2.5% (w/v) spot-on (Advocate®, Advantage® Multi) against *Dirofilaria repens* in experimentally infected dogs. Parasitol Res 114 (Suppl 1):S131–S144. https://doi.org/10.1007/s00436-015-4519-7

Savadelis MD, Ohmes CM, Hostetle JA, Settje TL, Zolynas R, Dzimianski MT, Moorhead AR (2017) Assessment of parasitological findings in heartworm-infected beagles treated with advantage Multi® for dogs (10% imidacloprid + 2.5% moxidectin) and doxycycline. Parasit Vectors 10(1):245. https://doi.org/10.1186/s13071-017-2190-9

Self SW, Pulaski CN, McMahan CS, Brown DA, Yabsley MJ, Gettings JR (2019) Regional and local temporal trends in the prevalence of canine heartworm infection in the contiguous United States: 2012-2018. Parasit Vectors 12(1):380. https://doi.org/10.1186/s13071-019-3633-2

Simón F, Siles-Lucas M, Morchón R, González-Miguel J, Mellado I, Carretón E, Montoya-Alonso JA (2012) Human and animal dirofilariosis: the emergence of a zoonotic mosaic. Clin Microbiol Rev 25(3):507–544. https://doi.org/10.1128/CMR.00012-12

Chapter 12
Guinea Worm Infection in Dogs: A Case of Reverse Zoonosis that Impedes *Dracunculus medinensis* Eradication

M. Teresa Galán-Puchades

Abstract *Dracunculus medinensis* or Guinea worm was largely considered an exclusive human parasite. The adult female *D. medinensis* (up to 100 cm long and 1.5–2.0 mm thick) inhabits and moves in the connective tissue, including the skin. Large females protrude from the skin causing unusual and unambiguous signs. Hosts become infected by drinking water containing the crustacean intermediate hosts (cyclopoid copepods known as water fleas) infected with Guinea worm L3 larvae. After years of a successful eradication campaign (focused mainly on preventing humans from drinking unfiltered or untreated water), Guinea worm transmission has been eliminated from most, but not all, countries. An unforeseeable high rate of dog dracunculiasis, mainly in Chad, has been detected as a result of the surveillance program started in 2012. This reverse zoonosis is preventing dracunculiasis from becoming the first infectious disease to be eradicated without a vaccine or specific medical treatment and the second human pathogen eradicated after smallpox. The alternative food-borne route of transmission, in which frogs and fish are involved, suggested by scientists of the Guinea Worm Eradication Program in an attempt to explain dracunculiasis transmission route in dogs, is also discussed.

Keywords *Dracunculus medinensis* · Dog dracunculiasis · Reverse zoonosis · Reservoirs · Guinea worm eradication program · Water-borne transmission · Food-borne transmission

---

M. T. Galán-Puchades (✉)
Faculty of Pharmacy, Parasite and Health Research Group, Department of Pharmacy, Pharmaceutical Technology and Parasitology, University of Valencia, Burjassot, Valencia, Spain
e-mail: mteresa.galan@uv.es

C. Strube, H. Mehlhorn (eds.), *Dog Parasites Endangering Human Health*, Parasitology Research Monographs 13,
https://doi.org/10.1007/978-3-030-53230-7_12

12.1 One Health Approach in Dracunculiasis

According to the World Health Organization (WHO), dracunculiasis, also known as Guinea worm disease (GWD), caused by the nematode *Dracunculus medinensis*, was the first parasitic disease set for eradication (https://www.who.int/dracunculiasis/eradication/en/). A fact that contributed to the idea of eradicability of the disease was the largely accepted belief that there was no known animal reservoir. However, nowadays, the zoonotic character of GWD has been accepted since several mammals, mainly dogs but also cats and baboons (*Papio anubis*) have been found infected with the same species affecting human beings (Galán-Puchades 2018; Thiele et al. 2018).

The One Health concept, introduced by USAID (United States Agency for International Development), WHO, FAO (Food and Agriculture Organization), and OIE (World Organization for Animal Health), recognizes that the health of humans is connected to the health of animals and the environment. Therefore, a clear example of a disease in which this concept is currently being applied is dracunculiasis.

The term "zoonosis" usually refers to a disease that is transmitted from animals to humans (also called "anthropozoonosis"), although actually zoonosis is any disease that is transmitted from animals to humans, or vice versa. The terms "reverse zoonosis" or "zooanthroponosis" refer to a human disease transferred to animals and then transferred back to humans (Messenger et al. 2014; Teshome 2019).

Due to the discovered high rate of canine dracunculiasis infection in Chad, dogs are being blamed for thwarting efforts to eradicate GWD (Callaway 2016; Molyneux and Sankara 2017). As a matter of fact, dogs are not, in this case, executioners but victims, since it was humans who initially transmitted the disease to their best friends, dogs. Now, dogs are only paying humans back with this reverse zoonosis.

But let us start from the beginning.

12.2 Guinea Worm Disease Throughout History

The best-documented parasitic disease known from ancient times seems to be caused by *D. medinensis* (Cox 2002). Given its large size, our ancestors are bound to have been aware of this worm.

The earliest descriptions are from the Egyptian medical Ebers papyrus (1550 BC), one of the oldest known medical works. Miller (1989) discusses the evidence for the occurrence, transmission, and treatment of dracunculiasis in ancient Egypt. The use of the word "dqr" in the papyrus seems to describe the gradual extraction of the worm from the body by winding it on a stick, which still is the only specific treatment of the disease nowadays (Fig. 12.1). Nevertheless, confirmation of the presence of Guinea worms in ancient Egypt comes from the finding of worms in Egyptian mummies (Nunn and Tapp 2000).

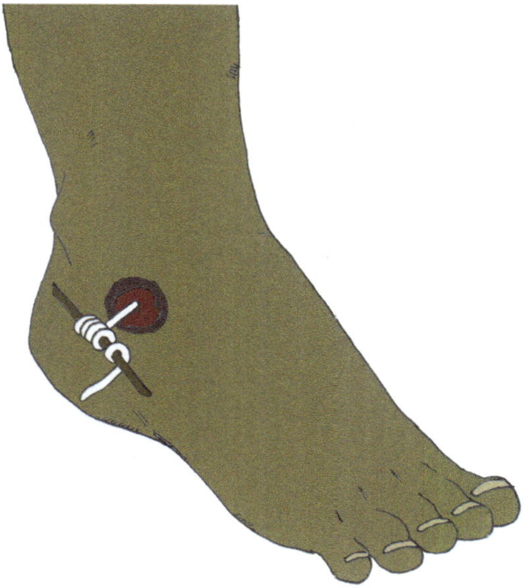

Fig. 12.1 Traditional method for removing *D. medinensis* female (illustration by Angela Debenedetti, PhD)

Apparently, some historians have identified references to this worm and its disease in the Bible. Most parasitologists accept that the "fiery serpent" that afflicted the Israelites in the Red Sea region after the Exodus from Egypt was actually Guinea worms (Bible, Numbers, 21,6) (Fig. 12.2).

Likewise, the rod of Asclepius, symbol of modern medicine, is depicted as a serpent coiled around a rod (Fig. 12.3). In ancient Greek mythology, Asclepius was the god of medicine and healing. The symbol might have its origin from the mentioned method of removal of *D. medinensis* (Fig. 12.1). This helminthic infection was so common then, that doctors advertised their skill by a graphical representation of a worm wound around a stick (Shetty et al. 2014).

Actually, the rod of Asclepius is the logo of WHO to symbolize its commitment to global health care (Fig. 12.4).

The Persian theologian and physician Avicenna, in around 1000 AD reported in his book *Liber canonis medicinae* on malaria and on many worms, especially on *Dracunculus*, which today in French is still called "fil d'Avicenne."

Explorers, navigators, travelers, physicians, epidemiologists, military officers, slave traders, administrators, among others, have contributed to the rich history of the Guinea worm (Edungbola 2019). In fact, Linschoten, a Dutch navigator, observed evidence of the disease at Hormuz (Persian Gulf) in 1584. Likewise, Sir James Emerson Tenner, who was the Colonial Secretary of the British Government in Ceylon from 1845 to 1850, also observed the Guinea worm and wrote that the natives attributed the disease to drinking water from particular wells (Gooneratne 1969). Actually, the name Guinea worm is attributed to Bruno or Braun, a traveler from Basle (Switzerland) who in about 1611 made several voyages to that part of the African coast, and on his return, published an account on local diseases (Gooneratne

Fig. 12.2 Israelites bitten by fiery serpents (Bible, Numbers, 21,6). (A print from the Phillip Medhurst Collection of Bible illustrations)

1969). In 1674, Georgius Hieronymus Velschius initiated the scientific study of the worm and the disease it caused (Velschius 1674).

European parasitologists remained ignorant of it until about the beginning of the nineteenth century, when British army medical officers began serving in India. In 1819, Carl Asmund Rudolphi discovered adult female worms containing the larvae, a discovery that was followed up in 1834 by a Dane known only as Jacobson. In 1836, D. Forbes, a British army officer serving in India, found and described the larvae of *D. medinensis* in water, and over the next few years several parasitologists, including George Busk, pursued the idea that humans became infected through the skin (Cox 2002). It was not until 1870 that the whole life cycle, including the stages in the crustacean intermediate host, was described by the Russian Alexei Pavlovitch Fedchenko (Fedchenko 1870a), when he was 27 years old (he died 2 years later in a climbing accident on Mont Blanc, France) (Litvinov 1991). A reproduction of his paper appeared in 1971 in the *American Journal of Tropical Medicine and Hygiene* (Fedchenko 1870b) on the occasion of the centenary of its publication (Anderson

Fig. 12.3 Statue of the rod of Asplecius (Archaeological Museum of Rhodes, Greece)

Fig. 12.4 World Health Organization logo

2000). His discovery that humans become infected by swallowing infected copepods paved the way to dracunculiasis control.

Dracunculiasis is likely to have originally infected monkeys and apes in the central part of Africa (Adamson 1988). Traders could then have brought infected animals to northern Africa. A close relationship between humans and apes, considered sacred animals, is known to have existed in Egypt during the period of the New Kingdom (1552–1069 BC) (Adamson 1988). The role of dogs in Guinea worm disease was also established at that time in Egypt. According to Adamson (1988), as Guinea worms are known to infect dogs, these animals may have become a reservoir of the disease in Lower Egypt, an area where dogs were numerous and lived in close association with humans.

Natural Guinea worm infections in dogs during the eighteenth to twentieth centuries were compiled by Muller (1971) and WHO (1996), including findings from Egypt, China, Kazakhstan, India, Ivory Coast, Timbuktu, and Tanzania. Therefore, apparently, Guinea worm infection in dogs has been a part of the history of dracunculiasis practically since its beginning.

Related to humans, dogs and dracunculiasis, curiously, in the art gallery of Bari (Italy) there is a painting from the sixteenth century that seems to be the only case of ancient dracunculiasis in the history of European art (Gaeta et al. 2017). The painting represents St. Roch, a French pilgrim who cured numerous people from the Black Death in the fourteenth century. In the painting, a kind of worm that protrudes several centimeters from the Saint's left leg can be clearly appreciated, a feature that is consistent with a case of dracunculiasis (Fig. 12.5, left). St. Roch, actually, is the Patron Saint of dogs and the typical iconography usually represents him beside a dog (Fig. 12.5, right), since a dog brought him food when he was sick.

12.3 *Dracunculus medinensis* Taxonomy

Though sometimes classed with the filarial worms, *Dracunculus* is not a filaria. *D. medinensis* current accepted taxonomy is:

Phylum: Nematoda

Class: Chromadorea

Order: Rhabditida

Suborder: Spirurina

Superfamify: Dracunculoidea

Family: Dracunculidae

Genus: *Dracunculus*

Species: *D. medinensis*

Adult dracunculoid nematodes occur in tissues and serous layers of mammals, fish, reptiles, amphibians, and birds. The extreme sexual dimorphism is a typical feature in Dracunculidae since females are far longer than males. This family has another genus, *Avioserpens*, a parasite of birds, while *Dracunculus* parasitizes

Fig. 12.5 St. Roch: Left, Anonymous painter (Archive of Pinacoteca Metropolitana di Bari *Corrado Giaquinto*, Italy. Tempera on canvas; fifteenth to sixteenth century); Right, Depicted by José de Ribera (Museo Nacional del Prado, Spain. Oil on canvas; 1631)

mammals and reptiles (snakes and turtles). The human Guinea worm, *D. medinensis*, is the best known and most studied species among the 14 recognized *Dracunculus* species, most of them present in the Old World (Cleveland et al. 2018).

Historically, as already pointed out, *D. medinensis* has sporadically been reported in dogs, with *D. insignis* being the species identified to infect domestic dogs and cats in the New World (the United States and Canada) (reviewed by Williams et al. 2018).

12.4 Dracunculiasis: Disease of Poverty that Causes Poverty

The female *D. medinensis* is a long, string-like worm. Dracunculiasis is the word in Latin for "affliction with little dragons," that, in addition to GWD, is also known as dracunculosis, dracontiasis, Medina worm disease, *Filaria medinensis* or Fil d'Avicenne. As a curiosity, among its local names are Farentit (Arabic), Mfa (Fanti and Twi), Kurkunu (Hausa), Naru (Hindi), Reshteh or Piyook (Persian), Rishta (Uzbek), Orok al Mai (Yemen), and Sobiya (Yoruba) (Muller 2002).

The disease causes functional disability with impaired mobility that leads to the inability to work or to attend school. The disability average lasts from 2 to 16 weeks. As a consequence, the economic losses reach millions of dollars annually and school absenteeism can exceed 60% (Smith et al. 1989; Ruiz-Tiben and Hopkins 2006). Those are the reasons why the disease was once called the "disease of empty granaries."

12.4.1 Morphology

Dracunculus medinensis is a thread-like worm. Considered one of the largest adult nematodes, the average elongated female measures 840 by 1.5 mm. The female possesses a prominent blunt, rounded anterior end. The rarely seen adult male is smaller than the female, measuring, on average, only 21 by 0.4 mm, not exceeding 40 mm. The anterior end of the male characteristically coils itself at least once.

The mouth is small and triangular, surrounded by a quadrangular, sclerotized plate. Lips are absent. Cephalic papillae are arranged in an outer circle of four double papillae, at about the same level as the amphids, and an inner circle of two double papillae, which are peculiar in that they are dorsal and ventral. Their esophagus has a large glandular portion that protrudes and lies alongside the thin muscular portion.

In young female worms, the vulva is placed in an equatorial position, being atrophied and nonfunctional in adults. The gravid uterus has an anterior and a posterior branch, each of which is filled with hundreds of thousands of embryos. The intestine becomes squashed and nonfunctional as a result of uterine pressure.

A major difficulty in taxonomy of dracunculids is the sparsity of discovered males. The length of the few specimens known ranges from 12 to 40 mm; spicules are unequal and 490–750 μm long. The gubernaculum ranges from 115 to 130 μm in length. Genital papillae vary considerably in published descriptions. Males remain in the connective tissues surrounding deeper muscles, usually in the thoracic region, and are either absorbed or eventually calcify.

First-stage larvae expelled by the adult female hatch from the eggshell just during the moment of birth. Larvae measure, on average, 653 by 23 μm, with a pointed tail that takes up one-third of the body and has a fully formed gut, although they do not feed.

12.4.2 Parasite Life Cycle: Water-Borne Transmission

Dracunculiasis belongs to the group of water-borne transmitted parasites that, in their life cycle, alternates between a mammal vertebrate definitive host and an arthropod intermediate host, depending, partially or completely, on water as their natural habitat. This group includes malaria, onchocerciasis, and lymphatic filariasis. These three parasitic diseases are transmitted by the bites of hematophagous arthropods with aquatic larval stages, as is the case of *Anopheles* mosquito in malaria, the black fly *Simulium* in onchocerciasis or Culicidae mosquitoes in lymphatic filariasis. However, the larval stages of *D. medinensis* become infective in tiny crustaceans belonging to the family Cyclopidae, genus *Cyclops* (known as water fleas), which develop their entire life cycle in water. Therefore, unlike the other mentioned parasites, dracunculiasis is transmitted only by means of drinking contaminated water with infected *Cyclops*. No other alternative route of infection had ever been considered until recently.

Definitive hosts become infected when, after drinking water containing infected *Cyclops*, the gastric juice effectively kills the water fleas and *D. medinensis* infective L3 larvae are able to escape through the hard *Cyclops* exoskeleton. According to the experiments carried out by Moorthy (1932), in cases of hyperchlorhydria, both, the water flea and the Guinea worm larvae, are killed by the gastric juice. In other conditions, such as hypochlorhydria or achylia gastrica, the acid in the gastric juice is not strong enough to kill the copepods and set the infective Guinea worm larvae free. Therefore, only those potential definitive hosts (humans, dogs, or others) with an adequate percentage of hydrochloric acid in the gastric juice, sufficient to kill only the water fleas, will become infected (Moorthy 1932). Since the lethal dose of hydrochloric acid for the copepods varies with different *Cyclops* species, the amount of infection in any place appears to depend, to a certain extent, on the particular *Cyclops* species. Mass infection, maturity, and *Cyclops* species as well as the concentration of the hydrochloric acid in the gastric juice appear to be the factors that ensure successful infection, and alterations in any of these factors may contribute to explain the variations in individual susceptibility (Moorthy 1932).

According to the National Research Council (1983), released larvae migrate quickly to the duodenal wall and proceed to the abdominal and thoracic cavities where they begin maturing in connective tissue. Male and female worms mate about 3 months after ingestion. The male worms die between the third and the seventh months of age and then become encysted, calcified, or are absorbed. The adult female worm lives in the connective tissues. At about 8 months, female worms usually move down to the lower limbs, where the uterus containing first-stage larvae develops to take up nearly the entire adult worm. Internal pressure and progressive senility cause the body wall and uterus of the parasite to burst, forcing a loop of the uterus through, freeing many juveniles. Each female worm releases about a million microscopic first-stage larvae into the water. Juveniles stimulate a violent allergic reaction that causes a blister in the skin of the host. Infected people frequently try to relieve the burning sensation by immersing the affected part in water. Contact with water causes the rupture of the uterus of the worm and stimulates the release of larvae into the water. The process is repeated intermittently over several weeks.

The released larvae remain active for 4–7 days in pond water or step wells, but are only able to infect an intermediate host for up to 72 h in a more favorable alkaline pH (Joshi 1992). However, their infectivity declines after only 12 h of free-living existence (Joshi et al. 1997). To develop further they must be eaten by copepods exceeding a certain minimum size. When the water temperature is above 21 °C, the larvae inside a *Cyclops* undergo second-stage molting and develop into the infective third stage in about 12–14 days. First-stage larvae swallowed directly by humans do not undergo further development and are probably killed immediately by gastric juices. *Dracunculus* larvae do not reach the third stage unless they enter the *Cyclops.*

According to the experimental studies conducted by Onabamiro (1951, 1954) and Joshi et al. (1997), *Cyclops* containing third-stage larvae tend to sink to the bottom of experimental jars and become active only when the bottom is stirred. Consequently, infected *Cyclops* would be more likely to be picked up during the dry season, when water levels in ponds or step wells are low. People drawing drinking water from stagnant surface-water sources during the height of the transmission season are exposed to higher rates of infected *Cyclops* (approximately 5 in 100 *Cyclops* may be infected). Dracunculiasis transmission is markedly seasonal and, in most parts of Africa (and formerly in India), the maximum incidence coincides with the planting season, resulting in great economic hardship. In semidesert areas of Africa drinking water is obtained from ponds during the rainy season but from deep wells for the rest of the year when the ponds are dry. However, in the humid savannah regions of West Africa where rainfall exceeds 150 cm per year, there is no transmission during the rainy season, when ponds turn into streams and *Cyclops* densities are low because of the large volume and turbidity of the water. Similarly, infection was highest during the dry season in step wells in India.

The life cycle of the parasite is well adapted to provide the maximum chances of transmission, as it takes the female almost a year to mature and release its larvae during the optimum period each year.

In 1982, WHO stated that although *Dracunculus* species are known to infect animals, the role of animal reservoir hosts in the transmission of *D. medinensis* to humans had not been clearly established. Likewise, at that time, WHO considered that even though there was no evidence to suggest that dracunculiasis was a zoonotic infection, and while the possibility of reintroduction of the parasite into unprotected human drinking water sources by possible reservoir hosts was very remote, it should not be entirely discounted (National Research Council 1983).

WHO practically anticipated what finally happened at the last stages of the dracunculiasis eradication campaign, that is, an unforeseeable Guinea worm infection in dogs is impeding that dracunculiasis could be the first infectious disease to be eradicated without a vaccine or specific medical treatment. However, in spite of the WHO recommendations, the role of reservoirs in the disease was apparently overlooked when the global campaign to eradicate the disease was initiated in the 1980s.

12.4.3 Symptoms and Treatment

The vast majority of infections involve a lower limb, but the females can emerge from other body parts, including the head, torso, upper extremities, buttocks, and genitalia (Ruiz-Tiben and Hopkins 2006).

People infected with Guinea worm are unaware of their infection for 10–14 months (average, 1-year incubation period) as there are usually no symptoms in the pre-patient period. The first sign appears a few days before the worm pierces the skin.

When the pregnant adult worm is ready to emerge, acute systemic symptoms develop (e.g., slight fever, an urticarial rash with intense itching, nausea, vomiting, diarrhea, dizziness), which are related to the formation of a blister. Patients experience a burning sensation and pruritus and they often place the affected part in water to relieve their discomfort. Infected people, if untreated, can develop chronic manifestations due to inflammation of the joints with clinical symptoms and signs of arthritis, synovitis, and muscle and tendon contractures with resultant ankylosis of the limb, and even tetanus (Ruiz-Tiben and Hopkins 2006). Aberrant (ectopic) locations of the worms such as the pancreas, lung, periorbital tissues, testis, pericardium, and the spinal cord have also been reported, but such migrations are rare.

Usually, only a single worm appears in a patient per year, but several (up to 20 or even more) can appear at the same time in one individual. Some female worms die before they emerge through the skin. Dead or ruptured worms can lead to the formation of sterile subcutaneous abscesses.

There is no curative drug or vaccine against dracunculiasis. The only treatment is the removal of the parasite by gentle traction with the aid of a stick (Fig. 12.1). The process can last for several days or even weeks due to the length of the female. Oral drugs to alleviate the associated pain, and topical antiseptics or antibiotic ointment to minimize the risk of secondary bacterial infections are recommended since they can also help in reducing inflammation, thus making the extraction of the worm easier (Ruiz-Tiben and Hopkins 2006). Infected hosts do not develop immunity and can be re-infected year after year.

12.5 Guinea Worm Eradication Program

In addition to the already mentioned largely accepted belief that there was no known animal reservoir, the simple Guinea worm water-borne life cycle and the easy recognition of infected people, are among the factors that made GWD to be considered an eradicable disease.

In 1980, US Centers for Disease Control and Prevention (CDC) suggested that GWD eradication would be the ideal indicator of the success of the United Nations 1981–1990 International Drinking Water Supply and Sanitation Decade (IDWSSD), as the disease is only transmitted via drinking water. In the following year, 1981,

GWD eradication was adopted as a subgoal of the IDWSSD. In 1984, the CDC created the WHO Collaborating Center for Research, Training, and Eradication of Dracunculiasis and subsequently, in 1986, the World Health Assembly adopted a resolution to eradicate GWD, defining eradication as the confirmed absence of clinical manifestations (interruption of transmission) for 3 or more years. At that time, there were 21 endemic countries, most of them in sub-Saharan Africa, but also Yemen, and Asian countries such as India and Pakistan, with a global incidence of GWD estimated to be 3.5 million cases annually and an additional 120 million people at risk.

However, the first major intervention supported by a government to eradicate the disease was carried out in the former Soviet Union between 1923 and 1931 by the Tropical Institute in Bukhara with the aim of eliminating the disease from the city and eight other permanently inhabited areas nearby, the only remaining foci of infection in the USSR (now Uzbekistan) at that time. Interestingly, in addition to protecting and cleaning water resources, draining ponds, and treating them with chemicals as a measure of prevention, dogs with dracunculiasis were destroyed (Litvinov 1991). Although the last human case of dracunculiasis in the former USSR was reported in 1931, *D. medinensis*, apparently, persisted in carnivorous animals in some areas (Litvinov 1991). Guinea worm was also eradicated in the south of Iran, where infection was confined to large covered cisterns (known as "birkehs"), in the 1970s. It was eradicated in Pakistan in 1994 and in India in 1997.

Due to the successful Guinea Worm Eradication Program (GWEP), WHO aimed at its eradication by 1995, later being postponed to 2009, and then to 2015 (WHO 2007). The Carter Center, the nonprofit organization established by former US President Jimmy Carter, took leadership of the project, integrating WHO, CDC, and UNICEF. By 2004, the Guinea worm was eradicated from Asia. In 2008, the Bill and Melinda Gates Foundation and the United Kingdom Department of International Development provided further financial support toward eradicating the disease. In 2012, the London Declaration for Neglected Tropical Diseases reaffirmed the commitment to eradicate the disease by 2020. During that period, the number of cases reported has been reduced by 99.99% to only 28 cases reported in 2018. However, this number increased to 53 in 2019. These human cases were detected in Chad (48), South Sudan (4), and Angola (1) (provisional figures, date January 13, 2020) (CDC 2020).

The GWEP strategy is based on the following interventions: (1) surveillance (including case management and containment); (2) provision of safe drinking-water sources; (3) vector control; (4) health education (personal prophylaxis); and (5) certification of eradication (Biswas et al. 2013).

Surveillance, a key element of disease control programs, is based on active house-to-house case searches. Communities were shown a Guinea worm photo identification card to assess whether anyone had seen a person with an emerging worm (Biswas et al. 2013), focusing surveillance only on humans, not on any other potential definitive hosts. As the number of cases declined, in order to increase the effectiveness of surveillance, the eradication program announced a reward (ranging from US$40 to US$160) for individuals reporting cases and also for infected people.

In spite of this successful campaign, events that have occurred in the last years related to *D. medinensis* epidemiology in certain African countries, are posing new challenges to its eradication (Dumiak 2018; LID 2019). In fact, as above-mentioned, WHO delayed GWD eradication again, this time to 2020 (Galán-Puchades 2017). However, this, apparently, unreachable deadline has recently been, once more, purportedly delayed to 2030 (Roberts 2019). The cause of these delays is the discovery, more than 30 years after the beginning of the eradication campaign, of nonhumans hosts, i.e., reservoirs of the disease, mainly dogs, but also cats and olive baboons, infected with the same *Dracunculus* species that infect humans (Eberhard et al. 2014; Thiele et al. 2018).

12.6 Dogs Burst into the Guinea Worm Life Cycle: The Scenario in Chad

For 144 years (since Fedchenko described the *D. medinensis* life cycle in 1870 until 2014), Guinea worm life cycle illustrations had only included humans as the only definitive host. However, based on the findings of infected dogs in Chad, Eberhard et al. (2014) upgraded dogs to the definitive host level, showing a dog, for the first time, besides a human in the figure that illustrated the *D. medinensis* life cycle (Fig. 9 in Eberhard et al. 2014).

Yet, what were the events that led to this novelty in the epidemiology of human dracunculiasis?

In fact, dogs have always been prowling the *D. medinensis* life cycle, but only when human cases of dracunculiasis dramatically decreased, did dog cases emerge from the darkness. According to Whitty (2015), the biggest challenges in eradication programs occur usually in their final stages. Consequently, there will always be the chance of reservoirs of the disease, not anticipated at the start of the eradication attempt. Reservoirs only become apparent when incident cases by the main route of transmission have been reduced (Whitty 2015); this being so in the case of dracunculiasis: at the final stage of Guinea worm eradication, an unanticipated (mainly) canine host emerged, just as the number of human cases hugely decreased.

In spite of the relevant fact that *D. medinensis*, the species affecting humans, had already been molecularly detected in a dog in Ghana in 2005 (Bimi et al. 2005), no epidemiological study of dogs was apparently conducted in any endemic country until 7 years later. Strangely, the authors of the finding of the human Guinea worm in a dog in Ghana stated that "in endemic communities, incidental infection of dogs, and perhaps other non-human mammals, with *D. medinensis* appears likely" (sic) (Bimi et al. 2005). To state, in 2005, that dog infections were *incidental* when at that time not a single effective surveillance of dog dracunculiasis had been carried out, was probably too hasty a conclusion, as demonstrated years later.

The first dog surveillance was implemented in Chad in 2012 (Eberhard et al. 2014). After 10 years without any human dracunculiasis case in that country, in 2010

Table 12.1 Human and dog dracunculiasis cases in Chad in the 2012–2019 period (CDC 2018a)

Year	Human cases	Dog cases
2012	10	27
2013	15	54
2014	13	113
2015	9	503
2016	19	1011
2017	15	817
2018	17	1040
2019[a]	48	1973

[a]Provisional figures

Table 12.2 Dog dracunculiasis cases in other countries than Chad in the 2015–2019 period (CDC 2016, 2017, 2018b, 2019a, 2020)

Year	Ethiopia	Mali	South Sudan
2015	13	1	1
2016	14	11	–
2017	11	9	–
2018	18	17	–
2019[a]	9	8	–

[a]Provisional figures

an outbreak was detected. Scientists did not know whether the disease was reintroduced or had continued to exist at very low levels (Eberhard et al. 2014). At the beginning of 2011, there were rumors of cases of worms in dogs that sounded like dracunculiasis. As a consequence, GWEP technical staff initiated the surveillance of dogs in Chad in April 2012. The results for the 2012–June 2013 period were 15 cases detected in humans and 56 in dogs. From that moment on, dog surveillance started to bring Guinea worm dog cases to light. Table 12.1 shows the figures of human and dog cases in Chad since 2012 (CDC 2018a). As reflected in the table, dog cases in Chad have considerably increased since 2015. Although this feature was considered a “mysterious epidemic” (Callaway 2016), it should be considered that since February 2015, a reward of US$20 has been offered for reporting infected dogs in Chad. Therefore, the high numbers of parasitized dogs since that year are probably a consequence of this incentive in such a poor country, where the monthly minimum wage is around US$110 (Galán-Puchades 2016).

Some scientists consider the presence of Guinea worms in dogs to be a recent phenomenon as a direct consequence of the eradication campaign, i.e., the parasite “jumps” to dogs to prevent being eradicated (Callaway 2016). However, an in-depth genetic study of human and nonhuman parasites, including those from dogs, revealed that dog dracunculiasis does not represent a novel host switch but represents a historically large and stable population (Thiele et al. 2018). Therefore, it can be concluded that, the almost decade-long period of zero cases reported in Chad prior to 2010 was, in fact, due to insufficient surveillance rather than the absence of infection (Thiele et al. 2018).

As a result of the epidemiological situation in Chad, other African countries have also been reporting canine dracunculiasis since 2015 (Table 12.2) although to a lesser extent. Nonetheless, a lack of reported cases in dogs does not necessarily mean absence of transmission, but absence or weakness of GWD surveillance (Sreenivasan et al. 2017). In fact, a total of 1881 villages were under active surveillance in Chad in 2018. However, in Ethiopia there were only 156, and 903 in Mali. In South Sudan 4046 villages were under surveillance, but no reward has been offered for finding dog cases, only for human cases and, curiously, only one dog case has been detected since 2015 (CDC 2018a).

Therefore, and in accordance with CDC, the steep annual increases in the total numbers of domestic dogs infected with Guinea worms in Chad after 2012 is due to the rapid expansion of villages under active surveillance. Likewise, the sensitivity of Chad's surveillance for Guinea worm infections also increased after the program launched a nationwide communication campaign in July 2017 to raise awareness of the cash rewards for reporting infected humans or animals. Consequently, the increase in dog dracunculiasis cases is not a result of an epidemic but the result of an intensified surveillance (CDC 2019b). The absence of any surveillance program before 2012 (like the one currently conducted) makes it impossible to know if the large numbers of infected dogs actually constitute an unusual situation in dog dracunculiasis epidemiology. Those who affirm that dog dracunculiasis is a recent and inexplicable phenomenon base their statement on the fact that interviewed elder residents could not recall dogs infected with Guinea worm (Eberhard et al. 2014; Callaway 2016). Given that dogs, and other animals, probably pull out the worms with their teeth (Muller 1971), unless skilled personnel had analyzed dog wounds, infected dog might have gone unnoticed.

12.7 On the Transmission Route of Guinea Worm Infection in Dogs

In addition to the first appearance of dogs in the Guinea worm life cycle in 2014, fish were also included for the first time as potential (transport) hosts in the *D. medinensis* life cycle by Eberhard et al. (2014). Currently, frogs are also considered potential paratenic hosts of the worm (Fig. 12.6). In fact, on the CDC website, a frog appears in the illustration that accompanies Guinea worm biology (https://www.cdc.gov/parasites/guineaworm/biology.html). Therefore, dracunculiasis, instead of being an exclusively water-borne anthroponosis, would also be a food-borne zoonosis (Galán-Puchades 2019).

This rather belated incorporation of aquatic animals in the dracunculiasis life cycle is a direct consequence of the discovery of Guinea worm reservoirs. According to the epidemiologists involved in GWEP, the explanation for finding numerous infected dogs but scattered human cases is due to a "turn" (sic.) of the worm that has led to this novel food-borne route of host infection (Enserink 2014). Scientists admitted not having any idea why this was happening only then (Enserink 2014). However, it should not be forgotten that there was no reliable dog surveillance in any

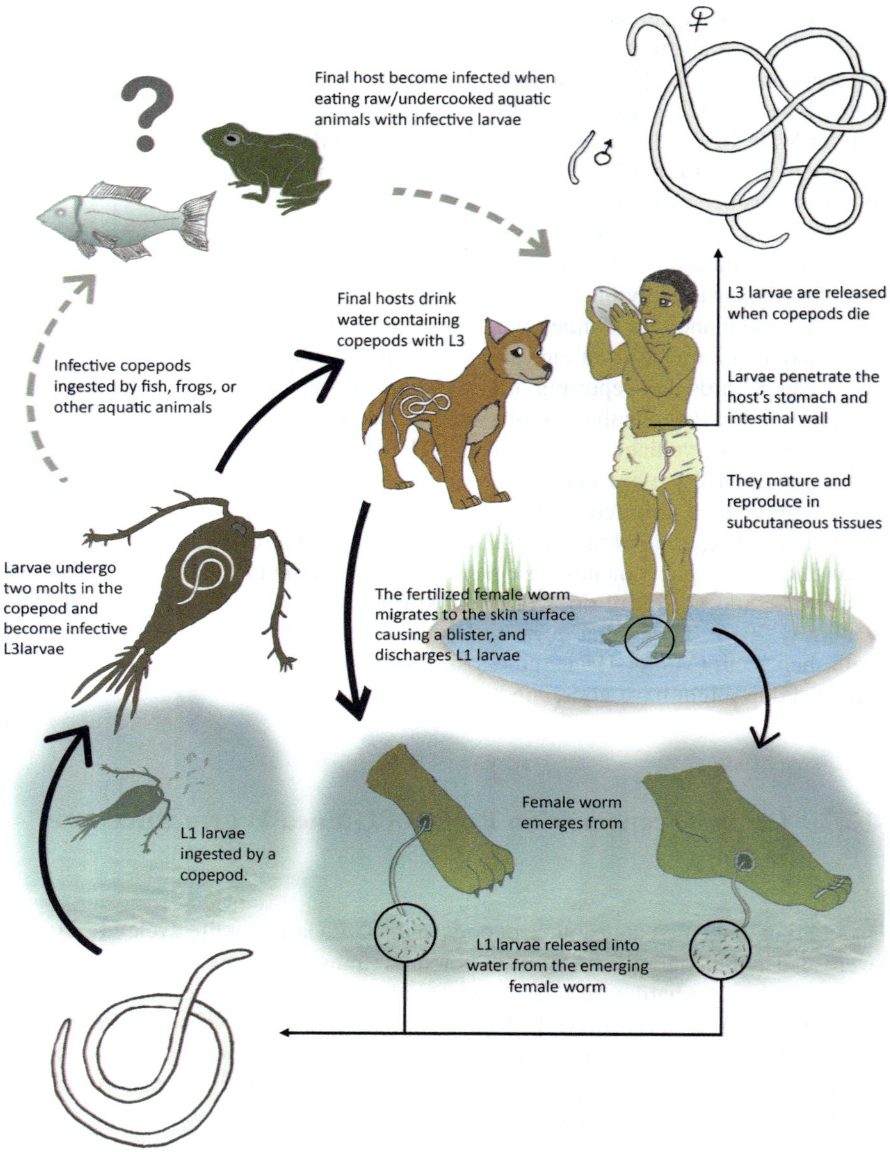

Fig. 12.6 *Dracunculus medinensis* life cycle (illustration by Angela Debenedetti, PhD)

endemic country prior to the current one which started in 2012. Therefore, it is difficult to establish the true degree of exceptionality of the Guinea worm prevalence found in dogs since there are no data for previous years.

Considering that *D. medinensis* is on the verge of eradication due to the control measures which, classically, have *exclusively* been aimed at the water-borne route, this suggested novel food-borne transmission has probably been of anecdotal importance, at least in humans. In fact, there were no differences in the consumption of aquatic animals (fish and frogs) between infected people and controls in Chad (Sreenivasan et al. 2017).

Epidemiologists are reluctant to accept the water-borne route of infection in the case of dogs since they lap water, which apparently is an unlikely method of acquiring infection as copepods scatter when being disturbed (Eberhard et al. 2014). However, a recent study has demonstrated that dogs normally ingest noninfected copepods while drinking (Garret et al. 2020). Infected copepods, not only with *D. medinensis* larvae (Onabamiro 1954), but also harboring larvae of other helminths (Pasternak et al. 1995; Franz and Kurtz 2002; Pulkkinen et al. 2000), decrease their motility, thus having reduced escaping ability, which, in turn, might favor their ingestion by dogs, and, thus, the completion of the life cycle.

The hypothesis of a food-borne route of dog dracunculiasis transmission, although possible, requires a concatenation of events, for paratenic as well as for transport hosts.

In the case of paratenic hosts, these animals accumulate the infective L3 larvae in their muscles. The parasite remains infective but no development takes place. To date, after several surveys, *D. medinensis* larvae have not been found in fish muscles (Cleveland et al. 2017, 2019). In addition, L3 larvae did apparently not disseminate in fish in experimental infections. Therefore, fish are not believed to be paratenic hosts (Cleveland et al. 2017). However, eight *D. medinensis* L3 larvae have been found in five frogs after studying 364 individuals (1.37% prevalence). The frogs harbored only one or two larvae, and only in one individual, three of them were found (Cleveland et al. 2017, 2019). Although it can be accepted that dogs eat frogs, considering the high numbers of infected dogs in Chad, it is hard to believe that the ingestion of frogs has any epidemiological importance given the minimum prevalence of *D. medinensis* infective larvae found in these aquatic animals.

Although no Guinea worm L3 larvae have been found in fish so far, fish have been proposed as possible transport hosts for these larvae. According to an experimental study conducted by Cleveland et al. (2017), when fish ate infected copepods, digested water fleas, and free larvae appeared in their intestines. *D. medinensis* larvae were not encapsulated, not hardy, highly susceptible to desiccation and exhibited a short life span in the fish intestinal tract (4 h or less). As a result, fish or their entrails must be consumed raw or undercooked within a short enough period of time. In Chad, dog dracunculiasis cases tend to spike in the summer during annual fish harvests (Callaway 2016). Considering that the incubation period of *D. medinensis* is about 10–14 months (Muller 1971), dogs would then probably have become infected in the summer of the previous year. Therefore, among the fish caught in summer, infected and non-infected, in order to get infected, dogs should preferentially ingest the infected ones, or their entrails, fast enough to prevent larvae desiccation/death in a country in which temperatures reach up to 43 °C (in the shade) in the summertime. Therefore, fish should be caught, taken to the village, eviscerated, and their entrails or the smallest individuals discarded in fewer than 4 h, considering the short life span of the larvae in the fish intestinal tract. Likelihood exists, of course, but the high dog prevalence requires a more likely route of infection. For dogs to become infected by eating transport hosts, a high prevalence of fish harboring infected larvae in the intestinal tract is required. Many transport fish mean, thus, many infected copepods in water, so that fish have access to them. Therefore, under these circumstances, dogs might become infected more easily drinking water with infected copepods.

Food consumed by dogs can vary daily. However, dogs drink unfiltered water several times a day every single day of their lives, logically in large amounts in summer, in the hottest months of the year, precisely when dogs become more frequently infected. Therefore, dogs are at risk of getting infected by the classic water-borne route on a daily basis, i.e., permanently. However, the ingestion of infected paratenic or transport hosts seems to be a rather accidental event that might, obviously, contribute to the infection of dogs, but probably to a lesser extent than the classic water-borne route.

12.8 A Reverse Zoonosis that Impedes *Dracunculus medinensis* Eradication

Although internal conflicts in Mali, Sudan, and South Sudan have hampered Guinea worm eradication efforts by preventing access to disease-endemic areas and disrupting in-country surveillance programs, the emergence of reservoirs, mainly dogs, seems to indicate that *D. medinensis* eradication will be extremely difficult—if not impossible (Roberts 2019). The erroneous expectation that dracunculiasis would be eradicated could even be detrimental to its control, as happened in the case of leprosy.

Smallpox was declared eradicated in 1980 after 3 years of global immunization led by WHO. Dracunculiasis was thought to be the second infectious disease, which was going to be eradicated due to its similar characteristics to smallpox, as both diseases do not have a mobile vector, the host has a limited carrier state, the disease can be easily diagnosed, also both have a limited area of endemicity as well as seasonal transmission, and prevention is inexpensive (Cairncross et al. 2002). However, and crucially, smallpox did not have reservoirs, but dracunculiasis does. And, it has happened before: in the 1930s, the drive to eradicate yellow fever died when scientists realized monkeys carried the virus.

Culling poultry to stop bird flu is routine. Likewise, Britain killed six million cows and sheep to stop a foot-and-mouth disease outbreak in 2001. However, killing dogs (as previously carried out in the former Soviet Union), although discussed, has, thus far, been rejected. People are obviously attached to their dogs, and they are needed for hunting and to protect huts against thieves, crops against baboons, and livestock against hyenas.

Not only in Africa, but also in other parts of the world, it is accepted that there are widespread but underreported animal dracunculiasis cycles completely independent of human infection (Muller 2002). However, it is unlikely that these zoonotic infections have any importance in human transmission in countries in which humans consume safe water. Yet, human cases sporadically emerge. For instance, though WHO declared India free of GWD in 2000, six cases were reported, from the north to the south of the country, during the 2002–2019 period (Pichakacheri 2019). These occasional human cases also complicate the global eradication scenario.

Unfortunately, failed or doomed eradication attempts outnumber successful ones (Whitty 2015). Therefore, from now on, and considering the important role of dogs in the maintenance of GWD, it would be more sensible to definitely change the approach of dracunculiasis eradication to human dracunculiasis elimination. According to WHO (https://www.who.int/bulletin/volumes/84/2/editorial10206html/en/), eradication is a permanent reduction to zero of the worldwide incidence of infection, while elimination is the reduction to zero of the incidence of the disease or infection in a defined geographical area. Eradication implies that routine intervention measures are no longer needed once interruption of transmission has been certified worldwide. However, elimination, as well as control, requires continued intervention measures and surveillance. If these intervention activities in dracunculiasis cease (with a concurrent decrease in financial resources), a re-emergence of the human disease is likely to occur as long as there are infected reservoirs.

Therefore, whether eradication of dracunculiasis by 2030 is realistic remains to be seen. Controlling transmission of Guinea worms in dogs (and other animals) will require a similar approach to that already in place for human transmission, albeit with greater emphasis on a One Health approach (LID 2019).

Regardless of cats and baboons, it is estimated that, only in the Chari river basin (not evaluated in other countries yet), there are around 60,000 dogs roaming (Roberts 2019). Consequently, how will it be possible to be sure that GWD has finally been vanquished in reservoirs?

References

Adamson PB (1988) Dracontiasis in antiquity. Med Hist 32:204–209

Anderson RC (2000) Nematode parasites of vertebrates. Their development and transmission, 2nd edn. CABI, Wallingford

Bimi L, Freeman AR, Eberhard ML et al (2005) Differentiating *Dracunculus medinensis* from *D. insignis*, by the sequence analysis of the 18S rRNA gene. Ann Trop Med Parasitol 99:511–517. https://doi.org/10.1179/136485905X51355

Biswas G, Sankara DP, Agua-Agum J et al (2013) Dracunculiasis (Guinea worm disease): eradication without a drug or a vaccine. Philos Trans R Soc B 368:20120146. https://doi.org/10.1098/rstb.2012.0146

Cairncross S, Muller R, Zagaria N (2002) Dracunculiasis (guinea worm disease) and the eradication initiative. Clin Microbiol Rev 15:223–246. https://doi.org/10.1128/CMR.15.2.223-246.2002

Callaway E (2016) Dogs thwart end to Guinea worm. Nature 529:10–11. https://doi.org/10.1038/529010a

CDC (2016) Guinea worm wrap-up #238. Public Health Service Centers for Disease Control and Prevention

CDC (2017) Guinea worm wrap-up #245. Public Health Service Centers for Disease Control and Prevention

CDC (2018a) Guinea worm wrap-up #258. Public Health Service Centers for Disease Control and Prevention

CDC (2018b) Guinea worm wrap-up #252. Public Health Service Centers for Disease Control and Prevention

CDC (2019a) Guinea worm wrap-up #259. Public Health Service Centers for Disease Control and Prevention

CDC (2019b) Guinea worm wrap-up #262. Public Health Service Centers for Disease Control and Prevention

CDC (2020) Guinea worm wrap-up #265. Public Health Service Centers for Disease Control and Prevention

Cleveland CA, Eberhard ML, Thompson AT et al (2017) Possible role of fish as transport hosts for *Dracunculus* spp. larvae. Emerg Infect Dis 23:1590–1592. https://doi.org/10.3201/eid2309.161931

Cleveland CA, Garrett KB, Cozad RA et al (2018) The wild world of Guinea worms: a review of the genus *Dracunculus* in wildlife. Int J Parasitol Parasit Wildl 7:289–300. https://doi.org/10.1016/j.ijppaw.2018.07.002

Cleveland CA, Eberhard ML, Thompson AT et al (2019) A search for tiny dragons (*Dracunculus medinensis* third-stage larvae) in aquatic animals in Chad, Africa. Sci Rep 9:375. https://doi.org/10.1038/s41598-018-37567-7

Cox FEG (2002) History of human parasitology. Clin Microbiol Rev 15:595–612. https://doi.org/10.1128/cmr.15.4.595-612.2002

Dumiak M (2018) New challenges to eradicating Guinea worm disease. Lancet Infect Dis 18:838. https://doi.org/10.1016/S1473-3099(18)30437-7

Eberhard ML, Ruiz-Tiben E, Hopkins DR et al (2014) The peculiar epidemiology of dracunculiasis in Chad. Am J Trop Med Hyg 90:61–70. https://doi.org/10.4269/ajtmh.13-0554

Edungbola LE (2019) The eradication of dracunculiasis (Guinea worm disease) in Nigeria. Academic, London

Enserink M (2014) Guinea worm eradication at risk in South Sudanese War. Science 343:236. https://doi.org/10.1126/science.343.6168.236

Fedchenko AP (1870a) Concerning the structure and reproduction of the Guinea worm (Filaria medinensis L.). Izvestiia Imperatorskago Obshchestva Liubitelei Estestvoznaniia Antropologii I Ethnografii, Moskova 8:71–81. (In Russian)

Fedchenko AP (1870b) Concerning the structure and reproduction of the Guinea worm (Filaria medinensis L.). Reprinted (1971) in: Am J Trop Med Hyg 20:511–523. https://doi.org/10.4269/ajtmh.1971.20.511

Franz K, Kurtz J (2002) Altered host behaviour: manipulation or energy depletion in tapeworm-infected copepods? Parasitology 125:187–196. https://doi.org/10.1017/s0031182002001932

Gaeta R, Bruschi F, Guiffra V (2017) The painting of St. Roch in the picture gallery of Bari (15th century): an ancient representation of dracunculiasis? J Infect 74:519–521. https://doi.org/10.1016/j.jinf.2017.02.002

Galán-Puchades MT (2016) Dogs and Guinea worm eradication. Lancet Infect Dis 16:770. https://doi.org/10.1016/S1473-3099(16)30080-9

Galán-Puchades MT (2017) WHO delays Guinea-worm disease eradication to 2020: are dogs the sole culprits? Lancet Infect Dis 17:1124–1125. https://doi.org/10.1016/S1473-3099(17)30565-0

Galán-Puchades MT (2018) The Guinea worm: a zoonotic parasite of dogs. Trends Parasitol 34:3–4. https://doi.org/10.1016/j.pt.2017.10.003

Galán-Puchades MT (2019) Dracunculiasis: water-borne anthroponosis vs. food-borne zoonosis. J Helminthol 94:e76. https://doi.org/10.1017/S0022149X19000713
Garret KB, Box EK, Cleveland CA et al (2020) Dogs and the classic route of Guinea worm transmission: an evaluation of copepod ingestion. Sci Rep 10:1430. https://doi.org/10.1038/s41598-020-58191-4
Gooneratne BW (1969) An additional historical note on the transmission of *Dracunculus medinensis*. Trans R Soc Trop Med Hyg 63:546
Joshi V (1992) Dracunculiasis in Jodhpur district: studies on some epidemiological and parasitological aspects. JoCD 24:191–193
Joshi V, Singhi M, Chaudhary RC (1997) Studies on dracunculiasis in the Indian Desert. J Arid Environ 37:181–191
LID (2019) Dracunculiasis—a case study for infection eradication. Lancet Infect Dis 19:1149. https://doi.org/10.1016/S1473-3099(19)30488-8
Litvinov SK (1991) How the USSR rid itself of dracunculiasis. World Health Forum 12:217–219
Messenger AM, Barnes AN, Gray GC (2014) Reverse zoonotic disease transmission (zooanthroponosis): a systematic review of seldom-documented human biological threats to animals. PLoS One 9:e89055. https://doi.org/10.1371/journal.pone.0089055
Miller RL (1989) Dqr, spinning and treatment of Guinea worm in P. Ebers 875. J Egypt Archaeol 75:249–254. http://www.jstor.org/stable/3821922
Molyneux D, Sankara DP (2017) Guinea worm eradication: progress and challenges-should we beware of the dog? PLoS Negl Trop Dis 11:e0005495. https://doi.org/10.1371/journal.pntd.0005495
Moorthy VN (1932) An epidemiological and experimental study of dracontiasis in Chitaldrug district. Indian Med Gaz 67:498–504
Muller R (1971) *Dracunculus* and dracunculiasis. Adv Parasitol 9:73–151. https://doi.org/10.1016/s0065-308x(08)60160-8
Muller R (2002) Worms and human disease. CABI, Wallingford. https://doi.org/10.1079/9780851995168.0000
National Research Council (1983) Opportunities for control of dracunculiasis. Report of a workshop. National Academic Press, Washington, DC. https://doi.org/10.17226/19479
Nunn JF, Tapp E (2000) Tropical diseases in ancient Egypt. Trans R Soc Trop Med Hyg 94:147–153. https://doi.org/10.1016/S0035-9203(00)90252-9
Onabamiro SD (1951) The transmission of *Dracunculus medinensis* by *Therocyclops nigarianusas* observed in a village in south-west Nigeria. West Afr Med J 1:159–160
Onabamiro SD (1954) The diurnal migration of cyclops infected with the larvae of *Dracrunculus medinensis* (Linnaeus), with some observations on the development of the larval worms. West Afr Med J 3:189–194
Pasternak AF, Huntingford FA, Crompton DW (1995) Changes in metabolism and behaviour of the freshwater copepod *Cyclops strenuus abyssorum* infected with *Diphyllobothrium* spp. Parasitology 110:395–399. https://doi.org/10.1017/s0031182000064738
Pichakacheri SK (2019) Guinea worm (*Dracunculus medinensis*) infection presenting as a diabetic foot abscess: a case report from Kerala. Natl Med J India 32:22–23. https://doi.org/10.4103/0970-258X.272111
Pulkkinen K, Pasternak AF, Hasu T et al (2000) Effect of *Triaenophorus crassus* (Cestoda) infection on behavior and susceptibility to predation of the first intermediate host *Cyclops strenuus* (Copepoda). J Parasitol 86:664–670. https://doi.org/10.1645/0022-3395(2000)086[0664:EOTCCI]2.0.CO;2
Roberts L (2019) Battle to wipe out Guinea worm stumbles. Nature 574:157–158. https://doi.org/10.1038/d41586-019-02921-w
Ruiz-Tiben E, Hopkins DH (2006) Dracunculiasis (Guinea worm disease) eradication. Adv Parasitol 61:275–309. https://doi.org/10.1016/S0065-308X(05)61007-X
Shetty A, Shetty S, Dsouza O (2014) Medical symbols in practice: myths vs reality. J Clin Diagn Res 8:12–14. https://doi.org/10.7860/JCDR/2014/10029.4730

Smith GS, Blum D, Huttly SRA et al (1989) Disability from dracunculiasis: effect on mobility. Ann Trop Med Parasitol 83:151–158. https://doi.org/10.1080/00034983.1989.11812323

Sreenivasan N, Weiss A, Djiatsa J-P et al (2017) Recurrence of Guinea worm disease in Chad after a 10-year absence: risk factors for human cases identified in 2010-2011. Am J Trop Med Hyg 97:575–582. https://doi.org/10.4269/ajtmh.16-1026

Teshome H (2019) Review on principles of zoonoses prevention, control and eradication. Am J Biomed Sci Res 3:188–197. https://doi.org/10.34297/AJBSR.2019.03.000660

Thiele EA, Eberhard ML, Cotton JA et al (2018) Population genetic analysis of Chadian Guinea worms reveals that human and non-human hosts share common parasite populations. PLoS Negl Trop Dis 12:e0006747. https://doi.org/10.1371/journal.pntd.0006747

Velschius GH (1674) Exercitationes de vena medinensis et de vermiculis capillaribus infantium. Auguste Vindel, Augsburg

Whitty CJ (2015) Political, social and technical risks in the last stages of disease eradication campaigns. Int Health 7:302–303. https://doi.org/10.1093/inthealth/ihv049

WHO (1996) Criteria for the certification of dracunculiasis eradication. WHO/FIL/96. 187 Rev. 1

WHO (2007) Global plan to combat neglected tropical diseases 2008–2015. World Health Organization, Geneva. WHO/CDS/NTD/2007.3

Williams BM, Cleveland CA, Verocai GG et al (2018) *Dracunculus* infections in domestic dogs and cats in North America; an under-recognized parasite? Vet Parasitol: Reg Stud Rep 13:148–155. https://doi.org/10.1016/j.vprsr.2018.05.005

Chapter 13
Ticks, Dogs, and Humans: An Endangered Community

Heinz Mehlhorn

Abstract Dogs and humans live in a close community and thus several parasites have developed methods to infect both host groups. This ability has increased considerably their chances of survival, since the vectors (leeches, mosquitoes, flies, ticks, etc.) live in different biotopes.

Keywords Mosquitoes · Ticks · Flies · Leeches

13.1 Introduction

Dogs accompany humans and their domestic animals since thousands of years. Therefore, it is not astonishing that humans might also become infected by parasites, which are mainly harbored by dogs. On the other side dogs had been and are of course also today very important for humans, since their presence offers personal safety and protection of homes besides important social contacts for persons being struggled by loneliness. Of course, dogs are also attacked by parasites and some of them may be spread also to humans eventually introducing severe diseases. Thus it is needed to minimize the transmission of agents of disease from dogs to humans and back (Bajer et al. 2013; Deplazes et al. 2012; Garcia Bocanegra et al. 2017; Imhoff et al. 2015; Jaenson et al. 2012; Levanov et al. 2016; Mehlhorn 2016a, b; Pfeffer and Dobler 2011; Randolph 2010; Roelandt et al. 2011; Yoshii 2019).

H. Mehlhorn (✉)
Department of Parasitology, Heinrich Heine University, Düsseldorf, Nordrhein-Westfalen, Germany
e-mail: mehlhorn@uni-duesseldorf.de

C. Strube, H. Mehlhorn (eds.), *Dog Parasites Endangering Human Health*, Parasitology Research Monographs 13,
https://doi.org/10.1007/978-3-030-53230-7_13

13.2 Symptoms Shown by (Potentially) Parasite-Diseased Dogs

The following symptoms might be seen at diseased dogs and can be induced by parasites, which—under certain conditions—might also become transmitted to humans (Tables 13.1, 13.2 and 13.3, Fig. 13.17).

Table 13.1 Symptoms of infections of dogs with parasites

Symptoms	Parasites
Itching, eczemas, red swellings, restlessness	*Leishmania* stages, worm larvae, mites, ticks, fleas, fly maggots, lice (e.g., *Linognathus setosus*)
Itching in the anal region, restlessness	Tapeworms, whipworms
Cough	*Pneumocystis carinii*, ascarids, heartworms, lungworms, lung mites
Loss of weight, apathia	Coccidia, *Hepatozoon canis*, *Babesia canis*, intestinal leishmaniasis, tapeworms, nematodes, liver flukes, ticks, mites, entamoebiasis
Vomiting	Liver flukes, ascarids, stomach worms
Diarrhea	*Giardia*, amoebae, liver flukes, hookworms
Blood in urine, kidney problems	Worms in the bladder and kidneys
Anemia	*Pneumocystis carinii*, *Leishmania* infections, liver flukes, fish tapeworms, ascarids, hookworms, lung worms, trichurids, ticks
Edemas	Liver flukes, heartworms, lungworms
Motility problems, paralysis	*Entamoeba infection*, *Babesia canis*, tapeworms, hookworms, heartworms, lungworms, huge amounts of ticks
Visual problems	*Leishmania* infection, *Toxoplasma gondii*, tapeworm cysts, heartworms, ascarids, fly larvae entering eyes

Table 13.2 Dog parasites and pathways of transmission

Pathways	Parasite species	Localization of infectious stages
Transmission via fecally contaminated food	*Giardia canis* (assemblages **C**, **D**)	Cysts in dog feces (Fig. 13.1)
	Entamoeba histolytica	Cysts in dog feces (Fig. 13.2)
	Cyclospora species	Oocysts in dog feces (Fig. 13.3)
	Toxoplasma gondii	Cysts in raw meat of animals (e.g., mice, pigs) (Fig. 13.4)
	Sarcocystis species	Raw meat of animals (Fig. 13.5)
	Intestinal and liver flukes	Raw meat (Fig. 13.6)
Transmission by eating raw or undercooked meat	Fish tapeworms, *Diphyllobothrium latum*	Raw fish meat (Fig. 13.7)
	Taenia tapeworms	Muscles of rabbits, hares, rodents, sheep (Fig. 13.8)
Transmission by oral uptake of worm eggs containing an oncosphaera larva	Dog and fox tapeworms (*Echinococcus* species)	Cyst in the liver of mammalia (Fig. 13.9)
Oral uptake of larva-containing raw fish	Kidney worm (*Dioctophyma renale*)	Oral uptake of larvae in raw meat of fish
Oral uptake of larva-containing meat	*Trichinella spiralis*	In raw meat of rodents in remnants of infected pork (Fig. 13.10)
Oral uptake of larva-containing eggs	Trichurids (*Trichuris* sp.), ascarids	Eggs in feces of final hosts, worms in caeca and colon of many mammals (Fig. 13.11)
Active skin penetration	Hookworms	Active skin penetration of free-living/ranging larvae (Figs. 13.12 and 13.13)
	Fly larvae	Being placed onto skin by female flies (Fig. 13.14a, b)
	Mites	In hair regions of other animals
Feeding of infected ectoparasites	*Dipylidium caninum* (Fig. 13.15)	Dogs become infected by feeding larvae in infected fleas (Fig. 13.15)
	Hepatozoon canis	Feeding of infected Mallophaga (e.g., *Trichodectes canis*) (Fig. 13.16)
Transmission inside dog mother, respectively, via mother's milk	(a) Ascarids (b) Lungworms (c) Hookworms (d) *Hammondia*/*Neospora* stages	Larval stages are found in organs, saliva, mother's milk, and/or in blood
Transmission during blood sucking of ectoparasites	(a) *Leishmania* species	Sandflies (*Phlebotomus* sp.)
	(b) *Babesia canis*	Ticks
	(c) Heartworm (*Dirofilaria immitis*)	Mosquitoes
	(d) *Borrelia* and *Anaplasma* stages	Ticks

Table 13.3 Important European hard ticks and pathogens transmitted to humans

Species	Distribution	Transmitted agents of disease
Ixodes ricinus	Europe until the Ural, North Africa	TBE (V), *Anaplasma*, *Borrelia*, *Rickettsia*, *Francisella* (all B), *Babesia* (P)
Ixodes hexagonus Hedgehog tick	Europe, North Africa	*Borrelia* (B)
Ixodes persulcatus Taiga tick	Eastern Europe, Russia, Asia	RTBE (V), *Anaplasma*, *Borrelia*, *Rickettsia*, *Francisella* (all B), *Babesia* (P)
Rhipicephalus sanguineus Brown dog tick	Mediterranean, tropical, and subtropical regions	*Rickettsia, Anaplasma, Ehrlichia* (all B)
Dermacentor reticulatus Forest tick	Focally in Germany, South- and Central Europe, Eastern Europe to Asia	*Rickettsia*, *Francisella* (B), *Borrelia* (?), *Coxiella* (B)
Dermacentor marginatus Sheep tick	Europe to Centra Asia	TBE (V), *Rickettsia* (B), *Francisella* (B), *Coxiella* (B)
Haemaphysalis punctata Red sheep tick	Europe to Central Asia	*Rickettsia* (B)
Hyalomma marginatum Brown tick	Mediterranean region, North Africa to Middle East	Crimean-Congo virus (V)

B bacteria, *P* protozoa, *V* virus, *RTBE* Russian tick-borne encephalitis

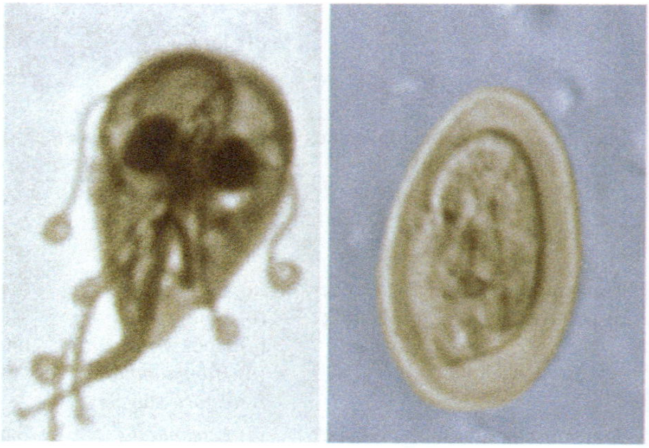

Fig. 13.1 Light micrograph of the flagellated stage (left) and the infectious cyst stage of *Giardia* sp.

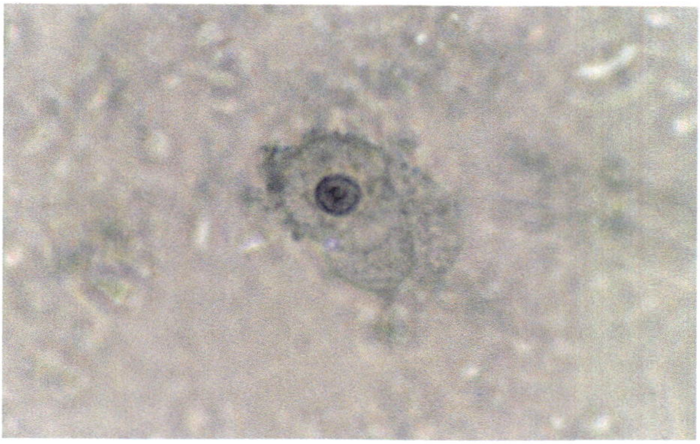

Fig. 13.2 Light micrograph of a minuta stage of *Entamoeba histolytica*, note the dense appearing nucleus and the central nucleolus

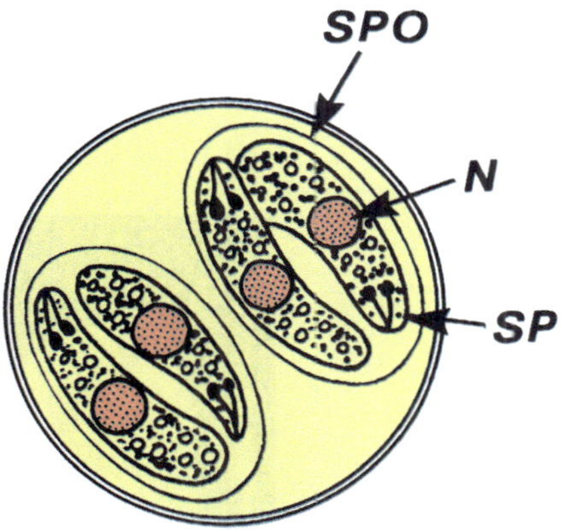

Fig. 13.3 Diagrammatic representation of an oocyst of *Cyclospora* sp. containing two sporocysts each with two sporozoites. N = nucleus; SP = sporozoite; SPO = sporocyst

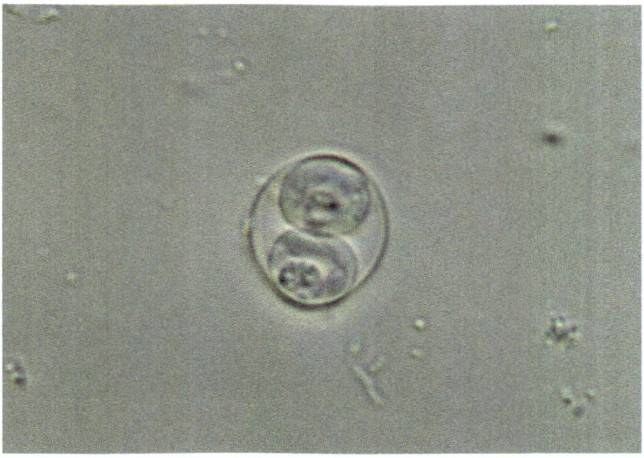

Fig. 13.4 Light micrograph of an oocyst of *Toxoplasma gondii* (containing two sporocysts) from cat feces. Each sporocyst contains four sporozoites

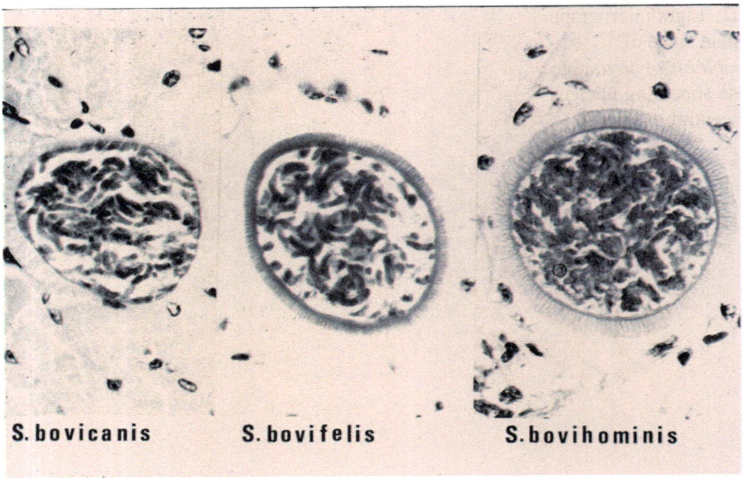

Fig. 13.5 Sections through sarcocysts in muscles of cattle that had been infected by ingestion of oocysts from the feces either of dogs, cats, or humans (*Sarcocystis bovihominis*)

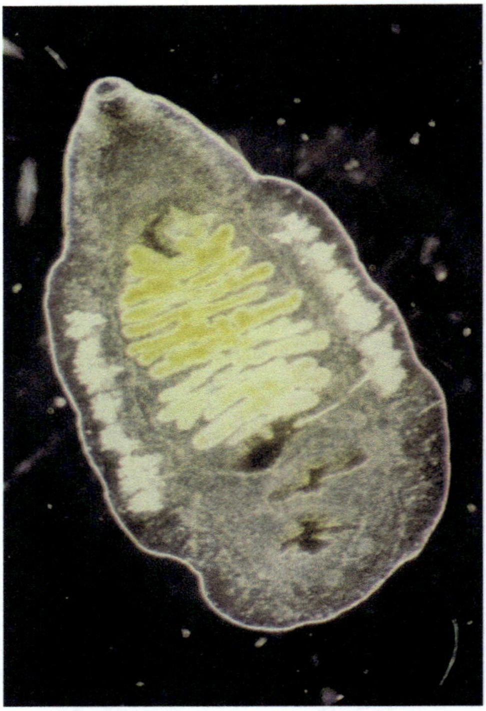

Fig. 13.6 Light micrograph of a colored so-called liver fluke (*Opisthorchis felineus*), which occurs in the small intestine of fish-eating hosts (dogs, cats, humans). The intestine is centrally situated and visible, since the body wall is transparent

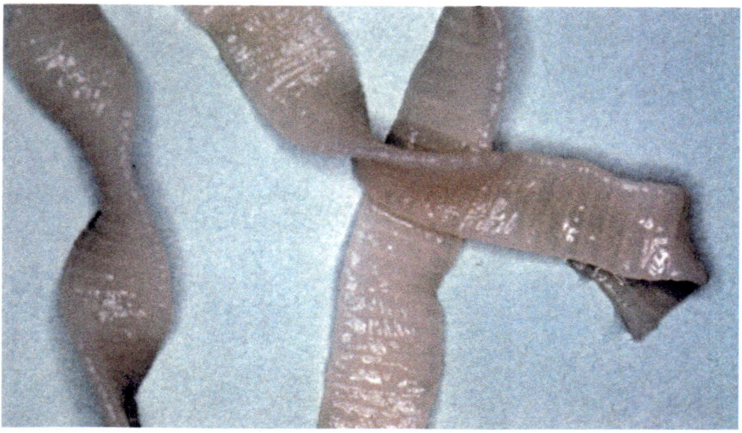

Fig. 13.7 Light micrograph of a portion of the band-like appearing tapeworm proglottids of *Diphyllobothrium latum*, which lives in the intestine of humans, dogs, and cats after having ingested raw, infected meat of freshwater fish

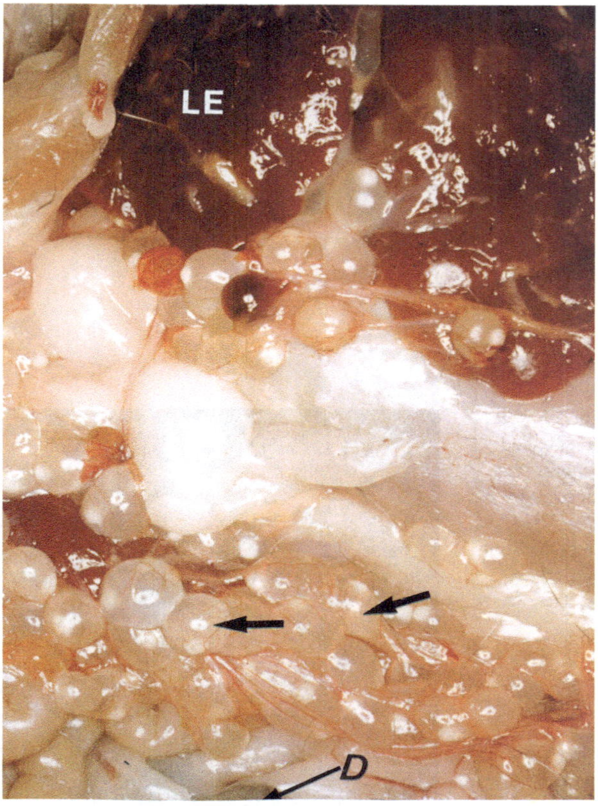

Fig. 13.8 Macrophoto of the omentum region of a rodent containing large numbers of the pea-sized, bladder-like cysticerci (arrows) of *Taenia pisiformis*. D = intestine; LE = liver

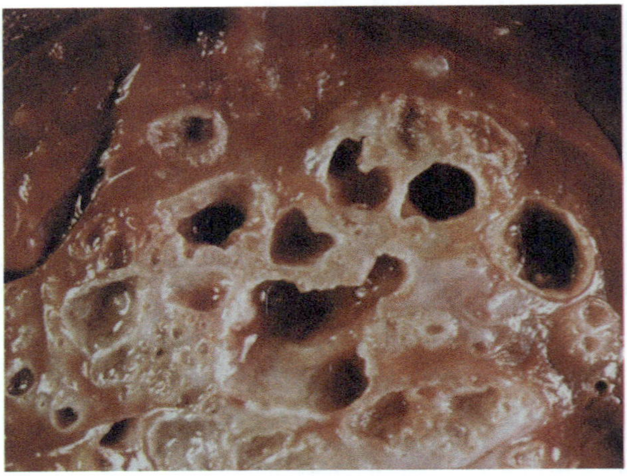

Fig. 13.9 Macrophoto of a diseased human liver after having ingested eggs of the fox tapeworm *Echinococcus multilocularis*

Fig. 13.10 Squeeze preparation of a muscle fiber-containing three larvae of the nematode *Trichinella spiralis*

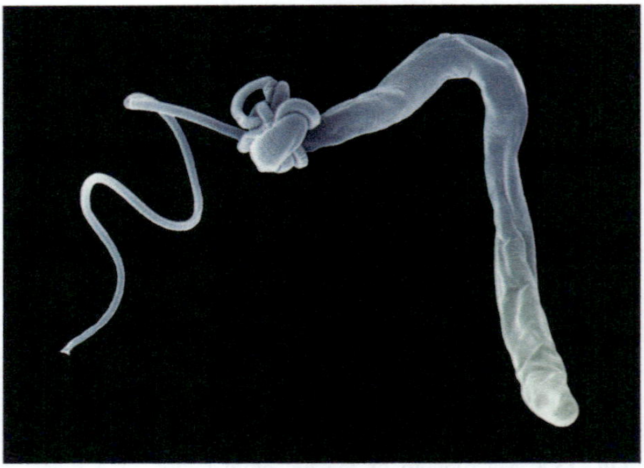

Fig. 13.11 Light micrograph of a so-called whipworm (*Trichuris* sp.), which is characterized by its whip-like body, which appears due to its flagellum-like anterior body

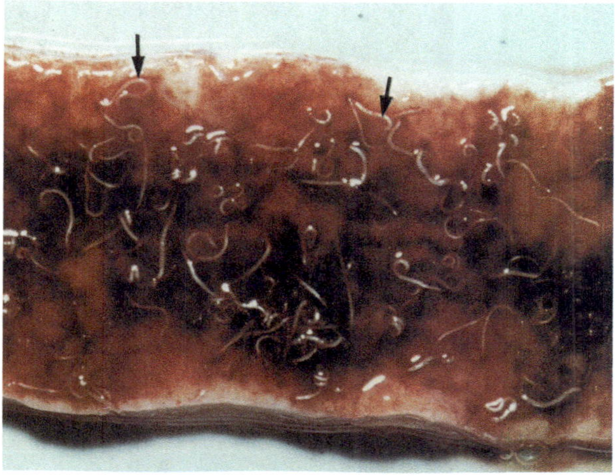

Fig. 13.12 Section of a blood-filled dog intestine containing numerous bloodsucking hookworms. Their eggs may also infect humans

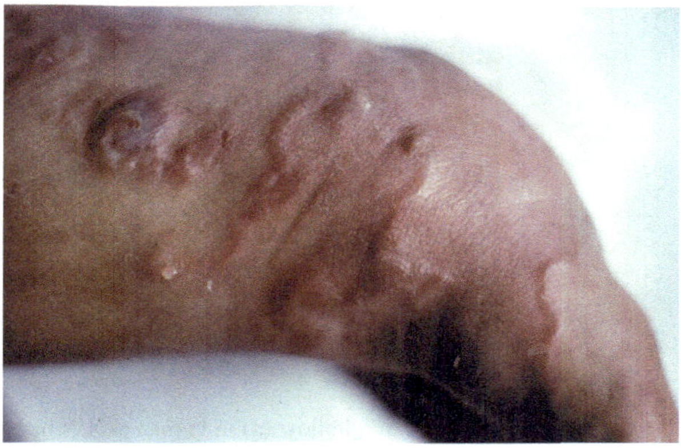

Fig. 13.13 Human skin showing wandering hookworm larvae

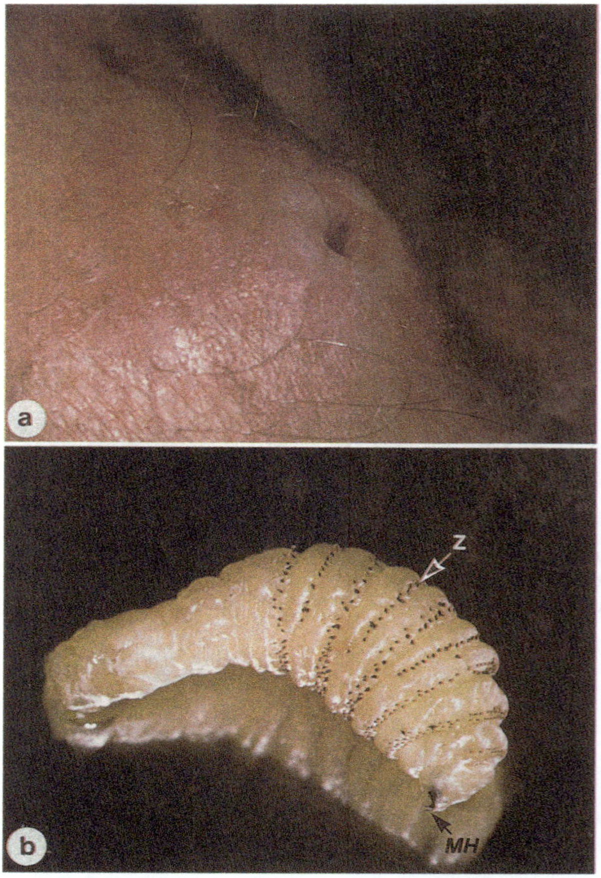

Fig. 13.14 Macrophotos of a hollow (**a**) in human skin, which had been filled by a fly larva (**b**). MH = mouth hooks; Z = small thorns

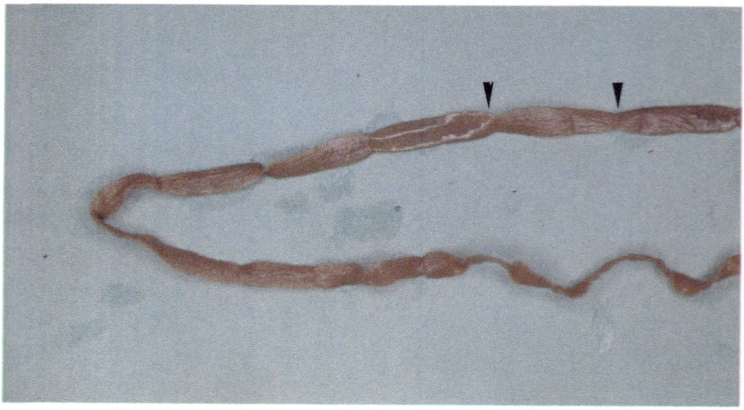

Fig. 13.15 Macrophoto of a strand of proglottids of the dog tapeworm *Dipylidium caninum* being excreted after use od praziquantel

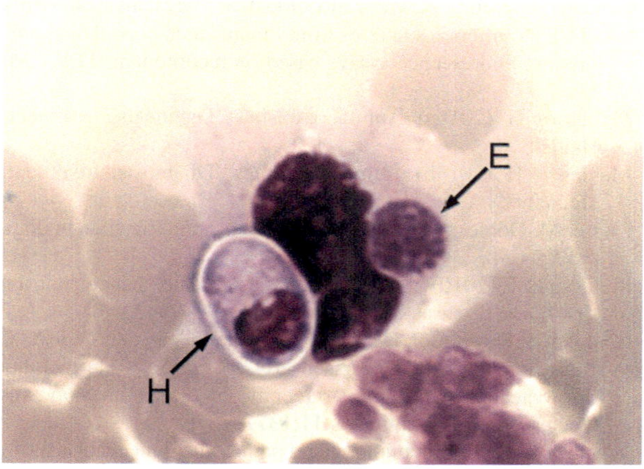

Fig. 13.16 Light micrograph of a gamont (H) of *Hepatozoon canis* in the remnants of a ruptured erythrocyte (E)

Fig. 13.17 Diagrammatic representation of the safe removal of attached ticks from the skin of hosts (humans, dogs, cats). It is recommended to use a forceps with very thin tips. The sucking apparatus, which has been entered into the skin, should be taken close to the skin so that the tick is not squeezed during removal thus avoiding mechanical transmission of agents of diseases

References

Bajer A, Rodo A, Bednarska M, Mierzejewska E, Welc-Falęciak R (2013) *Babesia canis* and tick-borne encephalitis virus (TBEV) co-infection in a sled dog. Ann Agric Environ Med 20:426–430

Deplazes P, Eckert J, von Samson-Himmelstjerna G, Zahner M (2012) Lehrbuch der Parasitologie, 3rd edn. Enke, Stuttgart

Garcia Bocanegra I, Jurada-Tarifa E, Cano-Terriza D et al (2017) Exposure to West Nile virus and tick-borne encephalitis. Transbound Emerg Dis 2018:1–8

Imhoff M, Hagedorn P, Schulze Y, Hellenbrand W, Pfeffer M, Niedrig M (2015) Review: sentinels of tick-borne encephalitis risk. Ticks Tick Borne Dis 6:592–600

Jaenson TG, Hjertqvist M, Bergström T, Lundkvist A (2012) Why is tick-borne encephalitis increasing? A review of the key factors causing the increasing incidence of human TBE in Sweden. Parasit Vectors 5:184

Levanov L, Vera CP, Vapalahti O (2016) Prevalence estimation of tick-borne encephalitis virus (TBEV) antibodies in dogs from Finland using novel dog anti-TBEV IgG MAb-capture and IgG immunofluorescence assays based on recombinant TBEV subviral particles. Ticks Tick Borne Dis 7:979–982

Mehlhorn H (2016a) Animal parasites. Diagnosis, treatment, prevention, 8th edn. Springer, Heidelberg

Mehlhorn H (2016b) Human parasites. Diagnosis, treatment, prevention, 8th edn. Springer, Heidelberg

Pfeffer M, Dobler G (2011) Tick-borne encephalitis virus in dogs—is this an issue? Parasit Vectors 4:59

Randolph SE (2010) To what extent has climate change contributed to the recent epidemiology of tick-borne diseases? Vet Parasitol 167:92–94

Roelandt S, Heyman P, De Filette M, Vene S, Van der Stede Y, Caij AB, Tavernier P, Dobly A, De Bosschere H, Vyt P, Meersschaert C, Roels S (2011) Tick-borne encephalitis virus seropositive dog detected in Belgium: screening of the canine population as sentinels for public health. Vector Borne Zoonotic Dis 11:1371–1376

Yoshii K (2019) Epidemiology and pathological mechanisms of tick-borne encephalitis. J Vet Med Sci 81:343–347

Chapter 14
Trematodes Attacking Dogs and Humans

Heinz Mehlhorn

Abstract Several trematode species run their life cycle using humans as final hosts and snails as intermediate hosts. Humans, however, are additionally attacked by trematode larvae, which may lead to severe skin inflammations.

Keywords *Bilharziella* species · *Opisthorchis felineus* · *O. viverrini* · *Heterophyes* species · Bird trematodes · Skin disease

Humans have adapted feeding activities like those of many vertebrate animals and live often in the same biotopes (Table 14.1). Thus it is not astonishing that humans are attacked by identical or at least very similar parasitic species.

These parasites are subdivided into two groups:

1. Parasites, which develop adult stages inside both humans and animals (Table 14.1)
2. Parasites, which stay as adults inside birds, while larvae remain in the skin of humans and animals

H. Mehlhorn (✉)
Department of Parasitology, Heinrich Heine University, Düsseldorf, Nordrhein-Westfalen, Germany
e-mail: mehlhorn@uni-duesseldorf.de

C. Strube, H. Mehlhorn (eds.), *Dog Parasites Endangering Human Health*, Parasitology Research Monographs 13,
https://doi.org/10.1007/978-3-030-53230-7_14

Table 14.1 Important flukes inside gallbladder/bile ducts

Name	Size of adults (mm)	Intermediate hosts 1 and 2	Final hosts Site of infection	Geographic distribution
Clonorchis sinensis	10–30 × 2–5	1. Snails 2. Fish	*Humans*, dogs, cats, pigs Bile ducts	East Asia, China, Korea
Opisthorchis felineus (syn. *O. tenuicillis*)	8–13 × 1.5–3	1. Snails 2. Fish	*Humans*, dogs, foxes pigs	Europe, Asia
Opisthorchis viverrini	6 × 2	1. Snails 2. Fish	*Humans*, dogs, cats	South East Asia
Pseudoamphistoma truncatum	2 × 0.5	1. Snails 2. Fish	*Humans*, dogs, cats, foxes, seals	Europe, Russia, India
Heterobilharzia regenti	9–18 × 2	1. Snails	*Dogs*, racoons, *humans* (?)	USA
Alaria alata	2.5 × 0.5–2	1. Snails 2. Tadpoles, frogs	Wolves, foxes, dogs, *humans* (?)	Europe, USA
Metorchis bilis	2.5–4.5 × 1.5	1. Snails	Dogs, cats	Europe, North USA

Note: Symptoms of disease start mainly (often) after return from trips in other countries

14.1 Trematodes (Flukes) Running Their Full Life Cycle also in Humans

1. *Name*: *Greek*: *opisthen* = behind; *orchis* = testis. *Latin*: *tenuis* = thin; *collum* = neck; *felis* = cat. The genus name *Opisthorchis* refers to the placement of the testes in the hind region of the worm. *Greek*: *heteros* = the other; *phyle* = origin
2. *Geographic distribution/epidemiology*: Worldwide; common in Asia, Europe; up to 15% of cats may be infected in Asia.
3. *Biology, morphology*: Most common are flukes of the species *Opisthorchis felineus* (syn. *tenuicollis*) and *O. viverrini*, which may occur in a broad spectrum of at least accidentally fish-eating hosts (cats, dogs, foxes, pigs, and humans). These flukes, which reach a length of 8–12 mm, parasitize inside the bile ducts of their hosts (Figs. 14.1a, b). A significant diagnostic criterion is the fact that their two testes have only slightly depressed lobes (Figs. 14.1a, b). The anterior testis has four lobes and the posterior one five. The eggs are shown in Fig. 14.2.
   Another group of worms = *Heterophyes* species, which are very small (~2 mm), is common in the small intestine of dogs and cats. For example, *H. heterophyes* has been found in Asia and in South European countries in up to 16% of cats and dogs (Figs. 14.3, 14.4, and 14.5). Their surface is closely covered with scales. In the case of *Opisthorchis* species as well as in these of the genus *Heterophyes*, water snails serve as first intermediate hosts and fresh and brackish water fishes

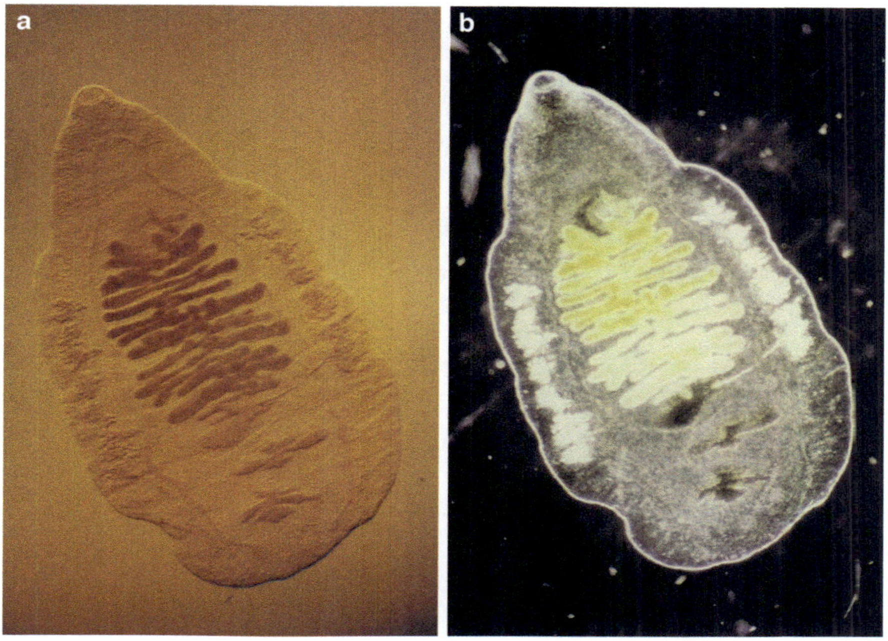

Fig. 14.1 (**a**, **b**) Light micrographs of unfixed adult worms of *Opisthorchis tenuicollis* (syn. *felineus*) using different techniques. Note that the internal organs shine through the surface of the worms. Peculiar is also the shape of the two testes in the terminal region of the worms. Reprinted by permission from Springer Nature: Springer International Publishing "Animal Parasites: Diagnosis, Treatment, Prevention" by Heinz Mehlhorn © 2016

as second intermediate hosts. In those of the latter group, the encysted metacercariae are situated in muscles and various other organs. In South-East Asia (Korea, China) also the species *Clonorchis sinensis* (10–30 mm × 2–5 mm) is common in dogs and cats as well as *Metorchisbilis* (2.5–4.5 × 1.5 mm) in Europe and North America and *Pseudoamphistomum truncatum* in Europe, Russia, and India.

4. *Symptoms of disease (Opisthorchiasis, Heterophyiasis)*:

   (a) *Opisthorchis tenuicollis*: Depending on the amount of ingested metacercariae, the symptoms of the disease may be absent or induce bad turns, showing intense inflammations of the bile ducts, enlargement of the liver and proliferation of the bile duct epithelia, which may induce carcinoma (especially common in humans!). *Clinical symptoms* are vomiting, loss of appetite, intestinal disorders, icterus, anemia, edemas, and/or ascites.
   (b) *Heterophyes species*: In cases of heavy infections: vomiting, disturbances in digestion, bloody diarrheas and hosts are easily susceptible to other infectious diseases.

Fig. 14.2 Light micrograph of an egg of *Opisthorchis felineus*. Note that the egg is covered at its top by a small lid. Reprinted by permission from Springer Nature: Springer International Publishing "Animal Parasites: Diagnosis, Treatment, Prevention" by Heinz Mehlhorn © 2016

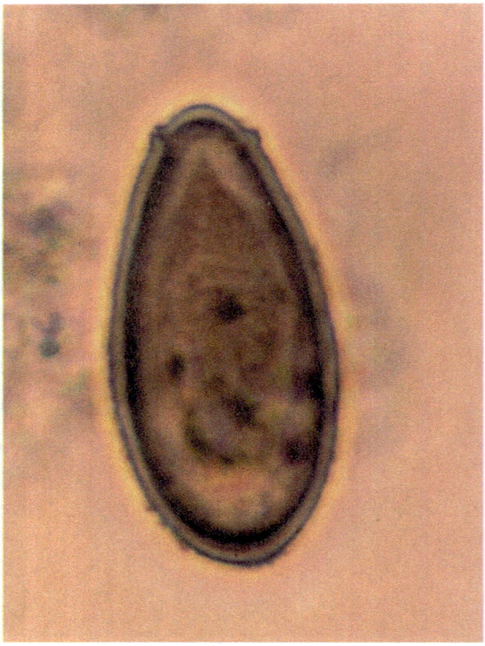

5. *Diagnosis*: Demonstration of the rather small eggs inside the feces by the help of the flotation method. These eggs possess in all species an operculum (Fig. 14.2). The eggs of *O. tenuicollis* measure 30 × 15 μm, while those of *Heterophyes* species are somewhat smaller and reach a mean size of 24 × 14 μm. However, other species of this group of parasites may reach larger sizes (up to 100 μm in length).
6. *Pathway of infection*: Cats, dogs, and humans are always infected by ingesting raw fish meat containing the 1–2 mm sized metacercariae of this species. In the case of *O. tenuicollis*, freshwater fishes are infected, while in the case of *Heterophyes* species, especially those fishes are infected which live in brackish or fully salty water.
7. *Prophylaxis*: Do not feed raw fish to home dogs or cats and avoid also consumption of such raw fish meat in restaurants.
8. *Incubation period*: About 1 week.
9. *Prepatent period*: 3–4 weeks in the case of *O. tenuicollis* and 7–9 days in the case of *Heterophyes* species.
10. *Patency*: It may take years in the case of *Opisthorchis tenuicollis* but only 1–3 months in the case of *Heterophyes* species.
11. *Therapy*: Oral uptake of praziquantel (3 days each 25 mg/kg body weight).

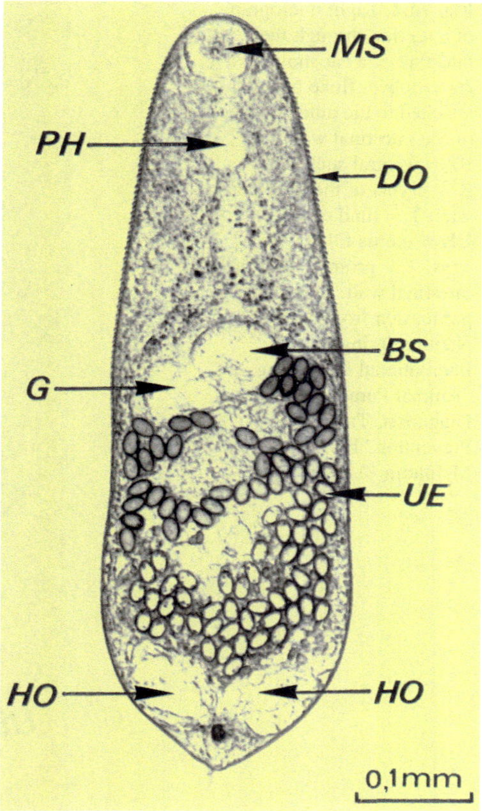

Fig. 14.3 Light micrograph of an adult worm of *Heterophyes heterophyes*. BS = ventral sucker; DO = thorns/scales; G = gonocotyl; HO = testis; MS = oral sucker; PH = pharynx; UE = uterus with eggs. Reprinted by permission from Springer Nature: Springer International Publishing "Animal Parasites: Diagnosis, Treatment, Prevention" by Heinz Mehlhorn © 2016

14.2 Schistosomatid Trematodes of birds

1. *Name*: Greek: *thrix, trichos* = tiny hair; *ornis, ornithos* = bird. Theodor Bilharz (1825–1862), a German physician, who in Cairo (Egypt) at first described these worms which today are called *Schistosoma* species. The genus *Diplostomum* was later renamed *Bilharzia*.
2. *Geographic distribution/epidemiology*: Worldwide.
3. *Biology, morphology*:

   (a) *Trichobilharzia species* (Table 14.2) occur in blood vessels close to the intestine in ducks and other water birds. The female worms reach a size of 3–5 mm, while the males are slightly larger. The latter bears the females in a species-specific hook containing ventral fold called *canalis gynaecophorous*.
   Inside the uterus of the females, only a single egg is located, which reaches a size of 140–210 μm × 50–70 μm and appears spindle like. As soon as it passes into the intestinal lumen, it already contains the miracidium, which

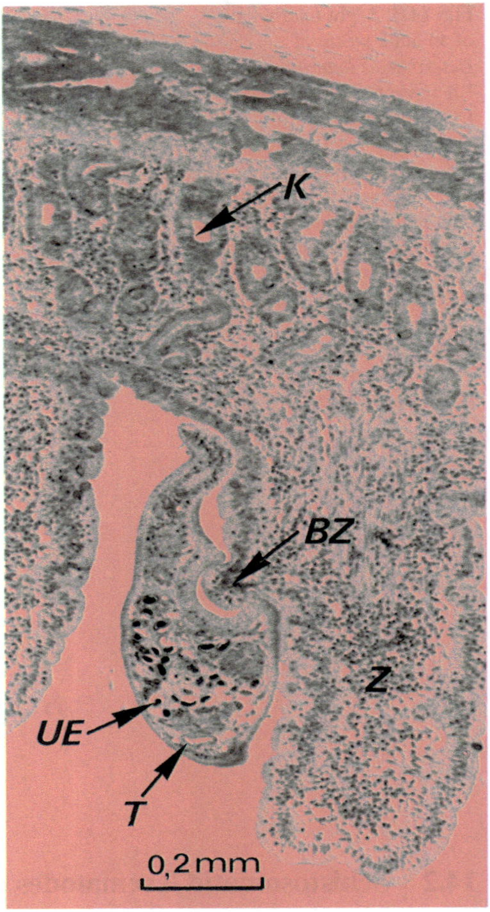

Fig. 14.4 Light micrograph of a section through the intestine of a cat showing a *Heterophyes* fluke being attached to the mucous layer of the intestinal wall. BZ = ventral sucker; K = crypts of the intestinal wall; T = hind end; UE = uterus filled with eggs; Z = protrusion of the intestinal wall. Reprinted by permission from Springer Nature: Springer International Publishing "Animal Parasites: Diagnosis, Treatment, Prevention" by Heinz Mehlhorn © 2016

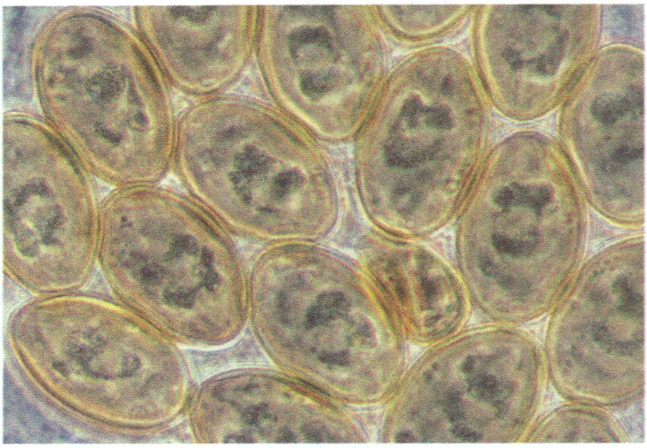

Fig. 14.5 Light micrograph of *Heterophyes* eggs (they already contain each a miracidium larva). Reprinted by permission from Springer Nature: Springer International Publishing "Animal Parasites: Diagnosis, Treatment, Prevention" by Heinz Mehlhorn © 2016

Table 14.2 Selection of species belonging to the subfamily *Trichobilharzia* and *Bilharziella* in Europe (E), Asia (A), Africa (AF), North America (NA) attacking humans and dogs

Species	Length	Intermediate hosts	Final hosts	Infested tissues	Prepatent period (days)
Trichobilharzia regenti (E)	~7 mm	*Radix* species	Ducks	Veins of nose mucosa	~20
Trichobilharzia szidati (E, A)	~3 mm	*Lymnaea* species	Ducks, other water birds	Intestinal wall, mesenterial veins	7–9
Trichobilharzia franki (E)	~4 mm	*Radix auriculata* and other *Radix* sp.	Ducks	Liver veins	10–13
Bilharziella polonica (E, A, AF, NA)	~2 mm	*Lymnaea* species, *Planorbarius*, *Planorbis*	Many water birds (geese, ducks, etc.)	Mesenterial and portal veins	80–90
Ornithobilharzia sp. (E, NA)	~3 mm	Snails in brackish water	Fresh and saltwater birds	Mesenterial veins	~50

may enter a broad spectrum of water snakes, when the host's feces have entered the water.

(b) *Bilharziella polonica* (Table 14.2): This species lives in the mesenteric blood vessels of *ducks*. The males reach a length of 4 mm, while the females are smaller (~2 mm). Both have a flattened body shape. The 400-μm long eggs excreted by the female have a typical shape being provided with a very long filament-like structure at one of the poles. They contain already a miracidium when they are deponed into water within the duck's feces. Intermediate hosts are snails of the genera *Lymnaea* and *Planorbis*.

(c) *Ornithobilharzia* species: These species parasitize in *seagulls* and several other water birds and occasionally in *geese*. Related species are found in Asia also in cattle, horses, and camels. The species inside birds reach a length of up to 8 mm. Their ovoid eggs measure 60–70 μm × 50 μm and contain already a miracidium when deponed in feces. The rather thick but smooth eggshell is provided with a terminal thorn. Intermediate hosts are several water snails (Fig. 14.6).

4. *Symptoms of disease*: Humans, dogs: When bathing in natural lakes, cercariae enter the skin of humans as they do in birds, but they are killed by the defense system of the skin. Then the remnants of these cercariae induce a peculiar inflammation called "swimmer's itch." The itching skin symptoms may stay for weeks and may reach high levels in sensitive persons (Fig. 14.7).
5. *Diagnosis*: Microscopical proof of the typical eggs in the feces with the help of the sedimentation method.

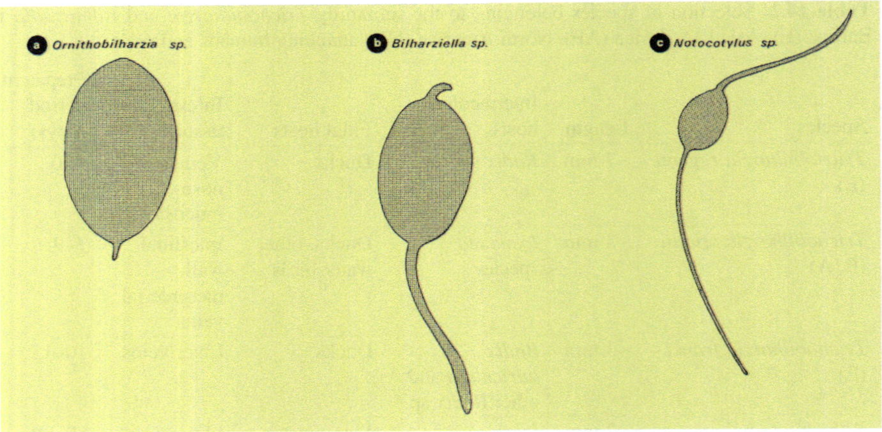

Fig. 14.6 Diagrammatic representation of the eggs of some schistosomatid trematodes of birds, the cercariae of which enter the skin of animals and humans leading to a so-called "bathing dermatitis" in humans, since no further development occurs in contrast to birds. Reprinted by permission from Springer Nature: Springer International Publishing "Animal Parasites: Diagnosis, Treatment, Prevention" by Heinz Mehlhorn © 2016

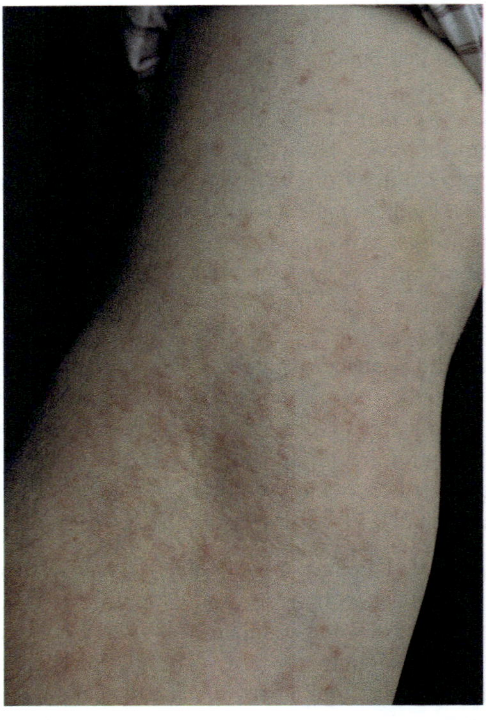

Fig. 14.7 Itching skin reactions due to penetrated schistosomal cercariae

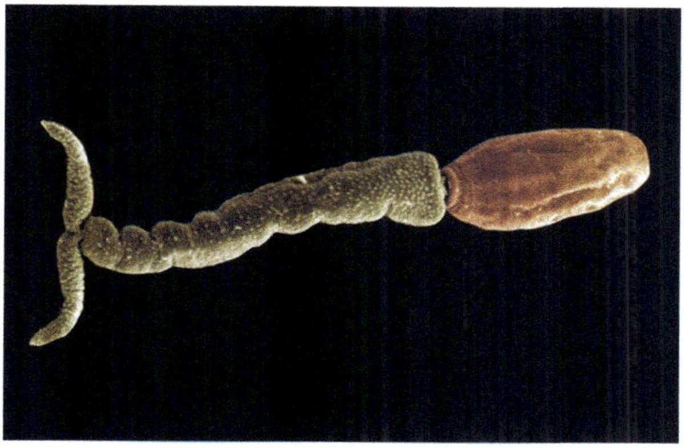

Fig. 14.8 Scanning electron micrograph showing the principal structure of a cercaria, which moves in the water by the help of its bifurcated motile tail

Table 14.3 Selection of schistosomatid species, the cercariae of which may enter the skin of humans, dogs, and other mammals and thus induce considerable skin-itching (Fig. 14.7)

Avian final hosts
• *Trichobilharzia regenti*
• *Trichobilharzia szidati*
• *Anserobilharzia brontae*
• *Dendritobilharzia pulverulenta*
• *Gigantobilharzia huronensis*
• *Bilharziella polonica*
Avian final hosts (exclusively marine)
• *Austrobilharzia* species
• *Ornithobilharzia canaliculata*
Mammalian final hosts (freshwater)
• *Heterobilharzia americana*
• *Schistosomatium douthitti*
• *Bivitellobilharzia nuiri*
• *Macrobilharzia macrobilharzia*

6. *Pathway of infection*: Percutaneously by penetrating cercariae (Fig. 14.8) during water contact (lakes, springs, rivers, etc.).
7. *Prophylaxis*: Practically impossible with respect to free-living birds.
8. *Humans*: Using the repellent spray Viticks (Fa. Alpha-Biocare, Neuss, Germany), which protects at least for 1 h when bathing in lakes.
9. *Incubation period*: Birds: 2 weeks; *humans*: in the case of bathing dermatitis: 24 h.
10. *Prepatent period*: 3–4 weeks until eggs appear first in feces.
11. *Patency*: At least 4–5 months.
12. *Therapy*: None in free-living birds. Cultured ducks and geese might be treated with praziquantel, if symptoms become visible. *Humans*: Use of desensitizing creams to overcome itching, which occurs as antigenic reactions (Wulff et al. 2007).

Parasite details have been revised and are fully discussed in Mehlhorn (2016a) (Table 14.3).

Further Reading

Ashour DS et al (2014) Insights into regulatory molecules of intestinal epithelial cell turnover during experimental infection by *Heterophyes heterophyes*. Exp Parasitol 143:48–54

Ashrafi K, Brant SV (2020) An efficient method for collecting the full-length adults, fragments, and eggs of *Trichobilharzia* spp. from the liver of definitive hosts. Parasitol Res 119:1167–1172

Boyce K et al (2012) Identification of a new species of digenean *Notocotylus malhamensis* from the bank vole *Myodes glareolus*. Parasitology 139:1630–1639

Dao TM et al (2014) *Opisthorchis viverrini*-like liver fluke in birds from Vietnam. J Helminthol 88:441–446

El-Seifi MA et al (2012) ELISA and some biochemical tests of Heterophyidae infection in laboratory animals. Parasitol Res 110:679–687

Elshazly AT et al (2008) Comparison of three immunodiagnostic tests for experimental *Heterophyes heterophyes* infection in dogs. Vet Parasitol 151:196–202

Gordy MA, Cobb TP, Hanington PC (2018) Swimmer's itch in Canada: a look at the past and a survey of the present to plan for the future. Environ Health 17:73

Haas W, Haeberlein S (2009) Penetration of cercariae into the living human skin: *Schistosoma mansoni* vs. *Trichobilharzia szidati*. Parasitol Res 105:1061–1066

Haas W, van de Roemer A (1998) Invasion of the vertebrate skin by cercariae of *Trichobilharzia ocellata*: penetration processes and stimulating host signals. Parasitol Res 84:787–795

Horák P, Mikeš L, Lichtenbergová L, Skála V, Soldánová M, Brant SV (2015) Avian schistosomes and outbreaks of cercarial dermatitis. Clin Microbiol Rev 28:165–190

Keiser J, Utzinger J (2009) Food-borne trematodiases. Clin Microbiol Rev 22:466–483

Korsunenko AV, Chrisanfova GG, Ryskov AP, Movsessian SO, Vasilyev VA, Semyenova SK (2010) Detection of European *Trichobilharzia* schistosomes (*T. franki*, *T. szidati*, and *T. regenti*) based on novel genome sequences. J Parasitol 96:802–806

Lichtenbergová L, Kolbeková P, Kourilová P, Kasný M, Mikes L, Haas H, Schramm G, Horák P, Kolárová L, Mountford AP (2008) Antibody responses induced by *Trichobilharzia regenti* antigens in murine and human hosts exhibiting cercarial dermatitis. Parasit Immunol 30:585–595

Macháček T, Turjanicová L, Bulantová J, Hrdý J, Horák P, Mikeš L (2018) Cercarial dermatitis: a systematic follow-up study of human cases with implications for diagnostics. Parasitol Res 117:3881–3895

Mahdavi SA et al (2013) Survey of migratory birds (Anatidae: *Anas* sp.) for schistosome parasites. Iran J Parasitol 8:333–336

Marszewska A, Strzała T, Cichy A, Dąbrowska GB, Żbikowska E (2018) Agents of swimmer's itch-dangerous minority in the Digenea invasion of Lymnaeidae in water bodies and the first report of *Trichobilharzia regenti* in Poland. Parasitol Res 117:3695–3704

Mehlhorn H (2016a) Animal parasites. Diagnosis, treatment, prevention, 1st edn. Springer International, Cham

Mehlhorn H (2016b) Human parasites. Diagnosis, treatment, prevention, 1st edn. Springer International, Cham

Morley NJ (2016) Cercarial dermatitis or cercariosis: defining the condition. Trends Parasitol 32:352–353

Pakharukova MY et al (2015) The first comprehensive study of praziquantel effects in vivo and in vitro on the European liver fluke *Opisthorchis felineus*. Int J Antimicrobiol Agents 46:94–100

Petney TN et al (2013) The zoonotic fish borne liver flukes *Clonorchis sinensis*, *Opisthorchis felineus* and *O. viverrini*. Int J Parasitol 43:1031–1046

Pomaznoy M et al (2013) Adult *Opisthorchis felineus* major protein fractions: comparison with liver flukes *O. viverrini* and *Clonorchis sinensis*. Exp Parasitol 135:297–306

Santoro M et al (2011) Helminth community structure of the Mediterranean gull (*Ichthyaetus melanocephalus*) in Southern Italy. J Protozool 97:364–366

Schuster RK et al (2007) Liver flukes in dogs and treatment with praziquantel. Vet Parasitol 150:362–365

Skala V et al (2014) Insights into the development of *Notocotylus attenuatus*. Parasitol Int 63:94–99

Southgate VR et al (1995) Mating behaviours in mixed infections of *Schistosoma haematobium* and *S. mattheei*. Parasitol Res 81:651–656

Thaenkhan U et al (2012) Families Opisthorchiidae and Heterophyidae: are they distinct? Parasitol Int 61:90–93

Tolstenkov OO et al (2012) The neuro-muscular system in fresh water furcocercaria from Belarus: I. Schistosomatidae. Parasitol Res 110:185–193

Traversa A et al (2012) A large outbreak of *Opisthorchis felineus* in Italy. Clin Microbiol Inf Dis 31:1089–1093

Vicente B et al (2014) A *Fasciola hepatica* derived fatty acid binding protein induces protection against schistosomiasis caused by *Schistosoma bovis*. Exp Parasitol 145:145–151

Vimalraj PG, Latchumikanthan A (2015) *Schistosoma spindale* infection in a captive jackal (*Canis aureus*). J Parasitol Dis 39:120–121

Weyher A et al (2010) Molecular identification of *Schistosoma mattheei* from feces of Kinda (*Papio cynocephalus kindae*) and Grayfoot baboon (*Papioursinus griseipes*) in Zambia. J Parasitol 96:184–190

Wulff C, Haeberlein S, Haas W (2007) Cream formulations protecting against cercarial dermatitis by *Trichobilharzia*. Parasitol Res 101:91–97

Yakhchali M et al (2013) Detection of infection with larval stages of *Ornithobilharzia turkestanicum* using PCR. Iran J Parastiol 8:627–633

[illegible] (2007) [illegible]
[illegible]
[illegible] Insights into the [illegible]
[illegible] (1976) Mating [illegible]
[illegible]
[illegible] (2012) Pandey [illegible]
[illegible]
[illegible] (2013) The [illegible]
[illegible] Research [illegible]
[illegible] (2017) A large [illegible]
[illegible]
[illegible] (2014) A [illegible]
[illegible]
[illegible]
[illegible]
[illegible] Molecular [illegible]
[illegible] and [illegible]
[illegible]
[illegible] (2016) [illegible]
[illegible]
[illegible] (2013) [illegible]
[illegible] PCR [illegible]

Index

A
Agamonematodum migrans, 153
Albendazole, 53, 163
Allopurinol, 38
Alpacas and llamas, 135
Alveococcus cyst, 116
Alveolar echinococcosis, 117, 119
Amastigotes, 35
Aminoglycoside paromomycin, 53
Ancylostoma braziliense, 172
Ancylostoma caninum, 148, 151, 169
  aberrant, 170
  anthelmintic resistance, 173
  anthelmintic resistant, 176
  buccal capsule, 151
  female, 148
  hookworm disease, 152
  hypobiosis, 151
  hypobiotic L3, 150
  infective larvae, 152
  life cycle stages, 148, 149
  molecular diagnostics, 171
  multidrug-resistant, 176
  muscular esophagus, 151
  prevalence, 152
  skin penetrating larvae, 150
  transmission, 153
  treatment guidelines, 153
  zoonotic hookworms, 171
Anthelmintic resistance, 173, 174
  development, 173, 176
  in greyhounds, 175
  in hookworms, 175
Anthelmintics, 152
Anthelmintic therapy, 17
Anti-*Leishmania* antibodies, 36
Anti-leishmanial treatment, 39
Auto infective cycle, 205

B
Baermann technique, 210, 211
Benzimidazoles, 53, 60, 63, 152
  febantel, 62
  fenbendazole, 60
Bilharziella polonica, 279
B lymphocytes, 31

C
Canine hookworms, 169
Canine leishmaniosis (CanL)
  clinical signs/laboratory findings, 37
  diagnosis, 34, 36
  endemic regions, 34
  ferritin, 34
  insecticides, 40
  and renal failure, 34
  resistance/susceptibility, 31
  stages, 39
  treatment, 38–40
  typical picture, 33
Canis lupus familiaris
  abilities, 1
  autosomes, 2
  breeds, 2

C. Strube, H. Mehlhorn (eds.), *Dog Parasites Endangering Human Health*, Parasitology Research Monographs 13,
https://doi.org/10.1007/978-3-030-53230-7

Canis lupus familiaris (*cont.*)
  chromosomes, 2
  genome, 2
  human cultures/religions, 2
  life spans, 2
  wolves, 1
Central nervous system (CNS), 10
Cephalic papillae, 246
Chagas disease, 74, 76, 79, 82
  nutritional and metabolic disorders, 74
Chari river basin, 257
Chloroquine, 66
Chronic strongyloidiasis, 209
Clinical staging system, 38
Colonoscopy, 169
Commercial vaccines, 40
Community-wide insecticide spraying campaigns, 94
Companion Animal Parasite Council (CAPC), 50, 51
Complete blood count (CBC), 37
cox1 sequences, 216
C-reactive protein (CRP), 34
Creeping eruption, 153, 160
Cryptosporidiosis, 65
Cucumber tape worm, 142
Culling poultry, 256
Cutaneous larva migrans (CLM)
  follicular cases, 164
  in oral mucosal, 163
  in Philippines, 164
  prevention, 164
  treatments, 163, 164
Cyclops, 247, 248
Cyst fertility, 132
Cysticercoid larvae, 142
Cytology, 35, 36

D
Demographic surveys, 82
Diffuse unilateral subacute neuroretinitis (DUSN)
  causative worm, 162
  characteristic, 161
  CLM, 162
  oral anthelmintics/laser mediated destruction, 161
  photocoagulation, 162
  size, 161
  SLO, 161
  subretinal space, 161
Dipylidium caninum, 143, 144
Direct–indirect switch, 206
Dirofilaria immitis, 228
  life cycle, 228
Dirofilaria repens
  life cycle, 229
Discrete typing units (DTUs), 75
Disseminated strongyloidiasis, 208
Dog dracunculiasis epidemiology, 253
Dog dracunculiasis transmission, 255
Dogs and humans
  European hard ticks and pathogens, 264
  parasites and pathways, 263
  symptoms of infections, 262
  transmission, 261
Domestic livestock, 136
Domestic transmission cycles, 80
Domestic vector species, 86
Dracunculiasis, 243, 247, 249, 256
  cases, countries, 252
  definitive hosts, 247
  eradication, 257
  human and dog, 252
  male and female worms, 247
  parasite, 248
  parasitic diseases, 247
  transmission, 248
Dracunculoid nematodes, 244
Dracunculosis, 246
Dracunculus medinensis
  life cycle, 254
  morphology, 246
  taxonomy, 244
Dracunculus species, 248
Drug treatment, 95

E
Echinococcus, 117
  diagnosis, 107
  drugs lethality, 117
  E. granulosus, 110, 114
  E. multilocularis, 115
  genotypes, 109
  geographic distribution, 108
  molecular biological details, 107
  multilocular cysts, 114
  prophylaxis, 117
  protoscolices, 117
  species distribution, 109
  specimens, 108
  symptoms of disease, 115
Echinococcus granulosus, 110–114, 123
  eggs, 110
  strains, 110
Echinococcus multilocularis, 114–116
Echinocuccus granulosus
  Australia eggs, 126

dingoes, 125
domestic animals
alpacas and llamas, 135
annual financial, 134
in Australia, 136
Australian sheep, 134
cattle, 133
domestic dogs, 133
GDR, 134
life cycle, 124
prevalence, 125
sheep herding dogs, 124
tapeworms, 124
transmission, 124, 125
Echocardiography, 79
Egg package (EP), 144
Electrocardiographic surveys, 78
Embryonation, 6
Entamoeba histolytica, 265
Enzyme-linked immunosorbent assay (ELISA), 16, 166
Eosinophilia, 209
Eosinophilic enteritis (EE), 165
A. caninum, 165, 168, 169
Australian, 168
cause, 165
diagnosing, 166, 167
hookworms, 170
percutaneous infection, 167
rarity, 165
symptoms, 166, 169
in the United States, 169
Eosinophilic gastroenteritis (EG), 165
Eosinophilic gastrointestinal disorders (EGID), 165
European Scientific Counsel Companion Animal Parasites (ESCCAP), 153
Excretory–secretory (ES), 166
Exfoliative dermatitis, 32, 33

F
Febantel, 62
Fenbendazole, 60
First-stage larvae (L1), 202
Food-borne transmission, 255
Free-living adult females, 202
Free-living adults, 202
Free-living males, 203

G
Gallbladder/bile ducts, 274
GenBank, 169
Giardia, 61
asymptomatic, 46
benzimidazoles, 60
CAPC map, 50
CDC data, 48
cysts, 47
diagnosis, 46
DNA identification, 46
drugs, 53
drug therapy, 52
gastrointestinal upset, 48
literature, 50
mortality, 47
nitroimidazoles, 54
oral vaccine, 52
outbreaks, 47
people and dogs, 46, 52
positive, 46, 50
prevalence, 48
sewage treatment, 47
subclinical, 45
treatment, 55, 63
ubiquitous nature, 50
Great Dividing Range (GDR), 134
Greyhound breeding farms, 175, 176
Guinea worm disease (GWD), 240
in ancient Egypt, 240
crustacean intermediate host, 242
D. medinensis, 240
D. medinensis life cycle, 251
dracunculiasis, 244
in Egypt, 244
GWEP technical staff, 252
infection, 249
maintenance, 257
WHO, 241, 257
zoonotic character, 240
Guinea worm eradication, 256
Guinea Worm Eradication Program (GWEP), 250, 251
Guinea worm infection, 253
reservoirs, 253
transport hosts, 256
Guinea worm water-borne life cycle, 249

H
Heartworm disease (HWD)
asymptomatic, 229
clinical presentations, 229
diagnosis, 230
prevention, 231
Heterophyes, 274, 278
Heterophyes heterophyes, 277
Heterophyes species, 274, 275
Homogonic cycle

Homogonic cycle (*cont.*)
  direct, 205
  indirect, 205
Hookworm disease, 151
Hookworm-related cutaneous larva margin (CLM/HrCLM), 154
  A. braziliense, 155
  autochthonous, 155
  chronic/persistent, 156
  creeping eruption, 154
  ELISA, 157
  enigmatic case, 160
  eosinophilia, 155
  follicular, 156
  hookworm larvae, 158
  hookworm species, 157
  optic disc edema, 160
  presentations, 156
  skin scraping, 157
  soil and/or sand, 154
  in United States, 154
Host infectiousness, 89
Host–parasite interactions, 198
Human dirofilariosis, 233
  pulmonary dirofilariosis, 233
  subcutaneous and ocular, 233
Human hydatid disease, 124
Human hydatidosis, 135
Human parasitic nematodes, 199
Human toxocarosis, 10
Hydatid disease, 124, 128, 134
Hydatid infection, 133
Hydatid larval stages, 110
Hyperinfection syndrome, 208, 209, 211
Hypobiosis, 150
Hypobiotic larvae, 150

I
Immune-mediated mechanisms, 31
Indirect hemagglutination assays, 79
Infectiousness, 88, 89
Infective third-stage larvae (iL3), 200, 202

K
kDNA PCR assays, 36

L
Lagomorphs, 132
Laparotomy, 165, 166
Leishmania donovani, 28
Leishmania infantum, 28, 29, 35
  clinical findings, 32
  detection, 35
  epidemiology, 29
  natural transmission, 30
  pathogenesis, 30
Leishmaniases, 28
Leishmaniosis, 28, 32, 34
  clinical signs, 29
  enzootic, 31
  visceral, 36
LeishVet association, 36
LeishVet stage I, 38
LeishVet stage II, 38
Life cycle, 110
Löffler's syndrome, 158, 159

M
Macropodid species, 127
Megacolon and megaesophagus, 78
Meglumine antimoniate, 38, 39
Meningoencephalitis, 78
Metronidazole, 53–55, 61, 66
Miltefosine, 38, 39
Mixed blood meals, 87
Molecular methods, 79, 80, 86
Moxidectin, 176
Myocardium, 78

N
Natural Guinea worm infections, 244
Neglected tropical diseases (NTDs), 196
Neurotoxocarosis (NT), 10, 13
  dopamine system, 13
  murine and human, 13
  parenchymal damage and haemorrhagic lesions, 14
  pathogenesis, 13
Nitazoxanide, 64
Nitroimidazoles, 54
  metronidazole, 54
  ronidazole, 60
  secnidazole, 59
  tinidazole and ornidazole, 54
Nonoverlapping transmission cycles, 92

O
Ocular *larva migrans* (OLM) syndrome, 10, 12
  clinical indication, 12
  symptomatology, 12

Oncosphaera larva, 142
Opisthorchis, 274
Opisthorchis felineus, 274, 276
Opisthorchis tenuicollis, 275
Ornidazole, 54
Ornithobilharzia species, 279
Oxfendazole, 63

P
Parasitic infections, 16
Parasitic nematodes, 173
Parastrongyloides, 204
Peracute hookworm disease, 151
Peridomestic transmission cycles, 80
Peripheral blood eosinophilia (PE), 165
Phlebotomine sand flies, 28
Pinworm, 196
Polymerase chain reaction (PCR) method, 29
Praziquantel-medicated baits, 136
Pre-molecular era, 213
Protoscolices, 110
Pseudorhabditis stercoralis, 196
Pulmonary dirofilariosis, 233
Pyrantel resistance, 174
Pyrethroids, 94

Q
Quantitative real-time PCR (qPCR), 88
Quantitative serological test, 39
Quinacrine, 53, 65
Quinolines, 65
  chloroquine, 66
  quinacrine, 65

R
Rabbits (*Oryctolagus cuniculus*), 132
Ravuconazole, 95
Reproduction
  in free-living generations, 204
  in parasitic adults, 203
Residual house spraying, 94
Reverse zoonosis, 240
Rhabditiform, 202
Rhabdonema strongyloides, 196
Ronidazole, 60

S
Sandworm eruption, 155
Scanning laser ophthalmoscope (SLO), 161
Schistosomatid species, 281
Schistosomatid trematodes, 277
  diagnosis, 279
SDS-PAGE analysis, 15
Secnidazole, 59
Serological cross-reactivity, 36
Serological tests, 16
Sex determination, 206
Sexual and asexual reproduction, 203
Small ribosomal SubUnit (*SSU*), 214
Strongylid nematodes, 173
Strongyloides, 204, 214
  life cycle, 196
Strongyloides intestinalis, 196
Strongyloides planiceps, 217
Strongyloides stercoralis
  detection, 209
  diagnosis, 209
  diagnostic methods, 197
  epidemiology, 198
  human-derived, 213
  infections, 199, 213
  life cycle, 200
  life history, 197
  medical importance, 198
  morphology, 197, 199
  parasitic adult females, 200
  parasitic worms, 201
  prevalence, 198, 212
  risk factors, 198
  sub-clades, 215
  symptoms, 209
  transplantation centers, 212
  treatment, 211
Strongyloidiasis
  hyperinfection syndrome, 208
  infection, 207
  parasitic adults, 196
  review articles, 197
Subcutaneous and ocular dirofilariosis, 233
Subcutaneous dirofilariosis (SCD), 231, 233
Sylvatic transmission cycles, 80

T
Tapeworms
  developmental cycle, 142
  diagnosis, 144
  geographic distribution/epidemiology, 141
  infectious cysticercoid larvae, 144
  Latin name, 141
  parasitizes, 141
  symptoms, 143

Tapeworms (*cont.*)
  terminal proglottids, 141
Tasmanian hydatid control campaign, 135
T-helper 1 (Th1) response, 30
Thiabendazole, 163
Thoracic radiography, 79
Threadworm, 196
Tinidazole
  and ornidazole, 54
Toxocara canis
  ascarid infection, 6
  biological characteristics, 6
  control and prevention, 18
  definitive hosts, 6
  development, 7
  diagnosis, 15
  distribution, 8
  embryonated, 6
  and immune cells/immune-related compounds, 14
  larvae, 6, 14
  life cycle, 9
  migration route, 10
  tracheal migration, 7
  transmission routes, 8
  treatment, 17
  VLM, 11
Toxocara eggs, 8
Toxocara-infected paratenic hosts, 10
Toxocarosis, 6, 11, 12, 18
Toxoplasma gondii, 265
Transmission
  eco-epidemiology
    demographic surveys, 82
    dog demography, 82
    host conditions, 82
    host-feeding patterns, 84
    prevalence, 87
Transmission cycles
  domestic, 80
  molecular epidemiology
    molecular methods, 92
    overlapping transmission cycle, 92
    sympatric transmission cycles, 93
  peridomestic, 80
  sylvatic, 80
Transplacental/intrauterine transmission, 7
Trans-sialidase inhibition assay, 79
Trematodes
  diagnosis, 276
  geographic distribution/epidemiology, 274
  prophylaxis, 276
Triatominae, 84
Triatomine host-feeding patterns, 84
Triatomine species, 85
Trichobilharzia species, 277, 279
Tropical Institute in Bukhara, 250
Tropical lifestyle, 167
Trypanosoma caninum, 77
Trypanosoma cruzi
  canine infections, 78
  chronic infections, 79
  consumption, 76
  diversity, 74
  dogs seropositive, 79
  domestic transmission, 90, 91
  DTU and clinical presentation, 75
  eco-epidemiology, 75
  genotypes, 89
  host infection, 75
  human infection, 74
  infection, 77, 87
  infectiousness, 88, 89
  insect vector and mammalian host, 75
  peridomestic transmission, 90
  prevalence, 87
  reservoir hosts, 74, 80, 81
  residual effects, 94
  sentinels of exposure, 90
  spatial distribution, 90
  sylvatic sources, 90
  transmission, 75, 81, 93
  transmission cycles, 74
  trypomastigotes, 75
  vaccine, 95
  vector-mediated and oral infections, 76
  zoonotic infection, 74
Trypanosomatids, 76

U
Urine protein/creatinine (UPC) ratio, 37

V
Vector-borne diseases, 235
Vector-borne infectious disease, 31
Vector control strategy, 94
Vector–host contact rates, 88
Vector-mediated zoonoses, 74
Visceral larva migrans (VLM), 11
  characteristics, 12
  classical, 11
  clinical picture, 11

syndrome, 10
Visceral leishmaniosis, 28

W
Water-borne transmission, 247, 248
Wild dogs, urban environments
in Australia, 126
camels, donkeys and horses, 131
deer, 130
feral cats, 127
feral pigs, 132
foxes, 126, 127
goats, 131
macropodid marsupials, 127, 128
residents living, 126
water buffalo, 130
wombats, 128
Wombats, 128

X
Xenodiagnosis
and hemoculture, 78
Xenointoxication, 94

Z
Zooanthroponosis, 240
Zoonosis, 240
Zoonotic infections, 248, 257

9783030532291